EARTH'S HAZARDS

Understanding Natural Disasters and Catastrophes

David M. Best
Northern Arizona University

David L. Hacker
Kent State Trumbull

Kendall Hunt
publishing company

Book Team

Chairman and Chief Executive Officer   Mark C. Falb
President and Chief Operating Officer   Chad M. Chandlee
Vice President, Higher Education   David L. Tart
Director of National Book Program   Paul B. Carty
Editorial Development Manager   Georgia Botsford
Senior Editor   Lynnette M. Rogers
Vice President, Operations   Timothy J. Beitzel
Assistant Vice President, Production Services   Christine E. O'Brien
Senior Project Editor   Charmayne McMurray
Permissions Editor   Renae Horstman
Cover Designer   Janell Edwards

Cover photos:

tornado: © Sean Martin, istockphoto.com
volcano: © james steidl, istockphoto.com
destruction: © JupiterImage Corporation

Chapter opener photos:

raging water: © Shutterstock, Inc.
tornado: © Shutterstock, Inc.
lava: © Shutterstock, Inc.

Brief Contents

Contents

vi   **Contents**

Chapter 13
Biological Hazards

Chapter 14
Human and Natural Interactions
on the Environment

Preface

The topic of natural hazards and disasters has always been of interest to the general populace. Each day on Earth brings some type of hazardous event that is part of the dynamic nature of the planet. This results in changes in the appearance of the landscape, and oftentimes the creation of a crisis for humans who are affected by the event. The broad range of disasters covered in this text addresses these events, from earthquakes, floods, and volcanoes to the biological hazards that often result indirectly from these events.

Normal geological processes have been studied and taught for more than two hundred years. The basics of river flow, landslides, and volcanoes were recognized early in the study of processes occurring on Earth, but geologists have developed a more detailed understanding of many others through the use of improved technology. By using satellite data and surface observations we are able to better explain how natural processes such as increased rainfall and river flow can generate floods in a given area. Monitoring movement in the subsurface gives insights into possible earthquake activity or volcanic eruptions. Examples of major disasters are numerous. During the past thirty years the eruption of Mount St. Helens in southwestern Washington, widespread flooding of the Midwest in the United States, the tsunami in South Asia and earthquakes, including the one in Haiti in January 2010, are just a few examples of geologically recent occurrences of life-changing events.

The dynamic nature of Earth became very evident in the latest stage of the preparation of this book. The following events occurred but are not addressed in the text:

- Magnitude 7.2 earthquake in Chile on March 11, 2010,
- Massive flooding in Rhode Island and Nashville, Tennessee in March and April 2010,
- Magnitude 6.9 earthquake in southern China on April 13, 2010 killing 2,200 people,
- Eruption of Eyjafjallajokull volcano in Iceland on April 14, 2010 disrupting air traffic across the Atlantic Ocean and on the European continent; activity continued for several weeks,
- Tornadoes strike Yazoo City, Mississippi on April 24, 2010 killing 12 people.

Featured Themes

The dynamic nature of the material covered in this text is presented through the use of carefully selected images and art work that reduce the amount of visual overload that can occur in such a course. Pictures are worth a thousand words, as we oftentimes hear; we have attempted to do this without offering too many images. Questions for Thought are at the end of each chapter as well as several web sites and brief lists of reference and reading material.

Topic Organization

Following an introduction to hazards in general and the development of the Earth system, volcanoes and earthquakes, which are many times interrelated, are presented. Discussions of mass movements and tsunami follow earthquakes, as the movement of land and water is often a result of earthquakes. Extraterrestrial hazards close out the first half of the material.

Almost 75 percent of Earth's surface is covered by water. Four chapters address the role that the atmosphere and water play in the creation of natural hazards. A general introduction to global climate, followed by streams and oceans, explains how the presence of water is a key force on Earth. The uncontrolled presence of cyclonic storms completes the section related to water. Wildfires, often the result of too little water in regions, have become an increasing problem in our country as well as in other parts of the globe. New topics provide material about biological hazards and the environment.

Several chapters include sections that address Lessons from the Geologic Past, as determined from the geologic record. The concepts and processes associated with the featured disasters have been stressed and the number of specific terms introduced with each topic has been held to a minimum.

Support Materials

An Instructor's Guide is available, in addition to a compact disc that contains images used in the text. Another resource is the availability of a web site that has links to current information on recent disasters (http://oak.ucc.nau.edu/dmb25). Numerous other sites exist on the Internet.

Acknowledgments

This work grew out an initial proposal by David B. Hacker, who contributed to the early stages of the text. I was asked by Tina Bower, Developmental Editor at Kendall Hunt at the time, to provide some material. My efforts evolved into writing a major portion of the text. Tina was succeeded by Lynne Rogers, Senior Development Editor, who has provided a continued high level of direction and encouragement. Challenges that always arise with the production of a textbook have been ably addressed by Charmayne McMurray, Senior Production Editor and Renae Horstman, Senior Permissions Editor. In somewhat quieter ways, but equally important, Paul Carty, Director of Publishing Partnerships, offered his high level support for completion of the project. Geology is an observational science which requires lucid artwork. This phase of the text was the charge of Craig White—job well done!

Throughout the writing of this book I often called on a cadre of colleagues to listen to ideas and provide me the impetus necessary to complete the work. The assistance of the following colleagues and professionals has proven to help me in the writing and presentation of material: from Northern Arizona University Lee Drickamer, Regents Professor Emeritus of Biology, James Wittke, School of Earth Sciences and Environmental Sustainability, and geology graduate students Chris Kassel, Mark Sutton, and Rachelle Wagner; Charles Denton, U. S. Forest Service (retired); Wendell Duffield, U. S. Geological Survey (retired) and for providing or suggesting image and image sources: James S. Best, Rich Giraud, Utah Geological Survey; Sara Jenkins, Desert Research Institute; Dave Norman, Washington State Department of Natural Resources; and Pat Stormer, Save the Light, Inc.. Discussions with Jeff Leid, Associate Professor of Biological Sciences and Alex Alvarez, Professor of Criminology and Criminal Justice at NAU improved my understanding of hazardous diseases. Special thanks go to Darrell Boomgaarden, who granted an interview relating his personal experiences during the Great Alaskan earthquake of March 1964.

I especially wish to recognize the help of my close friend and colleague, Syl Allred, Ph. D., Principal Lecturer in the Department of Biological Sciences at Northern Arizona University who collaborated with me in writing Chapter 14, Human and Natural Interactions on the Environment. He also provided support throughout the project.

Finally I wish to express sincere thanks to my wife Mary, who provided continued encouragement whenever challenges arose. Her persistence allowed the work to come to come to fruition. In addition her review of several chapters and her assistance with polishing the text made the material read much more clearly.

I also wish to express appreciation to the following reviewers who provided constructive and insightful comments.

Jean P. Kowal
University of Wisconsin, Whitewater

Martin Acaster
Portland Community College

Christine Aide
Southwest Missouri State University

Stephen T. Allard
Winona State University

Steven Altaner
University of Illinois

Dr. Thomas Bicki
Emerson College

Paul Bierman
University of Vermont

Susan Bilek
New Mexico Tech

Amy Bloom
Illinois State University

Jon Boothroyd
University of Rhode Island

Dr. Patricia Cashman
University of Nevada, Reno

Heather K. Conley
Illinois State University

Winton Cornell
University of Tulsa

Juliet Crider
Western Washington University

Charles Denton
U. S. Forest Service (retired)

David Dinter
University of Utah

Wendell Duffield
U. S. Geological Survey (retired)

Todd Feeley
Montana State University

Duncan Foley
Pacific Lutheran University

John Foster
California State University, Fullerton

Kevin Furlong
Penn State University

David Gillespie
Washington University, St. Louis

Luis Gonzalez
University of Kansas

Michael Hamburger
Indiana University, Bloomington

Linda Hand
College of San Mateo

Keith Henderson
Villanova University

Simon Katterhorn
University of Idaho, Moscow

Robert Kay
Cornell University

Chris Kent
Community College of Spokane

Attila Kilinc
Unviersity of Cincinnati

Jean P. Kowal
University of Wisconsin, Whitewater

Stephen D. Lewis
California State University, Fresno

Lawrence Malinconico
Lafayette College

Doug McKeever
Whatcom Community College

Sue Morgan
Utah State University

Ken G. Sutton
Portland Community College

Harold Tobin
University of Wisconsin, Madison

James Wittke
Northern Arizona University

I hope you find the text interesting and engaging and I ask that you make me aware of any errors, corrections, or additions you might have. The process of assembling a text is always difficult and oversights do occur. Thank you for your interest.

David M. Best
Flagstaff, Arizona
July 2010

About the Authors

David M. Best, Ph.D., grew up in North Carolina, where he graduated from the University of North Carolina at Chapel Hill with a B.S. in mathematics. He completed his M.S. in geology and then served in the U. S. Navy. After returning to UNC-Chapel Hill to obtain his Ph.D. in geology, he began his teaching career at Northern Arizona University in 1978. For more than three decades at NAU, David has taught classes in introductory and physical geology, geophysics, statistical methods, and the geology of Arizona. He has also served in various administrative roles, including department chair, associate dean, and dean of the College of Arts and Sciences. After returning to the teaching ranks in 2003, David began teaching geologic hazards along with the geocommunications course required of geology majors. Since 1998 he has team-taught a biology and geology field course in the national parks with Syl Allred.

David B. Hacker received his Ph.D. from Kent State University in 1998 and teaches geology on the Trumbull campus of Kent State University, where he is an associate professor of geology. His primary interests are in the overall field of structural geology, with emphasis on field mapping in the western United States. Presently David is working on analyzing structures and volcanism related to the growth of Miocene laccoliths in southwest Utah. He has completed field mapping in collaboration with government agencies such as the U.S. Geological Survey and the Utah Geological Survey.

Living with Earth's Natural Hazards

1

Key Terms

carrying capacity
closed system
dynamic equilibrium
Earth system science
exponential growth
frequency
geologic time scale
hypothesis
law (or principle)
magnitude
mitigation
natural catastrophe
natural disaster
natural hazard
rapid-onset hazard
recurrence interval
scientific method
slow-onset hazard
theory

Mount Shasta in northern California is the largest volcano in the region. Although the population is relatively sparse in the surrounding area, this volcano has the potential to erupt and produce massive landslides and lahars. Such an event occurred about 11,000 years ago, covering the landscape with rock and water from the melting of glaciers on the summit.

Does it seem like Earth is a dangerous or hazardous place to live? A *hazard* by definition is "a possible source of danger." Among some of Earth's natural hazards (or dangers) are hurricanes, tsunami, tornadoes, earthquakes, floods, mudslides, wildfires, and volcanic eruptions (**Figures 1.1a-c**). Through the media, we have seen these hazards manifest into disasters, like the 2004 tsunami in Asia, the 2005 earthquake in Pakistan, and the 2005 Hurricane Katrina in the United States, and wildfires across southern California in 2007, causing unprecedented devastation and loss of life (**Figures 1.2a-b**). These scenes of unprecedented devastation have been etched in our minds and have made us increasingly aware of our human vulnerability to the awesome power of nature. However, hazards such as these are simply *natural processes* involving physical, chemical, and biological mechanisms and forces that affect Earth's surface, and have been operating continuously for millions of years. They become hazards to humans only when people choose to live or work where these processes occur and become affected by them.

Affected we have been. Worldwide we are facing **natural disasters** on an unprecedented scale. Statistical data show that between 1974 and 2003 there were 6,367 natural disasters, not counting epidemics, that resulted in the reported deaths of slightly more than 2 million people. Cumulatively, about 5.1 billion people were affected in some way, 182 million people were made homeless, and estimated damages totaled $1.38 trillion. In the last decade alone, 86 percent of all disaster-related deaths were caused by natural hazards, with just 14 percent resulting from technological disasters such as transportation or industrial accidents. Asia alone suffered the most, with 75 percent of the deaths from natural disasters. Such large numbers may appear abstract and difficult to conceptualize, but they are a harsh reality for many families who have lost loved ones, had their homes reduced to rubble, or have watched their investments and economic future be destroyed by a natural disaster.

But what can we do? Natural hazards are inevitable natural phenomena, but their transition into disasters or catastrophes is often a result of the organization, distribution, and behavior of our society. Natural disasters know no political boundaries. Smoke from forest fires can make air unbreathable in neighboring nations, ash plumes from

natural disaster
The loss of life, injuries, or property damage as a result of a natural event or process, usually within a more local geographic area.

FEMA News Photo by Liz Roll.

Figure 1.1a The entire town of Grafton, IL, was under water in July 1993, when major flooding covered much of the Midwest. Waters did not recede in some areas until early in the fall.

NASA.

Figure 1.1b Hurricane Bonnie off the east coast of the United States in late August 1998.

USGS Photograph by Lyn Topinka.

Figure 1.1c A lahar from the eruption of Mount St. Helens damaged this home along the South Fork Toutle River, southwestern Washington.

Figure 1.2a Destroyed neighborhoods along the Gulf coast in Biloxi and Gulfport, Mississippi.

Figure 1.2b Neighborhoods in Rancho Bernardo, California, were destroyed by wildfires in late October 2007.

volcanoes can disrupt air traffic half a continent away, and emerging diseases and global warming can affect the entire planet. Whether an extreme hazard event becomes a natural disaster depends on our ability to predict, prepare, and mitigate. Most decision makers and policymakers agree that the key to reducing the vulnerability and risk of human populations to natural hazards is the integration of disaster preparedness, **mitigation**, and prevention measures into future policy development. Yet funding patterns, an indicator of real priorities, show that disaster relief—not reduction or prevention—tops the list of disaster-management funding, which leads to continued losses and suffering after we recover and a disaster strikes again.

Fortunately, public awareness of the losses in human lives and property from natural disasters is starting to grow, mostly through media images of devastation. Once we understand and accept the need for a change in our life styles, such as recycling or seat-belt use, our willingness to make the change increases. But we still have a long way to go if we are to focus on hazard preparedness and prevention rather than quick, band-aid solutions. Better information and public education about the natural environment and Earth processes are essential to reducing losses. Therefore, the focus of this book is to help develop an understanding of Earth's natural processes through the study of relationships between people and their environment. The information on natural hazards presented in this book is of practical value to people in making choices of where to live and understanding what hazards might be present so they may be avoided.

mitigation
The act of making less severe or intense; measures taken to reduce adverse impacts on humans or the environment.

From Natural Hazards, to Disasters, to Catastrophes: Interactions with Natural Processes

In describing our interaction with natural processes we use the terms natural hazard, disaster, and catastrophe. A **natural hazard** is an event that could have a negative impact on people and their property resulting from natural processes in the Earth's environment. Some of these events include earthquakes, hurricanes, tsunami, floods, volcanic eruptions, droughts, landslides, and coastal erosion. These events are the result of natural Earth processes that have been operating for several billion years;

natural hazard
An event or phenomenon that could have a negative impact on people and their property resulting from natural processes in the Earth's environment.

therefore they will continue to happen, regardless of whether or not humans are exposed to them. Only when these processes threaten humans and their property do they become hazards that may create a disaster. A natural disaster is the loss of life, injuries, or property damage as a result of a natural event or process, usually within a more local geographic area. A natural disaster could be as small as an individual's house damaged in a wildfire with losses handled through a private insurance carrier or a local community devastated by flooding that could require assistance from government agencies. A **natural catastrophe** is considered a massive disaster often affecting a larger region and requiring significant amounts of time and money for recovery. A catastrophe can be so large that it devastates a metropolis such as New Orleans that requires large amounts of disaster relief from outside the community, which can stifle the economy of an entire country.

natural catastrophe
A massive natural disaster often affecting a large region and requiring significant amounts of time and money for recovery.

Types and Characteristics of Natural Hazards

Natural hazards, and the disasters or catastrophes that result from them, come in a variety of forms. Natural hazards of terrestrial origin can be divided into three different categories: geologic, atmospheric (or hydro-meteorological), and environmental (or biological) (**Table 1.1**). These originate from the flow of energy and matter contained on or in our planet. Examples include geologic earthquakes and volcanic eruptions, atmospheric hurricanes and tornadoes, and environmental wildfires and diseases. Natural hazards of extraterrestrial origin include asteroid or comet impacts and solar-related events such as geomagnetic storms.

Hazards can also be categorized as **rapid-onset hazards** which expend their energy very quickly, such as volcanic eruptions, earthquakes, floods, landslides, thunderstorms, and lightning. These hazards can develop with little warning and strike rapidly. These are the hazards we hear about the most because of their violent effects and real-time media coverage. In contrast, **slow-onset hazards**, such as drought, insect infestations, disease epidemics, and global warming and climate change, take years to develop and are usually neglected by the media as not newsworthy until long after the effects have started.

Other types of hazards include man-made *technological hazards* that originate in accidental or intentional human activity, such as oil and chemical spills, building

rapid-onset hazard
A hazard that develops with little warning and strikes rapidly. They expend their energy very quickly, such as volcanic eruptions, earthquakes, floods, landslides, thunderstorms, and lightning.

slow-onset hazard
A hazard that takes years to develop such as drought, insect infestations, disease epidemics, and global warming and climate change.

TABLE 1.1	Categories of Natural Hazards and Examples	
TERRESTRIAL HAZARDS		
Geologic	**Atmospheric**	**Environmental**
• Volcanic eruptions	• Global warming	• Wildfires
• Earthquakes	• Tropical cyclones	• Biological diseases
• Landslides	• Storms	• Insect infestations
• Land subsidence	• Tornadoes	• Environmental pollution
• Tsunamis	• Droughts	
• Coastal erosion	• Lightning	
• Floods	• Blizzards	
EXTRATERRESTRIAL HAZARDS		
• Impacts from asteroids and comets		
• Solar flares		
• Geomagnetic storms		

fires, plane crashes, and acts of terrorism. *Anthropogenic hazards* are also caused by humans, but they affect our environment and ecosystem, which eventually affects us in the long run. These include things such as pollution, deforestation, and so on that lead to global warming, ozone layer destruction, increased magnitude of storms and landslides, and other effects. We will explore some of the anthropogenic hazards that are linked to certain natural hazards.

Often, several hazards and processes can be linked to a single event and occur at the same time or shortly after the main event. These multiple events must be taken into account in preparing in advance for a hazard. For example, along with strong winds, hurricanes are associated with intense precipitation that can cause flooding, erosion along the coast, and landslides on inland hill slopes. Volcanic eruptions can cause lahars (mudflows) and flooding, and if the volcano exploded or collapsed in the ocean, it can cause a tsunami.

Natural hazards also frequently occur in a series, meaning they follow an initial onset. Each subsequent hazard acts upon the effects of the preceding event. For instance, a hurricane can down enormous stands of timber, an event that could result in an disastrous outbreak of detrimental insect populations. Likewise, during long periods of drought, wildfires that burn standing or downed timbers can denude hillsides of vegetation and thereby trigger soil erosion, flooding, and landslides.

Natural Hazards in the United States: What's at Stake?

The extraordinary natural, climatic, and geographic diversity of the United States exposes people to a wide range of natural hazards throughout the nation. Natural hazards such as earthquakes, volcanic eruptions, wildfires, storms, and drought strike nearly every part of the United States, exacting an unacceptable toll of life, property, natural resources, and economic well-being (**Figure 1.3**). Each year, natural hazards cause hundreds of deaths and cost billions of dollars in disaster aid, disruption of commerce, and destruction of homes and critical infrastructure. The cost of major disaster response and recovery continues to rise, and property damage from natural hazards events doubles or triples each decade. The following discussion is a short summary of some common natural hazards in the United States intended to help illustrate the need for understanding natural hazards and reducing the occurrence and effects of natural disasters.

Severe Weather

Severe weather events such as tornadoes, hurricanes, storms, and heat waves strike many areas of the United States each year (**Figure 1.4**). Due to changes in population demographics and the establishment of more complex weather-sensitive infrastructures, the United States is more vulnerable to severe weather events today than in the past. It is now estimated that up to $2.2 trillion of the economy is affected annually by weather and climate events.

Over the past 30 years, coastal population growth has quadrupled, along with accompanying property and infrastructure development, to the extent that more than 69 million people now reside along hurricane-prone coastlines. At least 1,836 people lost their lives in Hurricane Katrina (**Figure 1.5**) and in the subsequent floods, making it the deadliest hurricane since the 1928 Okeechobee hurricane. The storm was also responsible for an estimated $81.2 billion (2005 U.S. dollars) in damage, making it the costliest natural catastrophe in our history.

Tornadoes are more common in the United States than anywhere else in the world, with an average of 1,000 reported annually nationwide, resulting in an average of 80 deaths and over 1,500 injuries every year. During May 2003, a total of 543 tornadoes hit the United States, breaking the previous record of 399 tornadoes recorded in 1992. The tornadoes that struck the Oklahoma City area on May 3, 1999, were

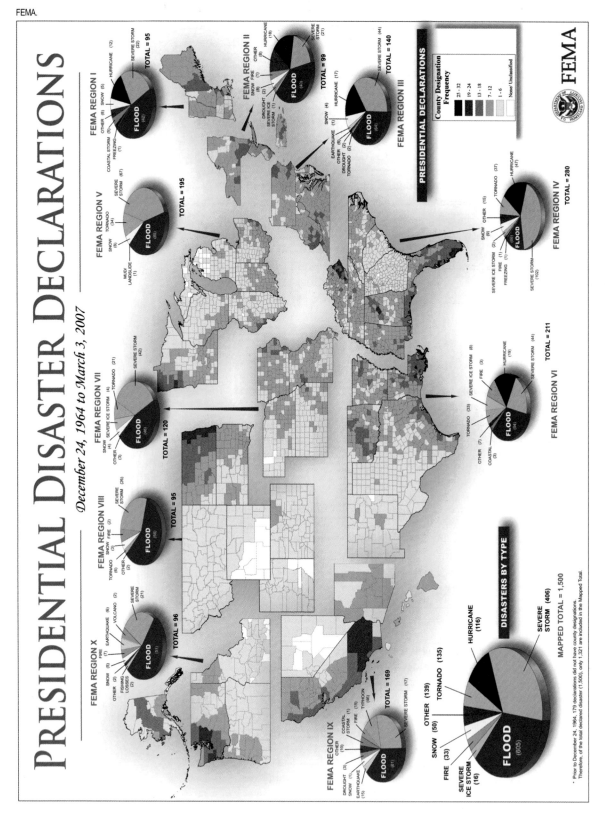

Figure 1.3 Presidential Disaster Declarations in the United States by county from 1964 to 2007 reflect the regional geographic distribution and human impacts of earthquakes, floods, hurricanes, tornadoes, severe storms, and wildfires.

Figure 1.4 Tornadoes struck the city of Lafayette, Tennessee, on February 5, 2008.

some of the most devastating in history, destroying over 2,500 structures and causing over $1 billion in damage (**Figure 1.6**).

Rapid-onset hazards such as hurricanes and tornadoes strike with deadly force, but slow-onset hazards can also be deadly. Such is the case with heat waves. In July 1995, a heat wave in Chicago killed 739 people.

Wildfires

Wildfires, also known as *wildland fires*, engulfing thousands of acres of land and encroaching on residential neighborhoods, have become all too familiar scenes (**Figure 1.7**). Wildfires, on average, burn 4.3 million acres (17,000 km^2) in the United States annually, and in recent years the nation has spent over $1 billion annually

Figure 1.5 Hurricane Katrina near peak strength on August 28, 2005. This developed into the most catastrophic storm to hit the United States in more than 80 years.

Figure 1.6 Several supercell thunderstorms produced more than 70 tornadoes that struck central Oklahoma, southern Kansas, and northern Texas on May 3, 1999. Forty-six people were killed and more than 8,000 homes and businesses were damaged or destroyed, at a cost of more than $1.2 billion.

Figure 1.7 Lake City, Florida, May 15, 2007. The Florida Bugaboo Fire raged out of control in some locations. The U.S. Department of Homeland Security's Federal Emergency Management Agency (FEMA) authorized five Fire Management Assistance Grants between March 27 and May 10, 2007, to help Florida fight fires in 16 counties.

NASA http://rapidfire.sci.gsfc.nasa.gov/subsets/?AERONET_La.Jolla/2007/297/AERONET_La.Jolla.2007297.aqua.250m.jpg

Figure 1.8 NASA satellite photo from October 24, 2007, showing the active fire zones and smoke plumes in southern California from Santa Barbara County to the U.S.–Mexico border.

for fire suppression. This does not include costs to communities and individuals in terms of structural losses or economic disruptions. The extreme fire seasons from 2000 to 2007 saw the largest areas burned by wildfires in the United States since the 1960s. The 2003 California fires burned over 743,000 acres (3007 km^2), destroyed 3,300 homes, and killed 26 people. The October 2007 California wildfires were a series of wildfires in which at least 1,500 homes were destroyed and over 500,000 acres (2,000 km^2) of land burned from Santa Barbara County to the U.S.–Mexico border (**Figure 1.8**). Two days into the fires, approximately 500,000 people from at least 346,000 homes were under mandatory orders to evacuate, the largest evacuation in the region's history. Nine people died as a direct result of the fire; 85 others were injured, including at least 61 firefighters.

Wildfires are not restricted to the drier climates of the western United States. The 2007 Bugaboo Scrub Fire, a wildfire in the southeastern corner of the country, raged from April to June, ultimately became the largest fire in the history of both Georgia and Florida. A thick, sultry smoke from the fires blanketed the city of Jacksonville, Florida and the entire area of northeast Florida, and southeast Georgia for many days, reducing visibility and causing many health concerns.

Drought

Drought is a complex, slow-onset, nonstructural-impact natural hazard that affects more people in the United States than any other hazard, with annual losses estimated at $6 to $8 billion. Compared to all natural hazards, droughts are considered the leading cause of economic losses. The 1988–1989 drought cost an estimated $39 billion nationwide and was at one time the greatest single-year natural disaster in U.S. history (Hurricane Katrina topped that amount in a single event). During 2002, over one-third of the nation experienced drought conditions. Growing population, a shift in population to drier regions of the country, urbanization, and changes in land and water use have increased the magnitude and complexity of drought hazards.

FEMA News Photo.

Figure 1.9 The January 17, 1994, Northridge, California, earthquake. Numerous highways were damaged and approximately 114,000 residential and commercial structures were damaged and 72 deaths were attributed to the earthquake. Damage costs were estimated at $25 billion.

Earthquakes

Each year the United States experiences thousands of earthquakes. An average of seven earthquakes with a magnitude of 6.0 or greater (enough to cause major damage) occur each year. About 75 million people in 39 states face significant risk from earthquakes, and earthquakes remain one of the nation's most significant natural hazard threats. The last major earthquake to strike a large urban area was the Northridge, California, earthquake that occurred on January 17, 1994 (**Figure 1.9**). Sixty-one people died as a result of the earthquake, and over 7,000 were injured. In addition, the earthquake caused an estimated over $15 billion in damage, making it one of the costliest natural disasters in U.S. history.

Volcanoes

The United States is among the most volcanically active counties in the world with about 170 active or dormant volcanoes (**Figure 1.10**). During the past century, volcanoes erupted in Washington, Oregon, California, Alaska, and Hawaii, devastating thousands of square kilometers and causing substantial economic and societal disruption and loss of life. Since 1980, 45 eruptions and 15 cases of notable volcanic unrest

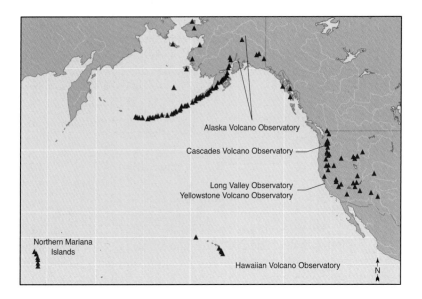

Alaska Volcano Observatory

Cascades Volcano Observatory

Long Valley Observatory
Yellowstone Volcano Observatory

Northern Mariana
Islands

Hawaiian Volcano Observatory

N

Figure 1.10 The U.S. Geological Survey (USGS) is responsible for monitoring the Nation's 170 active volcanoes (red triangles) for signs of unrest and for issuing timely warnings of hazardous activity to government officials and the public. This responsibility is carried out by scientists at the five volcano observatories operated by the USGS Volcano Hazards Program and also by state and university cooperators.

have occurred at 33 volcanoes, producing lava flows, debris avalanches, and explosive blasts that have invaded communities, swept people to their deaths, choked major rivers, destroyed bridges, and devastated huge tracts of timber forests. Volcanic ash plumes ejected into the atmosphere can be a costly and serious danger to aircraft. A Boeing 747 sustained $80 million in damages when it encountered ash from Mount Redoubt in Alaska during a 1989 eruption (**Figure 1.11**).

Floods

Floods are the most frequent natural disaster in the United States and can be caused by hurricanes, weather systems, and snowmelt (**Figure 1.12**). Failure of levees and dams and inadequate drainage in urban areas can also result in flooding. Nearly 75 percent of federal disaster declarations are related to flooding, which, on average, kills about 140 people each year and causes $6 billion in property damage. In 1993,

Alaska Volcano Observatory/United States Geological Survey, Photo by Cyrus Read.

FEMA, photo by Andrea Booher.

Figure 1.11 Redoubt Volcano has been active since a major eruption in 1989.

Figure 1.12 Volunteers filled more than 300,000 sandbags in one day in order to build levees along the Red River in Fargo, North Dakota, as the river crested at 41 feet in late March 2009.

flooding in the Mississippi Basin resulted in an estimated $12 to $16 billion in damages across nine states in the Midwest. An increase in population and development in river floodplains and an increase in heavy rain events over the past 50 years have gradually increased the economic losses from floods.

Landslides

Landslide hazards occur and cause damage in all 50 states (**Figure 1.13**). Severe storms, earthquakes, volcanic activity, coastal wave attacks, and wildfires can cause widespread slope instability. Each year landslides in the United States cause $1 to 2 billion in damage and more than 25 fatalities. The May 1980 eruption of Mount St. Helens caused the largest landslide in history—large enough to fill 250 million dump trucks. Human activities and population expansion, however, are major factors in increased landslide damage and costs.

Figure 1.13 Roads in Cougar, Washington, were destroyed in February 2009 by landslides and mudslides produced by a series of several winter storms that came off the Pacific Ocean.

Disease Epidemics

While disease outbreaks lack the rapid-onset aspect of other disaster discussed above, they potentially present an even greater threat to the United States population. The West Nile virus epidemic of 2002 demonstrates how outbreaks can become a public health emergency. In 2002, a total of 4,156 cases were reported, including 284 fatalities, not to mention the deaths of hundreds of thousands of birds and mammals. The cost of West Nile-related health issues was estimated at $200 million and exposed vulnerabilities in the U.S. health care system. It also served as both a warning and a wake-up call for the nation.

Human Population Growth: Moving Toward Disaster

Complicating all natural hazards issues is the rapid growth of our world population (**Figure 1.14**), along with the goal of having a better standard of living which is heavily tied to natural resource use. Remember that a natural hazard only becomes a

Figure 1.14 Population growth curve showing different parts of the world. North America and Europe maintain a fairly constant growth while Asia and Africa have much higher growth rates. Source: http://www.globalchange.umich.edu/globalchange2/current/lectures/human_pop/human_pop.html

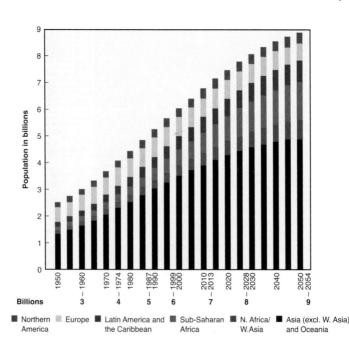

disaster when it affects people where they live or work. The increase we see in some natural disaster occurrences is due mostly to expansion of our population into hazardous areas which is occurring at a fast rate. The biggest population increases have been in cities and their suburbs (**Figure 1.15**). In the United States between 1950 and 2000, the urban population grew from 64 percent of the total population to 77 percent. Densely populated cities can be easy targets for natural catastrophes as witnessed in New Orleans in 2005. Mexico City, with over 20 million people, is a megacity in a region of earthquake and volcanic hazards. The southern and western regions of the United States (areas prone to drought, wildfires, hurricanes, earthquakes, and mudslides) are expected to grow by 32 and 51 percent, respectively, by the year 2050.

The increase in our world population by over 5 billion people during the past 200 years is without precedent in human history (**Figure 1.16**). The world population is estimated to have been only about 5 million people 10,000 years ago, but the nearly flat population curve began to rise about 8,000 years ago when agriculture and domestication of animals began to replace a hunter-gatherer culture. For thousands of years people lived in sparsely settled rural areas and growth rates were still very low, so population increased slowly but steadily to about 400 million by the Middle Ages (1100 to 1500). A sharp drop in population occurred when the Black Death struck Europe in the mid-1300s, but by 1700 the world population rebounded to about 650 million. Since that time, **exponential growth** in world population reached the world's first billion by the early 1800s, 2 billion by 1930, 3 billion by 1960, and 6 billion by 2000 (**Figure 1.17**). It is estimated that world population will reach 9 to 10 billion by 2050.

The rapid increase in human population is causing serious shortages of resources, including oil, food, and water, and is heavily degrading the environment. Some resources are renewable, such as water, whereas many others such as fuels (oil, natural gas, coal) and minerals (ores for copper, iron, etc.) are not. From population and natural resource data it becomes clear that it is impossible in the long run to support

© JupiterImages Corporation.

Figure 1.15 Cities throughout the world have grown rapidly in the past fifty years, using many more resources than in the past.

exponential growth

Growth in which some quantity, such as population size, increases by a constant percentage of the whole during each year or other time period; when the increase in quantity over time is plotted, this type of growth yields a curve shaped like the letter J.

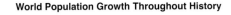

World Population Growth Throughout History

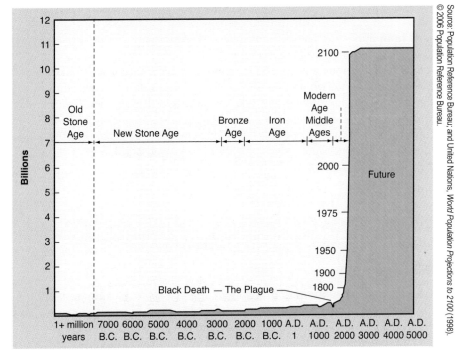

Source: Population Reference Bureau; and United Nations, *World Population Projections to 2100* (1998). © 2006 Population Reference Bureau.

Figure 1.16 Human population growth curve. Source: As shown and http://www.prb.org/Publications/GraphicsBank/PopulationTrends.aspx.

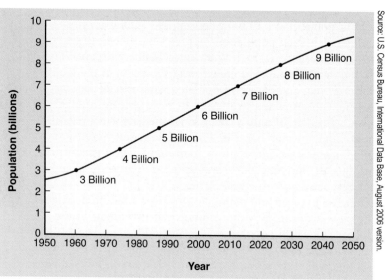

World Population: 1950–2050

Source: U.S. Census Bureau, International Data Base, August 2006 version.

Figure 1.17 Latest projections of world population from the U.S. Census Bureau showing that world population increased from 3 billion in 1959 to 6 billion by 1999, a doubling that occurred over 40 years. The Census Bureau's latest projections imply that population growth will continue into the twenty-first century, although more slowly. The world population is projected to grow from 6 billion in 1999 to 9 billion by 2042, an increase of 50 percent in 43 years.

carrying capacity
The maximum population size that can be regularly sustained by an environment.

exponential population growth with our finite resources. Scientists are worried that it will be impossible to supply resources and a high-quality environment for the billions of people who will be added to the planet in the twenty-first century. Some scientists suggest that the present population is already above our planet's **carrying capacity**, which is the maximum number of people the Earth can hold without causing environmental degradation that reduces the ability of the planet to support the population.

Compounding the problem is the fact that technologically advanced societies consume greater quantities of resources that are only partially deemed essential, with the remaining consumed for convenience or luxury. Therefore, to reach our goal of achieving a higher standard of living, our resource consumption increases substantially per person, not to mention the vast quantities of wastes generated. Human links to environmental degradation, due to our thirst for using more resources, can greatly increase the risks of natural hazards. For example, deforestation for wood production and land development (also a resource) can lead to more landslides and flooding. Therefore, population growth and resource consumption leads to more loss of life and property from individual hazard events, as well as more hazards or higher-magnitude hazards due to resource development.

One of the obvious ways to reduce natural disasters and catastrophes would be population control, for which there is no easy answer. Left unabated, some scientists predict that population growth will take care of itself through wars (mostly over resources such as land, fuel, minerals, water, etc.) and other catastrophes such as famine, disease, and ecosystem collapses. Others believe we will find better ways to control population through increased education, improved regulation of resources and space, and being environmentally friendly.

Impacts from Natural Hazards: Human Fatalities, Economic Losses, and Environmental Damage

Human Fatalities

Increasing world populations in urban areas and coastal regions has resulted in more people living in hazardous areas. As a result, more than 255 million people globally were affected by natural disasters each year between 1994 and 2003. During the same period, these disasters claimed an average of 58,000 lives annually, with a range of

TABLE 1.2	Natural Disasters, 1980–2006, 10 Deadliest Natural Disasters				
Date	Loss Event	Region	Overall Losses* U.S. $m	Insured Losses* U.S. $m	Fatalities
December 26, 2004	Earthquake, tsunami	South Asia	10,000	1,000	210,000
April 29-30, 1991	Cyclone, storm surge	Bangladesh	3,000	100	139,000
October 8, 2005	Earthquake	Pakistan, India	5,200	5	88,000
July-August 2003	Heat wave	Europe	13,000		50,000
June 21, 1990	Earthquake	Iran	7,100	100	40,000
December 8-19, 1999	Flash flood, landslides	Venezuela	3,200	220	30,000
December 26, 2003	Earthquake	Iran	500	19	26,200
November 13-14, 1985	Volcanic eruption	Colombia	230		25,000
December 7, 1988	Earthquake	Armenia	14,000		25,000
August 17, 1999	Earthquake	Turkey	12,000	600	15,000

© 2007 Münchener Rückversicherungs-Gesellschaft, Geo Risks Research, NatCatSERVICE.

* Original values as of December 2006.

10,000 to 123,000. In the year 2003 alone, 1 in 25 people worldwide was affected by natural disasters.

The 10 deadliest disasters between 1980 and 2006 are listed in Table 1.2 and the 10 deadliest historical disasters are listed in Table 1.3. Notice that as deadly as the 2004 Asian disaster was, it was not the deadliest in human history. Also notice that the greatest disasters occurred where human population density is high in a belt of poorer countries running from China and Bangladesh through India, northwestward into Iran and Turkey. Population growth, urbanization, and the inability of poor populations to escape from the vicious cycle of poverty make it more likely that there will be a continued, and increased, number of people who are vulnerable to natural hazards.

Economic Losses

The deaths and injuries caused by natural disasters are what grab our immediate attention, but there are economic losses as well that are increasing at a staggering rate (Figure 1.18). During the last decade, disasters caused an estimated $67 billion in damages per year on average, with a maximum of $230 billion and a minimum of $28 billion. The economic cost associated with natural disasters has increased 14-fold since the 1950s.

The list of most expensive events is dominated by hurricane storm events along coastal regions (Table 1.4). Tables 1.2 and 1.4 show that there is a disparity between the number of deaths in poorer countries and economic loss among the wealthy countries. The poorer, underdeveloped countries suffer increasing numbers of deaths, whereas developed countries suffer greater economic losses. In both the developed and undeveloped countries, population growth once again has led to the increases, with more people living in more dangerous places. However, in developed countries the number

TABLE 1.3	Ten Deadliest Natural Disasters			
Rank	Event	Location	Date	Death Toll (Estimate)
1.	1931 Yellow River flood	Yellow River, China	Summer 1931	850,000-4,000,000
2.	1887 Yellow River flood	Yellow River, China	September-October 1887	900,000-2,000,000
3.	1970 Bhola cyclone	Ganges Delta, East Pakistan	November 13, 1970	500,000-1,000,000
4.	1938 Huang He flood	China	1938	500,000-900,000
5.	Shaanxi earthquake	Shaanxi Province, China	January 23, 1556	830,000
6.	1839 India cyclone	Coringa, India	November 25, 1839	300,000+
7.	1642 Kaifeng flood	Kaifeng, Henan Province, China	1642	300,000
8.	2004 Indian Ocean earthquake/tsunami	Indian Ocean	December 26, 2004	225,000-275,000
9.	Tangshan earthquake	Tangshan, China	July 28, 1976	242,000*
10.	1138 Aleppo earthquake	Syria	1138	230,000

* Official government figure. Estimated death toll as high as 655,000.

TABLE 1.4	Natural Disasters 1980–2006, 10 Costliest Natural Disasters Ordered by Insured Losses				
Date	Loss Event	Region	Overall Losses* U.S. $m	Insured Losses* U.S. $m	Fatalities
August 25-30, 2005	Hurricane Katrina	USA	125,000	61,000	1,836
August 23-27, 1992	Hurricane Andrew	USA	26,500	17,000	62
January 17, 1994	Earthquake	USA: Northridge	44,000	15,300	61
September 7-21, 2004	Hurricane Ivan	USA, Caribbean	23,000	13,000	125
October 19-24, 2005	Hurricane Wilma	Mexico, USA	20,000	12,400	42
September 20-24, 2005	Hurricane Rita	USA	16,000	12,000	10
August 11-14, 2004	Hurricane Charley	USA, Caribbean	18,000	8,000	36
September 26-28, 1991	Typhoon Mireille	Japan	10,000	7,000	62
September 1-9, 2004	Hurricane Frances	USA, Caribbean	12,000	6,000	39
December 26, 1999	Winter storm Lothar	Europe	11,500	5,900	110

Source: Münchener Rückversicherrungs-Gesellschaft Geo Risks Research, NatCat SERVICE.
* Original values as of December 2006.

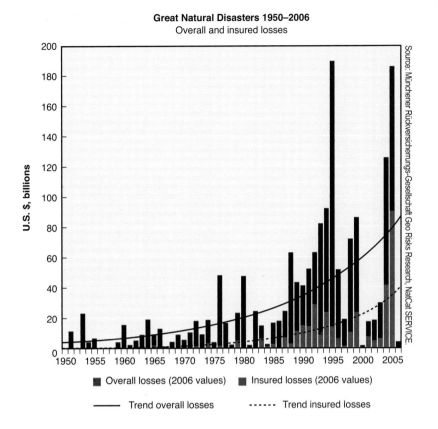

Great Natural Disasters 1950–2006
Overall and insured losses

Source: Münchener Rückversicherungs-Gesellschaft Geo Risks Research, NatCat SERVICE.

■ Overall losses (2006 values)   ■ Insured losses (2006 values)

—— Trend overall losses   - - - - Trend insured losses

Figure 1.18 Global overall (economic) and insured losses.

of deaths has not increased due to better prediction, forecasting, and warning systems, as well as safer buildings. Therefore, disasters cause the loss of money and lives wherever they occur, but wealthy countries always lose more money while poorer countries always lose more lives. Scientific predictions and evidence indicate that global climate change will increase the number of extreme events, creating more frequent and intensified natural hazards such as floods, storms, hurricanes, and droughts.

Environmental Damage

Natural hazards not only affect humans, but also affect and shape the environment and biodiversity. In August 1992 Hurricane Andrew had a strong economic effect not only on the city of Homestead, Florida, but also on the ecosystem of the nearby Everglades and other coastal waters. Despite the destruction in Homestead, timber losses during the winter storms of 1993–1994 cost insurers more than the total payout from Hurricane Andrew. The drought in 1988 in the Midwest and the 1993 floods both had impacts on riverine ecosystems as well as downstream coastal ecologies of the Gulf of Mexico, by changing water salinity and the addition of toxins and other pollutants washed out upstream by the floods. The floods also accelerated the spread of invasive species, like the Zebra mussels, in the Mississippi watershed. These few examples show that strategies for protecting renewable resources and biodiversity must be taken into account when looking at the effects of natural hazards.

Human Vulnerability and Natural Hazards

Each year natural hazards are responsible for causing significant loss of life and staggering amounts of property damage, and they continue to adversely affect millions of people every day. Statistics published by the International Strategy for Disaster Reduction, a subsidiary of the United Nations, show that there is an increasing trend

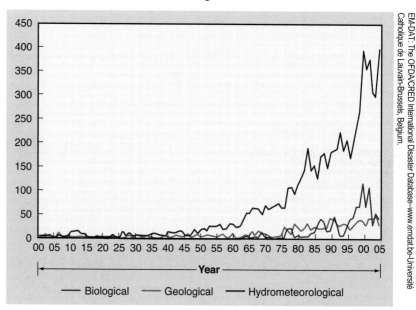

Number of Natural Disasters Registered in EMDAT 1900–2005

EM-DAT: The OFDA/CRED International Disaster Database–www.emdat.be–Université Catholique de Lauvain-Brussels, Belgium.

Figure 1.19 Number of natural disasters per year over the last 105 years. Landslides and floods are often associated with the atmospheric water cycle and are included in the hydrometeorlogical category.

of natural disasters, especially over the past decade (**Figure 1.19**). However, scientists don't see Earth as becoming more violent; instead, because of population and land management trends in our society, humans are increasingly becoming more *vulnerable* to the natural hazards, and more are therefore being reported. A summary of some reasons for the increase in disaster and catastrophe losses includes:

- Global population growth. In 1960, for example, there were 3 billion people living on Earth, and by 2000 there were 6 billion—a doubling in only 40 years.
- Increased settlement and industrialization of hazardous lands subject to floods, landslides, hurricanes, earthquakes, wildfire, volcanoes, and other hazards.
- Concentration of population and values in conurbations (when towns expand sufficiently that their urban areas join up with each other) and the emergence of numerous megacities—even in hazardous regions (e.g., Tokyo, with 30 million inhabitants).
- The increased demand for natural resources and rising standard of living in nearly all countries of the world produces growing accumulations of wealth and the need for increased insurance. Losses thus escalate in the event of a disaster or catastrophe.
- The vulnerability of modern societies and technologies in the form of structural engineering, lifelines (water, sewer, electrical, etc.), transportation services, and other networks that are both fragile and costly to repair when damaged by natural hazards. Such effects are mostly due to lack of scientific understanding or education and lack of awareness of the hazards.
- Global changes in environmental conditions, climate change, water scarcity, and loss of biodiversity, mostly due to human activities and interactions with natural hazards.
- Globalization of the world economy now makes us all vulnerable to disasters wherever they occur, for example, by shutting off supplies of goods.

Therefore, natural hazard events are only classified as disasters or catastrophes when people or their properties are adversely affected. An earthquake in the Gobi Desert or gales in the Antarctic are not natural catastrophes if they do not have any impact on human life or property. Conversely, a natural hazard event that is anything but extreme, like the natural flooding of a stream, can quickly become a disaster in a region that is densely populated and poorly prepared.

Studying Natural Hazards: Fundamental Principles for Understanding Natural Processes

People are usually surprised to learn that every day the Earth experiences earthquakes, volcanic eruptions, landslides, floods, fires, meteor impacts, extinctions, and storms. Natural disasters are often perceived as being "acts of God," with little causal relationship to human activities. However, a growing body of evidence points to the effects of human behavior on the global natural environment and on the possibility that certain types of natural disasters, such as floods, may be increasing as a direct consequence of human activity. Natural hazards, which the geologic record shows have been continuously shaping our planet for millions of years (and operating for billions of years), are not problems to be solved but are a critical part of how the Earth functions. Ecosystems and individual species have evolved to coexist with these hazards. Natural hazards become disasters only when natural forces collide with people, structures, or other property. Understanding how the Earth works and our relationship with it is the first step in reducing the impacts of natural hazards. The following concepts serve as a basic framework for our study and understanding of different natural hazards discussed in later chapters.

The Earth System and the Human Connection

Earth itself is a complex and dynamic planet, always in motion, and not static or unchanging. Earth's major components of land, water, air, and life do not exist as isolated entities, but are interconnected to form a unified dynamic whole. Even though we can study each component individually, each is related to the continuously interacting entity that we call the *Earth system*. We observe the most obvious connection between Earth's components when natural disasters strike. Events such as destructive volcanic eruptions, disastrous landslides, devastating earthquakes, tsunamis, hurricanes, and floods affect people in obvious ways. Even though we cannot prevent these natural hazards from happening, the more we study and understand how they operate, the more we will be able to predict them and mitigate their effects. Scientists who study the interconnections of Earth's components use an interdisciplinary approach known as **Earth system science**, which aims to study our planet as a *system* composed of numerous interconnecting subsystems governed by natural laws.

Earth system science
Study of our planet as a system composed of numerous interconnecting subsystems governed by natural laws.

A system is any assemblage or combination of interacting components that form a complex whole. A system is a group of any size composed of interacting parts within which we are interested in studying the flow of energy and matter. In our home we can talk about the heating system, electrical system, or plumbing system. Each functions as an independent unit, transferring material and energy from one place to another, and each is driven by a force that makes the system operate. A good example of a system is an automobile with its various components (or subsystems), such as the engine, transmission, steering, brakes, and radiator, that all function together as a whole when energy is supplied (gasoline in this case). The many different component parts mutually adjust to function as a whole, with changes in one part bringing changes in the others. For example, if the oil pump in the car malfunctions and causes less oil lubricant to enter the cylinders, then increased friction will cause more heat to be generated in the engine, in turn causing the cooling system to adjust to the need for additional heat removal.

The overall nature of the Earth system is similar to an open system because it exchanges both energy and matter with its surroundings. Earth receives energy from the sun and radiates it back into space, along with energy that leaves the Earth's interior. However, most of the Earth's materials (such as rocks and water) are neither gained nor lost to space but are continuously transferred and recycled within the Earth system. Thus, we most often view Earth as a self-contained **closed system** with regard

closed system
A system in which no matter or energy can leave or enter from the outside.

to matter (even though some meteors fall to Earth and small amounts of gas escape into space). The solar energy that enters this closed system causes the air and water to move and flow in patterns that can be hazards, such as hurricanes and wind storms, while heat energy from within the Earth causes motions that result in earthquakes, volcanic eruptions, and drifting continents.

Human Activities and Hazards

dynamic equilibrium
The state in which the action of multiple forces produces a steady balance, resulting in no change over time.

Natural systems tend toward balance among opposing factors or forces. The interactions of these energy sources produce and maintain a balanced state called **dynamic equilibrium**. A change in one part of the system will tend to become balanced by a change or changes in another part of the system to maintain overall equilibrium. Therefore, this is not a static equilibrium but a dynamic one, and human activities can cause or accelerate changes in natural systems.

The Earth's land surface is an important resource for people, plants, and the other animals with which we share the planet. Humans use land for housing, transportation, agriculture, manufacturing, mineral and energy production, deposition of waste products, and aesthetics. The increase in human population has increased the demand for land use, such as for urban and agricultural purposes. This has led to an increasing rate of human-induced change to our planets surface. Human activity, including agriculture, urbanization, and mining, is now at the level where we move as much soil and rocks on an annual basis as do natural processes such as river transport. It appears that human activities are the most significant processes shaping the surface which in turn upsets the dynamic equilibrium of Earth's system and produces undesired hazards. For example, changes such as altering the steepness of hillslopes and removing vegetation can lead to unwanted increases in landslide and flooding events. Deforestation can lead to soil erosion (loss of a natural resource) and increase landslides and flooding.

frequency
Occurrence of specific events. Used in interpreting the past record of events to predict occurrences of that event in the future.

The impact of humans on the environment is broadly proportional to the size of the population as well as technological advances. For example, air pollution can lead to the slow onset of hazards. The exhaust from one automobile pollutes only the air in its immediate vicinity, which can be dispersed through the atmosphere with negligible global impact. However, the collective exhaust from millions of automobiles has a global impact. Burning of fossil fuels over the past 250 years of increasing industrialization has caused measurable increases in atmospheric pollutants, such as the greenhouse gas CO_2, leading to global climate change. Next to human population growth, climate change has the greatest impact on the environment. Global warming will cause sea levels to rise as glaciers melt, leading to increased coastal erosion. Warming of the oceans will produce a warmer atmosphere and increase the **frequency** and severity of weather-related thunderstorms, tornadoes, and hurricanes. Changes in climate patterns can affect food-producing areas and increase precipitation in some areas (producing flooding and landslides), while expanding desert areas in others (causing droughts). Many of these changes could lead to population shifts, which could initiate wars or social and political upheavals. We must not forget that humans are part of the Earth system and that our activities can produce changes with potential consequences for us and our planet.

Time Perspectives: Geological and Historical Records

geologic time scale
A relative time scale based upon fossil content. Geological time is divided into eons, eras, periods, and epochs.

Geology is the scientific study of the Earth and its history. One of the important concepts that sets geology apart from most other sciences is the consideration of vast amounts of time. Geologists use the **geologic time scale** in describing the timing and relationships between events that have occurred during the history of the Earth (**Figure 1.20**). Many of the natural hazard processes have been operating on Earth long

Era	Period		Epoch	Approximate Ages (in millions of years)
Cenozoic	Quaternary		Recent Pleistocene	
				— 16 —
	Tertiary		Pliocene Miocene Oligocene Eocene Paleocene	
				— 66 —
Mesozoic	Cretaceous		Late Early	
				—144—
	Jurassic		Late Middle Early	
				—208—
	Triassic		Late Middle Early	
				—245—
Paleozoic	Permian		Late Early	
				—286—
	Carboniferous	Pennsylvanian	Late Middle Early	
				—320—
		Mississippian	Late Early	
				—360—
	Devonian		Late Middle Early	
				—408—
	Sllurian		Late Middle Early	
				—438—
	Ordovician		Late Middle Early	
				—505—
	Cambrian		Late Middle Early	
				—545—
Precambrian	Locally divided into Early Middle and Late			
				— 3,900 + —

Figure 1.20 The Geologic time scale.

before humans populated the planet. Geologically, humans have been here for a very short period of time compared to the 4.6-billion-year-old Earth. The most primitive human-type fossil remains are no older that 3 to 4 million years. Modern humans (*Homo sapiens*) evolved during the Pleistocene Epoch less than half a million years ago and migrated out of their African birthplace as recently as 80 thousand years ago to occupy Europe and beyond. On a geologic time scale, this is a very short time for humans to populate the Earth (less than 0.06 percent of the age of the Earth). Needless to say, humans have had an enormous impact on the surface of the Earth—an impact that is far out of proportion with the length of time we have occupied the planet.

The geologic record is full of examples of disasters that humans have never witnessed during historical times—for example, asteroid collisions large enough to cause the extinction of the dinosaurs, or violent caldera eruptions such as that at Yellowstone. Knowing that these hazards happened in the past means they certainly will happen again, and we should not be surprised when they do. Studying the events of the geologic past is vital to understanding the frequency with which we can expect hazardous events to occur.

The Scientific Method in Studying Earth's Hazards

Natural hazards, as the term implies, are a natural phenomenon and so can be studied using scientific principles. The Earth system is governed by natural laws that provide the keys to understanding the landscapes we live on and the processes that formed them. Our understanding of natural processes operating on Earth today and in the past has been influenced by the use of the scientific method for studying natural phenomenon.

The **scientific method** is based on logical analysis of collected data and observations to solve problems. From initial observations, scientists develop a tentative explanation to explain the cause, or "why," of the phenomenon being studied, which is known as a **hypothesis**. To see if the hypothesis is an explanation of the observations, it must be tested with new observations and experiments. A stated hypothesis should always be testable since the science of discovery evolves through continual testing with new observations and experiments. Alternate hypotheses are usually developed to test other potential explanations for the observed phenomenon, which are also tested. If the observations and experiments are inconsistent with a hypothesis, then they are rejected as an explanation for the phenomenon, or the hypothesis is revised and new tests are developed. If a hypothesis continues to be supported by continued testing over a long period of time, it becomes known as a **theory**. A theory, then, is an explanation for some natural phenomenon that has a large body of supporting evidence and has been tested for validity through the scientific method.

Continued testing of a theory that is irrefutably correct leads to the statement of a **law** (or **principle**). Most disciplines have laws, such as those in biology, chemistry, economics, geology, and physics.

scientific method
A systematic way of studying and learning about a problem by developing knowledge through making empirical observations, proposing hypotheses to explain those observations, and testing those hypotheses in valid and reliable ways.

hypothesis
A tentative explanation to explain the cause, or why, of the phenomenon being studied.

theory
A comprehensive explanation of a given set of data that has been repeatedly confirmed by observation and experimentation and has gained general acceptance within the scientific community.

law (or principle)
The unvarying sequence of a set of naturally occurring events.

Hazard Mitigation and Management: Reducing the Effects from Natural Hazards

The effects of natural hazards will continue to increase unless concerted efforts are made to reduce them. At stake is our physical, economic, and social well-being. But these are increasingly at risk as the world population grows and concentrates in hazard-prone areas. Large numbers of buildings, critical facilities, and lifelines (utilities) remain or are still being constructed that are vulnerable to natural hazards. Natural disasters and catastrophes from Earth's natural hazards cannot always be prevented, although some are avoidable with foresight and planning. Most of the planet's seemingly deadly natural forces are largely the result of people making bad choices concerning where to live, deliberate placement in harm's way, or failure to see and/or act on warning signs. The main goal of preventing or avoiding a natural disaster or catastrophe is to understand and assess the natural processes at work that lead to natural hazards. Human awareness through education is a vital step toward knowing and understanding the risks involved.

To address the increasing number of losses (both to life and property) from natural hazards and foster an understanding of natural hazard processes, the United Nations (UN) declared 1990 through 2000 as the International Decade for Natural Disaster Reduction. The UN goals were to:

- Improve the capacity of each country to mitigate (which means to reduce the effects of something) the effects of national disasters.
- Apply existing scientific and technological knowledge.
- Foster advances in science and technology.
- Disseminate new and existing technical information.
- Develop measures for the assessment, prediction, prevention, and mitigation of natural disasters through technological assistance and technology transfer, demonstration projects, education and training, and evaluation of program effectiveness.

To meet this goal and develop a strategy for the United States, the U.S. Subcommittee on Disaster Reduction (under the Office of Science and Technology Policy) defined and grouped hazard reduction and disaster management activities under nine broad categories:

1. Research and development
2. Hazard identification
3. Risk assessment
4. Risk communication
5. Prediction
6. Mitigation
7. Preparedness
8. Response
9. Recovery

This strategy builds on existing programs and activities within many federal agencies, such as the Federal Emergency Management Agency (FEMA), the United States Geological Survey (USGS), and the National Oceanic and Atmospheric Administration (NOAA). Even though we are interested here in natural hazards, these concepts can be used in dealing with technological hazards as well. The reader will find, and should keep in mind, that these topics are recurring themes throughout the chapters of this book on the natural hazards. We can group these nine concepts into three groups: *before, during,* and *after natural hazards strike.* It is now recognized that more concerted efforts must be placed on the activities before a natural hazard strikes, even though the actions during and after a hazard strikes remain important.

Before a Natural Hazard Strikes

Time and again, when hazards strike, we find that areas that had properly prepared for the hazard suffered fewer losses and recovered more quickly than those that did not have preparedness programs in place. Being prepared requires seven steps: (1) hazard process research and development, (2) hazard identification, (3) risk assessment, (4) risk communication, (5) mitigation; (6) prediction, and (7) preparedness, described as follows:

1. Hazard process research and development. The science activities dedicated to improving understanding of the underlying processes and dynamics of each type of hazard. This includes fundamental and applied research on geologic, meteorological,

epidemiological, and fire hazards; development and application of remote-sensing technologies, software models, infrastructure models, and organizational and social behavior models; emergency medical techniques; and many other science disciplines applicable to all facets of disasters and disaster management.

Knowledge of how hazards operate is the fundamental first step in understanding how to reduce impacts to humans and the environment. Basic science research leads to fundamental breakthroughs in understanding natural processes, such as the theory of plate tectonics that greatly improved our knowledge of volcanic and earthquake hazards. Long-term scientific studies of coupled ocean–atmosphere oscillations enabled NOAA in 1997 to predict a powerful El Niño during 1997–1998 that produced large rainfall-induced landslide events in California. NOAA and USGS worked with the California Office of Emergency Services to prepare landslide hazard maps showing former landslide deposits and debris-flow source areas along with dangerous rainfall thresholds for the San Francisco Bay area. This prediction allowed individuals and emergency responders to prepare for the extreme hazard event. No loss of life occurred from landslides in this area, as compared to 25 deaths in comparable storms from an El Niño event in 1982.

2. **Hazard identification.** Determining which hazards threaten a given area. This includes understanding an area's history of hazard events and the range of severity of those events. The continuous study of the nation's active faults, seismic risks, and volcanoes is included in this category, as are efforts to understand the dynamics of hurricanes, tornadoes, floods, droughts, and other extreme weather events.

Since natural disasters are not random "acts of God" but are naturally repetitive events caused by natural processes, we can identify potential hazards and assess their probability of occurring in the area. Some events may not occur more than once in a person's lifetime or even for several generations. Also, many events such as volcanic eruptions—with hundreds to thousands of years elapsing between eruptions—are not on the same timeline as humans (average life expectancy of 70 years), thus we often do not see the danger. Therefore we become complacent by not knowing there is a hazard until disaster strikes. It is important, therefore, to take our working knowledge of natural hazards (from hazard process research) and identify their potential occurrences. Also known as *hazard assessment*, hazard identification consists of determining *where* and *when* hazardous events occurred in the past and how they affected the area.

Hazards vary widely in their power, and the extent to which hazards impact an area is partly a function of their **magnitude**, the energy released, and the interval between occurrences, or frequency. All other factors being equal, larger-magnitude events occur less frequently but cause more damage than those of a smaller magnitude. The **recurrence interval** is how frequently, on average, a hazard event of a certain magnitude occurs. This information is important and useful to planners and public officials responsible for making decisions in the event of a possible disaster.

Geologists are well trained to consider the long- and short-term history of an area and assess natural hazards. Geologists understand that if a destructive event occurred in this area before, then it could possibly happen again. Therefore, hazard assessment studies usually try to: (1) quantify the recurrence of each hazard; (2) increase the amount of detail available about the location, frequency, and magnitude of the physical and biological effects on maps; and (3) ensure that assessments consider possible links between hazards (for example, the triggering of floods by hurricanes or an increase in wildfires during a drought).

3. **Risk assessment.** Determining the impact of a hazard or hazard event on a given area. This includes advanced scientific modeling to estimate loss of life, threat to public health, structural damage, environmental damage, and economic disruption that could result from specific hazard event scenarios. Risk assessment takes place both before and during disaster events.

magnitude
The size or scale of an event, such as an earthquake.

recurrence interval
(1) A statistical expression of the average time between events equaling or exceeding a given magnitude. (2) The average time interval, usually in years, between the occurrence of an event of a given magnitude or larger.

A risk assessment tells us something about the loss expected from each hazard, which elements specifically are at risk (population, property, natural resources, environment), and how easily the various elements can be damaged by each hazard. Risk assessments include estimates of the overlapping vulnerability of new technologies and the way and extent to which human actions may influence the risk. Depending on the vulnerability of individuals or a community to a hazard of a certain magnitude, risk is essentially a hazard considered in terms of its recurrence interval and expected costs. Thus the greater the hazard, the shorter its recurrence interval; the more vulnerability, then the greater the risk. The higher the risk, the more critical that the hazard-specific vulnerabilities should be targeted by mitigation and preparedness efforts. However, if there are no vulnerabilities there will be no risk—for example, an earthquake occurring in a desert where nobody lives.

4. Risk communication. Public outreach, communication, and warning at every stage of hazard management. Risk communication includes raising public awareness and instigating behavioral change in the areas of mitigation and preparedness; the deployment of stable, reliable, and effective warning systems; and the development of effective messaging for inducing favorable community response to mitigation, preparedness, and warning communications.

Once a risk has been identified, people who live and work in hazard-prone areas need to know the nature and probable impact of natural hazards and what they can do to protect themselves before as well as during and after a natural disaster occurs. This information is conveyed through schools, special television programs, publications, workshops, and many other channels. The challenge here is to make sure that accurate information is presented by highly credible sources and understood by the public leading to preventative and precautionary actions. Federal agencies, working with state and local governments, usually try to promote the importance of disaster prevention and mitigation programs.

5. Mitigation. Sustained actions taken to reduce or eliminate the long-term risk to human life and property from hazards based on hazard identification and risk assessment. Examples of mitigation actions include planning and zoning to manage development in hazard zones, stormwater management, fire fuel reduction, acquisition and relocation of flood-prone structures, seismic retrofit of bridges and buildings, installation of hurricane straps, construction of tornado-safe rooms, and flood-proofing of commercial structures.

Once the risk for each hazard is known, mitigation efforts attempt to prevent hazards from developing into disasters or to reduce the effects of disasters when they occur. Thus they are designed to reduce our vulnerability to a natural hazard. The mitigation phase differs from the other phases of response and recovery because it focuses on long-term measures for reducing or eliminating risk. From disaster to another, we are now seeing that the benefits of mitigation before an event greatly affect the losses after the event, and that we need to move more in this direction to reduce the impacts of hazards. Like the adage "an ounce of prevention is worth a pound of cure," *mitigation is the most cost-efficient method for reducing the impact of natural hazards.* However, progress toward this end has been slow. The implementation of mitigation strategies can also be considered a part of the recovery phase if applied after a disaster occurs; however, any actions that reduce or eliminate risk over time are still considered mitigation efforts.

Mitigation measures can be divided into structural methods (engineering solutions) or non-structural (land use management) methods depending on the type of hazard risk. *Structural methods* include building codes that specify materials, construction techniques that minimize damage to structures during a hazard event, and construction of large-scale structural interventions such as flood levees and seawalls. Historically, we have mostly employed large engineering projects in order to try to control natural processes for the protection of lives and property. *Nonstructural methods* include restrictive zoning, relocation, and land abandonment. Zoning laws can greatly reduce loss of lives and structures if properly implemented and enforced. Zoning can limit population density in potentially hazardous areas as well as specify the kind of structures permitted in a hazardous area. However, in many hazard-prone areas, housing developments, utilities, schools, fuel storage tanks, and hospitals still sprawl across the land (for example, across active fault lines in earthquake-prone parts of California). This construction occurred because the hazard was not realized accurately at the time, or the potential risk was glossed over. Now, corrective measures are too late and will have to wait until the recovery phase after an event, so in the meantime structural measures are usually employed.

In the long run, the best solution is to abandon the use of certain hazardous areas and adapt them for public use, by constructing parks and athletic fields on river floodplains. Examples include Santa Barbara, California, Rapid City, South Dakota, and Scottsdale, Arizona (**Figure 1.21**). Unfortunately, these actions are usually taken after disasters have struck, but still are examples of proactive land-use management. The designation of public recreation, wildlife, or wilderness areas with minimal permanent structures is a prudent use of land in hazardous areas. This method greatly reduces the risk to people, avoids the expense of evacuations before an event, avoids emergency response expenses, and minimizes reconstruction costs afterward. An added benefit of this approach is that it affords our growing population of people beautiful recreation areas and needed open spaces.

Courtesy of David M. Best.

Figure 1.21 Indian Bend Wash Greenbelt in Scottsdale, Arizona, lies within the drainage of an area that floods every year from thunderstorm activity. Minimal cleanup is required following a storm to return the area to its normal use and appearance.

Land-use planning methods are an obvious and cost-effective way of avoiding many disasters or catastrophes. However, it is sometimes difficult to impose land-use restrictions because areas are already heavily populated, or because people want to live as close as possible to scenic areas such as coasts, rivers, and mountains. Citizens often resent being told they cannot live in a certain area and oppose attempts at restrictions because they feel it infringes on their property rights, and attempts at regulations in the public interest often are met with political and legal opposition.

6. **Prediction.** Predicting, detecting, and monitoring the onset of a hazard event. Federal agencies utilize weather forecast models, earthquake and volcano monitoring systems, remote-sensing applications, and other scientific techniques and devices to gather as much information as possible about the what, when, and where of a potential hazard, as well as the severity of each threat.

Predicting the time, place, and severity of a hazard event saves lives and reduces losses when it is followed by timely and effective communication of warnings to the public. Some disaster events can be predicted fairly accurately, such as the arrival of coastal hurricanes, tsunami, certain storms, and some river floods; some precursor events have led to better volcanic eruption predictions. Our capability to predict natural hazards has grown considerably as a result of scientific and technological advances in recent decades. However, variability still exists from hazard to hazard, and we still need to monitor and observe hazards and possible precursors to close the gaps and translate predictions into easily understood warnings that can be communicated quickly to the public.

7. **Preparedness.** The advance capacity to respond to the consequences of a hazard event. This means having plans in place concerning what to do and where to go if a warning is received or a hazard is observed. Communities, businesses, schools, public facilities, families, and individuals should have preparedness plans.

Emergency plans must be prepared and tested many times over before a natural hazard strikes. Even the most comprehensive and well-constructed contingency plans should be tested and evaluated through field exercises and study so that planning activity voids can be filled. Although planning for small, frequent natural hazards seems challenging and expensive at the time, planning for infrequent disasters is more difficult but equally important.

During a Natural Hazard Strike

Even while a natural hazard is in progress, actions can be taken to reduce the impact; the response phase takes place during a natural hazard event.

8. Response. The act of responding to a hazard event. Hazard response activities include evacuation, damage assessment, public health risk assessment, search and rescue, fire suppression, flood control, and emergency medical response. Each of these response activities relies heavily on information and communication technologies.

During the initial response following the occurrence of a hazard first responders arrive and perform their acts of saving lives and stabilizing the injured. A great deal of training goes into preparing all the various emergency services that are required to operate in a coordinated, efficient manner. Search and rescue teams move into high gear in order to reach injured people in the shortest amount of time. The role of individual citizens is also important as they can provide assistance with many tasks that do not require specific technical knowledge. Sandbagging swollen rivers requires the help of many people, but a smaller number of technical assistants are needed to oversee the task.

After a Natural Hazard Strikes

As the emergency response period following a disaster wanes, individuals and communities enter a long period of recovery, redevelopment, and restoration.

9. Recovery. Activities designed to restore normalcy to the community in the aftermath of a hazard event. Recovery activities include restoring power lines, removing debris, draining floodwater, rebuilding, and providing economic assistance programs for disaster victims. As with response, the recovery process relies heavily on the availability of up-to-date data and information about the various community sectors, and on the technology to obtain and communicate that information.

The primary goal of the recovery phase is to restore an area affected by a disaster to its pre-disaster condition. Once the initial emergencies are dealt with, a more calculated response is used to make decisions that involve the clean-up and possible reconstruction of damaged or destroyed property. Undoubtedly infrastructure such as electrical distribution, water systems, and transportation need to be restored to their normal operational levels (**Figure 1.22**). Federal and state financial aid are often available to help complete these needs.

Figure 1.22 Union Pacific Railroad workers repairing an area washed out by flooding near a street crossing on Galveston Island, Texas. Tidal surge and hurricane force winds brought by Hurricane Ike in September 2008 damaged infrastructure across the region.

Disasters are sometimes seen as opportunities to rebuild a community or relocate it so that it will not suffer great losses in the future from a repeat performance of a natural hazard. The challenge is that there must be a strong commitment from citizens and the local government of the affected area to rebuild in such a manner that the best hazard mitigation methods are employed.

However, the window of opportunity for change is very narrow following a disaster. Following the devastation of New Orleans by Hurricane Katrina, scientists studying the circumstances surrounding the catastrophe suggested it was a good time to relocate the city instead of rebuilding it. New Orleans is built on a floodplain area of the Mississippi River, and levees were constructed along the river to protect it from flooding. The levees stopped the floods as well as the deposition of sediments that built up the delta to begin with. This resulted in the city's land sinking below sea level, which required the building of more levees around the entire city and pumping of water to keep the city dry. In the meantime, the Mississippi River—which is the lifeblood of the city—naturally changes its course over time across the delta and tends to move westward away from the city. To prevent this from happening, the U.S. Army Corps of Engineers spends billions of dollars in attempts to prevent this natural event from happening and thus keep the river flowing toward the city. The city continues to sink below sea level. Higher levees must be built and massive pumps must keep operating (using natural resources) to keep it dry. Is it time to move to higher ground? The citizens and local government decided to stay and rebuild in hopes that another massive hurricane is not on the horizon. Studies have shown that major hurricanes hit the central Gulf Coast region every 100 to 200 years, a time span long enough to not be a concern to most people.

Lessons and Challenges

Today we hear more about natural hazards and have seen more vividly the negative effects these hazards have on people, property, and the environment. We cannot totally avoid natural hazards, but how we perceive and react to them determines our destiny. We can act to minimize and reduce their impacts. After all, natural hazards do not become disasters or catastrophes unless the people and communities they touch are not prepared to deal with them. The previously described strategy of reducing the impacts from inevitable natural hazards can be strengthened as we learn more about natural processes and work together to mitigate their hazards. However, best laid plans are not always easily followed and we are not out of the woods yet, so to speak. We still have a long way to go if we are to focus on preparedness and prevention rather than the usual bandage solutions.

Our general approach to disaster management has remained reactive, focusing on relief, followed by recovery and reconstruction. Prevention planning or community preparedness has been poorly funded and has not been a major policy priority. Relief remains media-friendly, action oriented, easy to quantify (e.g. tons of food distributed, number of family shelters shipped), and readily accountable to donors as concrete actions in response to a disaster. With the increase in magnitude of disaster impacts, mostly in poorer developing countries, concern is rising over inadequate preparedness and prevention. Natural disasters create serious setbacks to the economic development of countries. This has been proven time and time again, particularly in the last decade with the devastation caused by Hurricane Mitch in Central America; the Yangtze River floods in China; earthquakes in Turkey, Iran, and Indonesia; and with Hurricane Katrina in the United States. All of these events diverted economic development funds toward reconstruction.

One of the biggest challenges, then, is to change our culture from one of continually responding to and recovering from a natural disaster to one of learning to live with natural hazards through mitigation efforts, which are more cost efficient. The USGS and the World Bank calculated the worldwide economic losses from natural disasters in the 1990s could have been reduced by $280 billion if $40 billion had been invested in disaster preparedness, mitigation, and prevention strategies. Unfortunately, many policies concerning hazards are driven by politics and special interests.

Policymakers must understand that mitigating hazards is prudent both politically and economically. Developers, companies, and local governments often allow or encourage people to move into known hazardous areas. Developers, real estate companies, and some corporations are reluctant to admit to the presence of hazards for fear of reducing the value of land and scaring off potential clients. So the adage "buyer beware" applies here. Many local governments consider any news about hazards to be bad for growth and economy and resist restrictive zoning for fear of losing their tax base. Even developers and private individuals view restrictive zoning as infringement on their property rights and doing as they please with their property. But having good land-use practices can lead to an improved quality of life, new recreation opportunities using open-space lands, and a stronger tax base in the long run.

Instead of mitigating hazards, natural disasters continue to kill and inflict human suffering as well as destroy property, economic productivity, and natural resources. Financial assets are diverted from much needed investments in our future—such as research, education, and the reduction of crime, disease, and poverty. Inaction today regarding natural hazards compromises our safety, economic growth, and environmental quality for future generations. Natural disasters greatly jeopardize our goal of sustainable development, which meets the needs of the present without compromising the ability of future generations to meet their needs. Sustainable development is usually understood to require (1) economic growth, (2) protection of the environment, and (3) sustainable use of ecological systems. However, a fourth criterion of equal importance is that sustainable development must be resilient to the natural variability of the Earth and the solar system. By variability, we mean natural hazards events such as earthquakes, tsunamis, hurricanes, and so on that brutally interrupt our societies and ecosystems. These events have not merely punctuated Earth's history as much as they have defined it, and so understanding them is the first step in our success in coping with them.

We have opportunities to prevent or reduce future natural disasters rather than just picking up the pieces after a hazard unleashes its energy upon us. Just like the adage, "an ounce of prevention is worth a pound of cure," we know that a habit of a good diet and exercise is a first defense against heart disease and stems off the higher cost of medical expenses associated with hospitalization or surgery. The need is the same for us as individuals to develop good habits about how we live and build on this planet. It is hoped that this book will help readers to understand natural hazards and what might occur in the event of a disaster, and to make sound decisions about them.

Summary

Natural hazards are the result of natural processes involving physical, chemical, and biological mechanisms and forces that affect Earth's surface, and they have been operating continuously for millions of years. These processes become hazards to humans when people live or work where these processes occur and become affected by them. Each year, natural hazards become disasters or catastrophes that are responsible for thousands of deaths and billions of dollars in disaster aid, disruption of commerce, and destruction of homes and critical infrastructure.

Hazards can also be categorized as rapid-onset hazards, which expend their energy very quickly, such as volcanic eruptions, earthquakes, floods, landslides, thunderstorms, and lightning. These hazards can develop with little warning and strike rapidly. In contrast, slow-onset hazards, such as drought, insect infestations, disease epidemics, and global warming and climate change, take years to develop. Rapid population growth is forcing development in hazardous areas and has increased the demand for natural resources such as land, water, petroleum, metals, and timber.

Living with natural hazards requires us to keep some basic principles in mind. The dynamic nature of the Earth produces many types of hazards. We must recognize where these events can occur and alter our life styles in order to co-exist with nature. The effects of natural disasters will continue to increase as the population expands into more inhospitable regions on Earth.

References and Suggested Readings

Brown, L. R., C. Flavin, and S. Postel. 1991. *Saving the planet*: New York, W. W. Norton & Co.

National Science and Technology Council Committee on Environment and Natural Resources. Grand challenges for disaster Reduction. A report of the subcommittee on disaster reduction, June 2005.

National Science and Technology Council Committee on Environment and Natural Resources. Natural disaster reduction, a plan for the nation. A report by the committee on Earth and Environmental Sciences, December 1996.

National Science and Technology Council Committee on Environment and Natural Resources. Reducing disaster vulnerability through science and technology. An interim report of the subcommittee on disaster reduction, July 2003.

National Science and Technology Council Committee on Environment and Natural Resources. Reducing impacts of natural hazards. A report by the committee on Earth and Environmental Sciences, May 1992.

Parfit, M. 1998. Living with natural hazards. *National Geographic* 194 (1): 2–39.

Smith, Keith. 2001. *Environmental Hazards: Assessing risk and reducing disaster*, 3rd ed. Routledge, London.

Tobin, Graham A. and Burrell E. Montz. 1997. *Natural Hazards: Explanation and Integration:* The Guilford Press, New York.

Web Sites for Further Reference

http://www.census.gov
http://www.earthobservatory.nasa.gov
http://www.fema.gov
http://www.prb.org/DataFinder.aspx
http://www.un.org/popin/
http://www.usgs.gov

Questions for Thought

1. Explain the differences between natural hazards, disasters, and catastrophes.

2. What natural hazards are most common in the United States?

3. What are the nine hazard reduction and disaster management activities?

4. Contrast and explain the general nature of disaster losses in developed countries versus poor countries.

5. What is the difference between scientific hypotheses and scientific theories?

6. What is the world's population today, and how do growth rates compare to historical trends?

7. Explain the concept of carrying capacity for a species of this planet.

8. What is the difference between rapid-onset and slow-onset hazards?

9. Explain the concept of a system and how humans are a part of the Earth system.

10. How do human activities upset the dynamic equilibrium of the Earth's system?

The Dynamic Earth System

Earth has a wide range of environments that allow life to exist. The interaction of water, the atmosphere, and land generate natural hazards that affect both people and the land.

2

Key Terms

asthenosphere
atmosphere
biogeochemical cycle
biosphere
carbon cycle
conduction
convection
convection cell
convergent plate boundary
core
crust
divergent plate boundary
Earth system
ecosystem
geosphere
gravitational energy
heat transfer
hydrologic cycle
hydrosphere
igneous rock
impact energy
inner core

(*Continued*)

island arc	oceanic trench	sedimentary rock
kinetic energy	outer core	silicate
lithosphere	plate	stratosphere
lithospheric plate	plate boundary	subduction
mantle	plate tectonic	subduction zone
mesosphere	potential energy	transform fault
metamorphic rock	radiation	transform plate boundary
mineral	radioactive decay	troposphere
oceanic crust	rock	volcanic arc
oceanic ridge	rock cycle	weathering

Earth system
Composed of the geosphere, hydrosphere, atmosphere, and biosphere, and all of their components, continuously interacting as a whole.

As mentioned in the first chapter, we live on a planet that is complex and dynamic, and has been in a continuous state of change since its origin about 4.6 billion years ago. The present day landscapes and features we observe and live with are still changing as the result of interactions among Earth's many internal and external components. In studying the Earth, it is clear that the planet can be viewed as a system of interacting parts or spheres (or layers). The geosphere, hydrosphere, atmosphere, and biosphere, and all of their components, can be defined separately, but each is continuously interacting as a whole called the **Earth system** (**Figure 2.1**). Geologists attempt to understand the past, present, and future behavior of the whole Earth system by taking a planetary approach to the study of global change.

The Earth's surface is a major boundary in the Earth system where the four spheres intersect, where energy is transferred, and therefore, where we find volcanoes, landslides, earthquakes, severe weather, and long-term climate changes that impact humans. The dynamic Earth system is powered by energy from two sources, one external and the other internal. Energy from the Sun drives the external processes that occur in the atmosphere, hydrosphere, and at Earth's surface of the geosphere. Processes such as weather, climate, ocean circulation, and erosion of the land are driven by the Sun's energy. Earth's interior portion of the geosphere is the second source of energy. Heat remaining from when the planet formed along with heat produced continuously from decay of radioactive elements, powers the internal processes to produce volcanoes, earthquakes, and mountains. This constant motion of the Earth's spheres will continue as long as the energy sources remain.

Figure 2.1 View of the Great Lakes region of the United States, where the four spheres on Earth interact constantly.

Courtesy NASA.

Earth's Spheres

Atmosphere

Earth is surrounded by a thin envelope of gases called the **atmosphere**. The gases consist of a mixture of primarily nitrogen (78 percent), oxygen (21 percent) and minor amounts of other gases, such as argon (less than 1 percent), carbon dioxide (0.035 percent) and water vapor. These gases are held near Earth by gravity but thin rapidly with altitude. Almost 99 percent of the atmosphere is concentrated within 30 kilometers of Earth's surface and forms an insignificant fraction of the planet's mass (less than 0.01 percent). Despite its modest dimensions, the atmosphere plays an integral role in planet dynamics. It supports life for both animals that need oxygen and plants that need carbon dioxide. It supports life indirectly by controlling climate, acting as a filter, and being a blanket that helps retain heat at night and shield us from the Sun's intense heat and dangerous ultraviolet radiation. The atmosphere is also responsible for transporting heat and water vapor from place to place on Earth, with solar heat being the driving force of atmospheric circulation. Energy is continuously being exchanged between the atmosphere and the surface. This interchange influences the different effects of weather and climate. The atmosphere is divided into five layers (**Figure 2.2**) in which different factors control the temperature.

atmosphere
The air surrounding the Earth, from sea level to outer space.

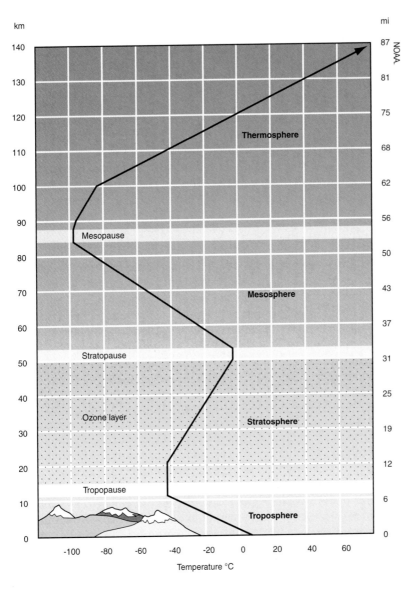

Figure 2.2 Earth's atmosphere consists of five layers. The troposphere has the greatest effect on activity at the surface. The yellow line shows the average temperature; note the variation of temperature with increased altitude. The ozone layer shields the surface from incoming ultraviolet radiation.

Layers of the Atmosphere

troposphere
The lowest part of the atmosphere that is in contact with the surface of the Earth. It ranges in altitude above the surface up to 10 or 12 kilometers.

stratosphere
The level of the atmosphere above the troposphere, extending to about 50–55 kilometers above the Earth's surface.

hydrosphere
The part of the Earth composed of water including clouds, oceans, seas, ice caps, glaciers, lakes, rivers, underground water supplies, and atmospheric water vapor.

The **troposphere** is the first layer above the surface and contains half of the Earth's atmosphere and about 75 percent of the total mass. This is where all plants and animals live and breathe. Virtually all of the water vapor and clouds exist in this layer and almost all weather occurs here. The air is very well mixed and the temperature decreases with altitude. Air in the troposphere is heated primarily from the ground up because the surface of the Earth absorbs energy and heats up faster than the air does. The heat is spread through the troposphere because the air is slightly unstable. At the tropopause the steady decline in temperature with altitude ceases abruptly and the cold air of the upper troposphere is too dense to rise above this point. As a result, little mixing occurs between the troposphere and the stratosphere above. The tropopause and the troposphere are known as the lower atmosphere.

The next layer is the **stratosphere** where many jet aircrafts fly because it is very stable. It extends to about 55 km above the Earth and this layer plus the troposphere make up 99 percent of the total mass of the atmosphere. In the Earth's stratosphere, the temperature remains constant to about 20 km and then the temperature increases with altitude to the stratopause (where temperature changes again) at 50 km. The stratosphere is heated primarily from above by solar radiation instead of from below as in the troposphere. The presence of ozone (O_3) causes the increasing temperature in the stratosphere by absorbing ultraviolet rays which causes the temperature in the upper stratosphere to reach those found near the Earth's surface. Ozone is more concentrated around an altitude of 25 kilometers and is important in protecting life on Earth by absorbing much of the harmful high-energy ultraviolet rays before they reach the surface of the planet.

Ozone concentrations decline in the upper portion of the stratosphere, and, at about 50 km above Earth, temperature decreases once again with altitude. This second zone of declining temperature is the mesosphere where the atmosphere reaches its coldest temperature of as low as around −93°C due to very little radiation absorption. The mesosphere starts just above the stratosphere and extends to 85 km high until reaching the mesopause. The gases in the mesosphere are thick enough to slow down meteorites hurtling into the atmosphere, where they burn up by friction, leaving fiery trails, "shooting stars," in the night sky. The regions of the stratosphere and the mesosphere, along with the stratopause and mesopause, are called the middle atmosphere.

The thermosphere starts just above the mesosphere and extends to 600 km high. The thermosphere contains auroras and is where the space shuttle orbits. The temperature increases with altitude due to the Sun's energy and can reach as high as 1,727°C. The air is very thin in the thermosphere and small changes in energy can cause large changes in temperature. The thermosphere also includes the region of the atmosphere called the ionosphere, which is a region of the atmosphere that is filled with charged particles. The thermosphere layer is known as the upper atmosphere.

Hydrosphere

The **hydrosphere** includes all of Earth's water which continually circulates among the oceans, continents, and atmosphere. The most prominent feature of the hydrosphere is the global ocean, which covers 71 percent of the Earth's surface and contains about 97 percent of its water (**Figure 2.3**). Ocean currents transfer heat from place to place in the atmosphere, and the oceans also alter global climate. The remaining three percent of Earth's water is found as freshwater in ice caps, glaciers, lakes, streams and groundwater. This freshwater is responsible for creating many varied landforms found on the surface of our planet. Water evaporates from the oceans and moves through the atmosphere, precipitating as rain and snow, and returns to the oceans in streams, groundwater, and glaciers. This movement of water over the Earth's surface erodes and transports weathered rock material and deposits it as sediment thus constantly modifying the landscape.

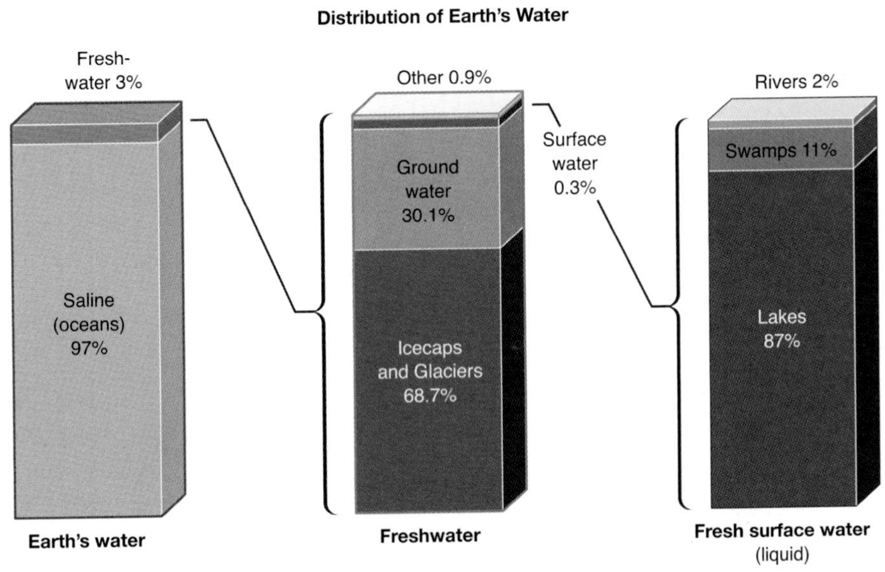

Distribution of Earth's Water

Figure 2.3 Most of Earth's water is found in the oceans. About two-thirds of the relatively small amount of freshwater is confined to cold environments.

Biosphere

The **biosphere** includes all life on Earth in a thin layer that includes the uppermost geosphere, the hydrosphere, and the lower atmosphere (**Figure 2.4**). Plants and animals depend on the physical environment for the basics of life and are clearly affected by Earth's environment. Organisms breathe air, need water, and live in a relatively narrow temperature range. Land organisms depend on soil, which is part of the geosphere. Ultimately, the biosphere strongly influences the other three spheres. Essentially, the present atmosphere has been produced by chemical activity of the biosphere, mainly through photosynthesis by plants. Organisms control some of the composition of the oceans; for example, many organisms extract calcium carbonate from seawater for their bones and shells, and when they die, this material settles to the seafloor and accumulates as sediment that lithifies to limestone. Also, the biosphere formed Earth's natural resources of coal, oil, and natural gas. Therefore, many of the rocks in the Earth's crust can have their origins traced back to some form of biological activity.

Geosphere

Beneath our feet are the rocks and soil that make up the solid part of our planet known as the **geosphere**, or solid Earth. The solid materials of Earth are separated into layers based on chemical composition and physical properties. The compositional layers consist of the hot, central core, surrounded by a large mantle that comprises the majority of Earth's volume, and a thin, cool **crust** at the surface. The layers separated by different physical properties include the **inner core, outer core, mesosphere, asthenosphere,** and **lithosphere** (**Figure 2.5**). The characteristics of these layers are controlled by an increase in density, temperature, and pressure with depth.

Layers of the Geosphere

The crust is the outermost and thinnest layer of the Earth and consists of two types. The continental crust is thick (20–70 km) and consists of a variety of rock types but has a gross overall composition similar to granite, with an average density of 2.7 g/cm^3,

biosphere
The living and dead organisms found near the Earth's surface in parts of the lithosphere, atmosphere, and hydrosphere. The part of the global carbon cycle that includes living organisms and biogenic organic matter.

geosphere
The soils, sediments, and rock layers of the Earth including the crust, both continental and beneath the ocean floors.

crust
The rocky, relatively low density, outermost layer of the Earth.

inner core
The solid central part of Earth's core.

outer core
The liquid outer layer of the core that lies directly beneath the mantle.

mesosphere
(1) (Geosphere) The division of the mantle between the asthenosphere and the outer core (2) (Atmosphere) The division of the atmosphere above the stratosphere that begins about 50 kilometers (31 miles) in altitude and extends to about 80 kilometers (50 miles).

asthenosphere
The uppermost layer of the mantle, located below the lithosphere. This zone of soft (plastic), easily deformed rock exists at depths of 100 kilometers to as deep as 700 kilometers.

lithosphere
The outer layer of solid rock that includes the crust and uppermost mantle. This layer, up to 100 kilometers (60 miles) thick, forms the Earth's tectonic plates. Tectonic plates float above the more dense, flowing layer of mantle called the asthenosphere.

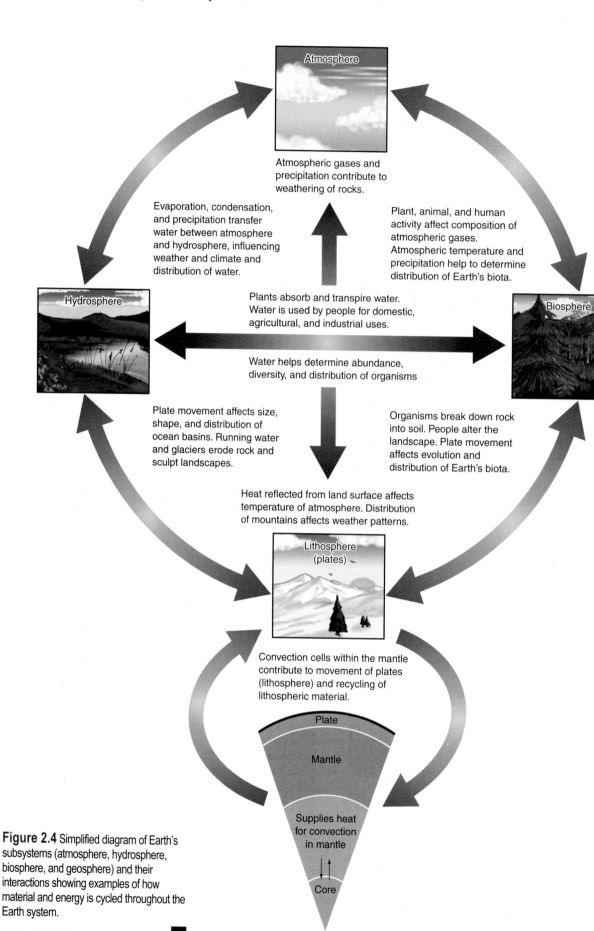

Figure 2.4 Simplified diagram of Earth's subsystems (atmosphere, hydrosphere, biosphere, and geosphere) and their interactions showing examples of how material and energy is cycled throughout the Earth system.

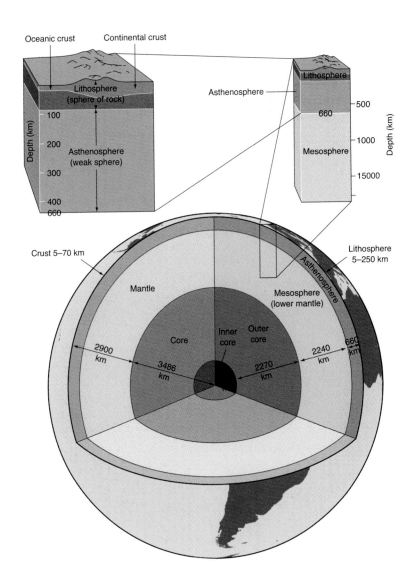

Figure 2.5 Internal structure of the geosphere. Notice that the lithosphere includes both the crust and the uppermost layer of mantle that overlies the asthenosphere.

and contains considerable amounts of silicon and aluminum. The **oceanic crust** is thin (5–10 km) and consists of the dark, igneous basalt that contains more iron and magnesium than granite, thus making oceanic crust denser (3.0 g/cm³) than continental crust. Both continental and oceanic crusts are cooler relative to the layers below, and thus consist of hard, strong rock that is physically rigid. The chemical makeup of the crust is dominated by eight elements (Table 2.1).

The **mantle** is the middle layer between the crust and core and occupies about 83 percent of Earth's volume. It is denser than the crust with a calculated density ranging from about 3.3 g/cm³ in its upper part to about 5.7 g/cm³ near the contact with the core. The composition is mostly of the rock peridotite which is a dark, dense igneous rock containing **silicate** minerals (compounds of silicon and oxygen atoms) with abundant iron and magnesium. Although the chemical composition is similar throughout the mantle, the temperature and pressure increase with depth. Temperatures near the top of the mantle approach 1000°C, increasing to about 3,300°C near the mantle-core boundary. At these temperatures the mantle should be molten rock. However, the high pressure from the thickness of overlying rocks prevents the mantle rocks from melting. This is because rocks will expand as much as 10 percent when they melt and the high pressures make it difficult for rock to expand and therefore inhibit melting. If the combination of temperature and pressure effects is close to the rock's melting point, the rock remains solid but loses its strength which makes it weak and plastic. If the temperature rises, or the pressure decreases, the rock will begin to

oceanic crust

That part of the Earth's crust of the geosphere underlying the ocean basins. It is composed of basalt and has a thickness of about 5 km.

mantle

The layer of the Earth below the crust and above the core. The uppermost part of the mantle is rigid and, along with the crust, forms the 'plates' of plate tectonics. The mantle is made up of dense iron and magnesium rich (ultramafic) rock such as peridotite.

silicate

Refers to the chemical unit silicon tetrahedron, SiO_4, the fundamental building block of silicate minerals. Silicate minerals represent about one third of all minerals and hence make up most rocks we see at the Earth's surface.

TABLE 2.1	Chemical Composition of Earth's Crust. Data are normalized to 100 percent on the basis of the first eight elements.	
Element	Percent by Weight	Percent by Volume
Oxygen (O)	46.6	93.8
Silicon (Si)	27.7	0.9
Aluminum (Al)	8.1	0.5
Iron (Fe)	5.0	0.4
Calcium (Ca)	3.6	1.0
Sodium (Na)	2.8	1.3
Potassium (K)	2.6	1.8
Magnesium (Mg)	2.1	0.3
All other elements	1.5	0.3
TOTAL	100.0	100.0

melt and form magma (molten rock). These temperature and pressure changes with depth cause the strength of the rocks to vary with depth and create layering within the mantle, as well as in the core below.

The uppermost mantle below the crust is relatively cool and its pressure low, which are similar physical conditions to the crust and produce hard, strong rocks as well. The rigid physical properties of both the crust and upper mantle cause the two to combine and form an important outer layer called the lithosphere (rock sphere). The lithosphere, consisting of the crust (either continental or oceanic) and upper mantle, averages 100 km thick but is as much as 200 km thick in some continental regions (**Figure 2.6**).

The asthenosphere (weak sphere) is the zone in the upper part of the mantle, about 250 km thick, that is more ductile or plastic than the rest of the mantle. The

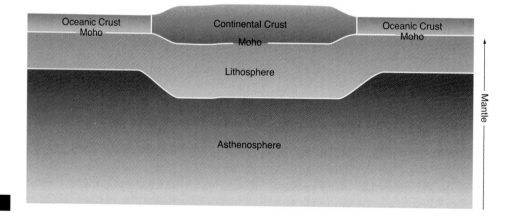

Figure 2.6 The outer portion of Earth's layers includes oceanic and continental crustal material, along with a rigid lithosphere and a hotter, plastic asthenosphere.

temperatures and pressures in this zone are at the right combinations (approaching the melting point of mantle rock) where the rocks lose much of their strength and become soft and plastic, like warm putty or hot road tar, and can flow slowly. Partial melting within the asthenosphere generates magma (molten rock) which can rise sometimes to the surface to form volcanic eruptions. This is because the magma is molten and less dense than the mantle rock from which it was derived.

The lower mantle, or mesosphere ("middle sphere"), is a solid and forms most of the volume of the Earth's interior. Because the high pressure at these depths offsets the effect of high temperature, the rocks are stronger and more rigid than the overlying asthenosphere. Despite the stronger nature of the mesosphere, it never gets as strong as the lithosphere and remains plastic and capable of flowing slowly over geologic time.

The **core** is the inner most region of Earth and has an average calculated density of 11 g/cm³ and comprises about 16 percent of the planet's total volume. The composition is considered to be predominantly metallic iron and some nickel. The core is subdivided into two zones based on very different properties: the inner core is solid and the outer core is a molten liquid. Near the center, the core's temperature is estimated to be nearly 7000°C, hotter than the Sun's surface, and pressures are 3.5 million times that of Earth's atmosphere at sea level. This extreme pressure is responsible for keeping the inner core solid despite being hotter than the outer core.

The combination of the rigid lithosphere lying over the non-rigid asthenosphere has extremely important geologic implications for natural hazards. The lithosphere is broken into numerous individual pieces called **plates** that move over the asthenosphere. This motion is the basis for the plate tectonic theory. Movements of the plates create earthquakes, volcanism, and mountain uplifts.

core
The innermost layer of the Earth, made up of mostly of iron and nickel. The core is divided into a liquid outer core and a solid inner core. The core is the densest of the Earth's layers.

plate
Slab of rigid lithosphere (crust and uppermost mantle) that moves over the asthenosphere.

Earth Processes and Cycles

As a system, remember that Earth is an assemblage or combination of interacting components that form a complex whole. The system is driven by the flow of matter and energy through the components. Each of Earth's major spheres, the atmosphere, hydrosphere, biosphere, geosphere, are usually described and studied as separate entities, each containing numerous interacting smaller systems; however, each sphere and their components interact with each other continuously, exchanging both matter and energy.

Several energy and material (matter) cycles are fundamental to our understanding and study of natural processes and hazards connected to the dynamic movements of matter and energy. These include the **rock cycle**, the **hydrologic cycle**, and the **biogeochemical cycle**. Keep in mind that during the course of these cycles, matter is always conserved, that is, it is neither created nor destroyed, but continuously changes form. The cyclical exchange of matter and energy occurs between various storage reservoirs within the spheres and on various time scales. This cyclic movement of matter and energy plays a large role in producing natural disasters when we find ourselves in the path of their motions.

rock cycle
The sequence of events in which rocks are formed, destroyed, and reformed by geological processes. Provides a way of viewing the interrelationship of internal and external processes and how the three rock groups relate to each other

hydrologic cycle
The cyclic transfer of water in the hydrosphere by water movement from the oceans to the atmosphere and to the Earth and return to the atmosphere through various stages or processes such as precipitation, interception, runoff, infiltration, percolation, storage, evaporation, and transportation.

Material for the Cycles

The materials that make up the different Earth spheres, and what energy performs work on, is known as matter, which is defined as anything that has mass and occupies space. Therefore, the atmosphere, water, plants, animals, minerals, and rocks are composed of matter made up of atoms of different elements. Matter exists in three states: solids, liquids, and gases. The cycling of this matter is important in Earth's processes.

biogeochemical cycle
Natural processes that recycle nutrients in various chemical forms from the environment, to organisms, and then back to the environment. Examples are the carbon, oxygen, nitrogen, phosphorus, and hydrologic cycles.

Forms of Energy

Natural processes responsible for natural disasters (such as erosion, atmospheric circulation, deformation, melting, etc.) occur on or within the Earth and require energy for their operation of performing work on matter. Energy is defined as "the ability to do work," which can be constructive or destructive. Work itself can mean moving something, lifting something, warming something, or lighting something. Energy exists in two states: stored and waiting to do work (**potential energy**), or actively moving (**kinetic energy**). Potential energy is so named because this stored energy or work has the *potential* to change the state of other objects when released. The kinetic energy of an object is the energy it possesses because of its motion and is an expression of the fact that a moving object can do work on anything it hits and thus it quantifies the amount of work the object could do as a result of its motion. There are many different forms of energy that can be transformed from one to the other:

- **Heat energy** is exhibited by moving atoms and the more heat energy an object has, the higher its temperature, which is a measure of its kinetic energy. When gasoline is burned in a car's engine, the heat energy is converted to kinetic energy that moves the car.
- **Radiant energy** is energy carried in the form of electromagnetic waves such as visible light, ultraviolet rays, and radio waves. Most of the Sun's energy reaches Earth in this form (especially light) and is converted to heat energy.
- **Nuclear energy** is energy released from an atom's nucleus. Energy is released by several processes including radioactive decay, where an unstable radioactive nucleus decays spontaneously into a new nucleus, and nuclear fusion where multiple nuclei join together to form a new element nucleus and heat-producing nuclear reactions where two nuclei merge to produce two different nuclei.
- **Elastic energy** is potential energy formed through deformation (stretching, bending, or compressing) of an elastic material like a rubber band. When energy is released, it is converted to kinetic energy and heat (by friction). Earthquakes exhibit a release of energy stored in elastically deformed rocks that eventually break.
- **Electrical energy** is produced by moving electrons through material and is often converted into heat energy. Lighting storms are good examples.
- **Chemical energy** is produced by breaking or forming chemical bonds and this type of energy is usually converted to heat.
- **Gravitational energy** is produced when an object falls from higher to lower elevations. As the object falls, the energy can be converted to kinetic energy or heat.

Energy Systems for the Cycles

Earth's major cycles and processes are powered by energy from two sources. The Sun drives external processes that occur in the atmosphere, hydrosphere, and surface of the lithosphere. Climate, weather, ocean circulation, and erosional processes are the result of energy from the Sun being transformed into heat energy which powers these external processes. The second source of energy is derived from the Earth's interior. Heat left over from the Earth's formation along with heat produced from the constant decay of radioactive elements powers the internal processes forming volcanoes, earthquakes, and mountains.

Energy from the Sun

The Sun is the most prominent feature in our solar system and contains approximately 98 percent of the total solar system mass (**Figure 2.7**). The Sun is composed of hydrogen (about 74 percent of its mass, or 92 percent of its volume), helium (about

potential energy
The energy available in a substance because of position (e.g., water held behind a dam) or chemical composition (hydrocarbons). This form of energy can be converted to other, more useful forms (for example, hydroelectric energy from falling water).

kinetic energy
The energy inherent in a substance because of its motion, expressed as a function of its velocity and mass, or $MV^2/2$.

Figure 2.7 Earth is the third planet from the Sun. When the size of Earth is compared to that of the Sun, more than one million Earths could fit inside the Sun and 109 Earths could be placed across its equator.

25 percent of mass, 7 percent of volume), and trace quantities of other elements. The Sun's outer visible layer is called the photosphere and has a temperature of 6,000°C and a mottled appearance due to the turbulent eruptions of energy at the surface. Solar energy is created deep within the core of the Sun, where the temperature (15,000,000°C) and pressure (340 billion times Earth's air pressure at sea level) are so intense that nuclear reactions take place. Here, nuclear fusion produces energy by fusing hydrogen atoms together to form atoms of helium. During nuclear fusion, some of the mass from the hydrogen atoms is expelled as energy and carried to the surface of the Sun by convection (see following discussion), where it is released as radiant energy. Energy generated in the Sun's core takes about a million years to reach its surface, but every second, 700 million tons of hydrogen are converted into helium and 5 million tons of pure energy are released. The Sun has been producing energy since the solar system formed 4.6 billion years ago, and it is estimated that solar energy production will continue at this level for another 5 billion years.

The radiant energy from the Sun travels in many different wavelengths (**Figure 2.8**) through space, with about 43 percent being visible light, 49 percent near-infrared, and 7 percent ultraviolet (UV) solar radiation received on Earth. Some of the energy is immediately reflected back into space (about 40 percent) by the atmosphere and the Earth's surface (mostly the oceans). Some of the energy is converted to heat and is absorbed by the atmosphere, hydrosphere, and lithosphere. Heat drives the hydrologic cycle, causing evaporation of the oceans and circulation of the atmosphere, and, with gravity, rain to fall on the land and run downhill. Thus energy from the Sun is responsible for such natural disasters as severe weather and floods.

In the biosphere, some energy is absorbed and stored by plants during photosynthesis and used by other organisms, or is stored in fossil fuels such as coal and petroleum. This stored energy in the biosphere is released through burning to form heat energy, which also releases the stored carbon that produces carbon dioxide—a greenhouse gas that is partially responsible for global warming issues.

Energy from Earth's Interior

Energy in the form of heat is generated within the Earth's interior and contributes only about 0.013 percent of the total energy reaching the Earth's surface (the other 99.987 percent from the Sun). However, this smaller amount of energy that flows to the Earth's surface is responsible for deformational events that build mountains and cause earthquakes, and for melting of rocks to create magmas that result in volcanism. Heat energy is produced from three energy sources within the Earth's interior:

Radioactive decay continuously produces heat energy from radioactive isotopes such as ^{235}U (Uranium), ^{232}U, ^{232}Th (Thorium), and ^{40}K (potassium) (see Appendix A). Radioactive isotopes have unstable nuclei that break down (decay) to

radioactive decay
Natural spontaneous decay of the nucleus of an atom where alpha or beta and/or gamma rays are released at a fixed rate.

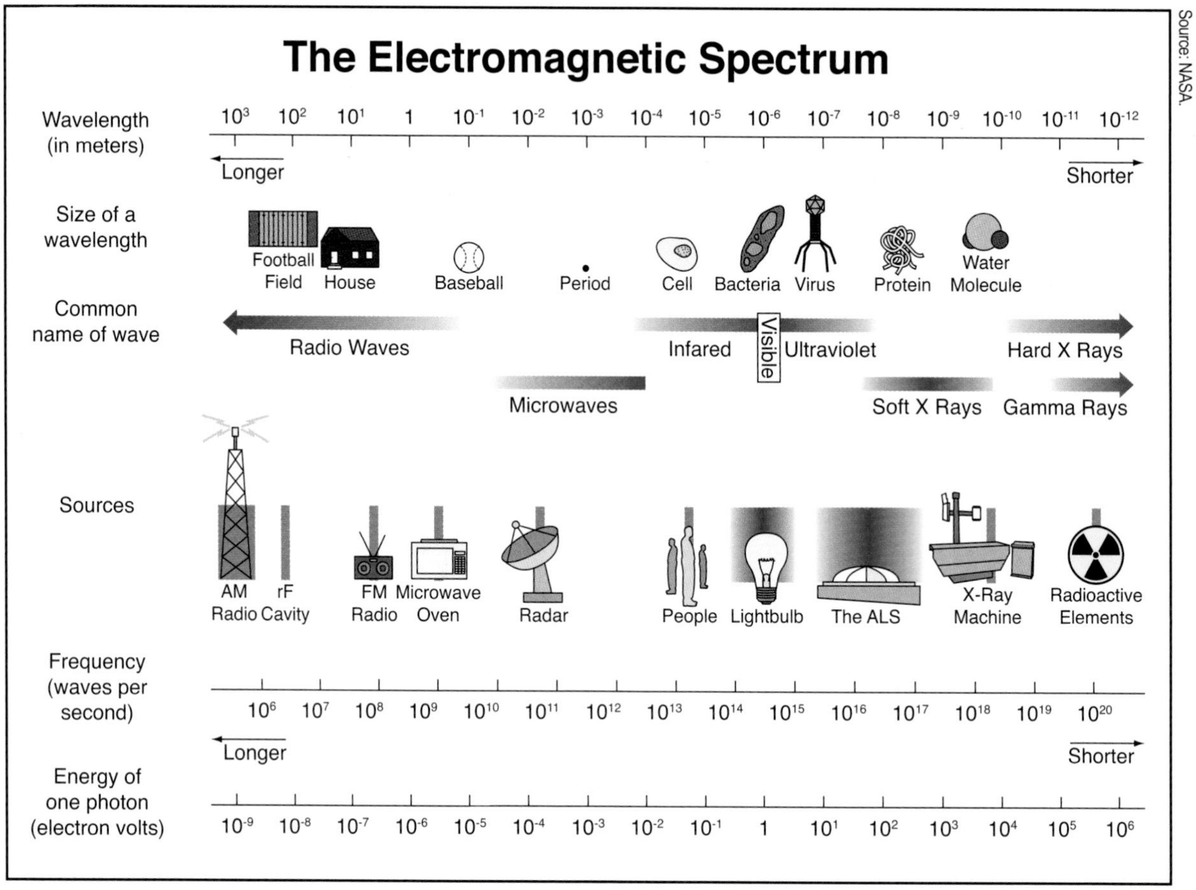

The Electromagnetic Spectrum

| Wavelength (in meters) | 10^3 | 10^2 | 10^1 | 1 | 10^{-1} | 10^{-2} | 10^{-3} | 10^{-4} | 10^{-5} | 10^{-6} | 10^{-7} | 10^{-8} | 10^{-9} | 10^{-10} | 10^{-11} | 10^{-12} |

Longer ←   → Shorter

Size of a wavelength: Football Field, House, Baseball, Period, Cell, Bacteria, Virus, Protein, Water Molecule

Common name of wave: Radio Waves, Microwaves, Infared, Visible, Ultraviolet, Soft X Rays, Hard X Rays, Gamma Rays

Sources: AM Radio, rF Cavity, FM Radio, Microwave Oven, Radar, People, Lightbulb, The ALS, X-Ray Machine, Radioactive Elements

| Frequency (waves per second) | 10^6 | 10^7 | 10^8 | 10^9 | 10^{10} | 10^{11} | 10^{12} | 10^{13} | 10^{14} | 10^{15} | 10^{16} | 10^{17} | 10^{18} | 10^{19} | 10^{20} |

Longer ←   → Shorter

| Energy of one photon (electron volts) | 10^{-9} | 10^{-8} | 10^{-7} | 10^{-6} | 10^{-5} | 10^{-4} | 10^{-3} | 10^{-2} | 10^{-1} | 1 | 10^1 | 10^2 | 10^3 | 10^4 | 10^5 | 10^6 |

Figure 2.8 The electromagnetic spectrum has an extremely wide range of wavelengths. Visible light is only a small portion of the spectrum.

---

gravitational energy
The force of attraction between objects due to their mass and is produced when an object falls from higher to lower elevations.

impact energy
Cosmic impacts with a larger body convert their energy of motion (kinetic energy) to heat.

---

a more stable isotope and expel subatomic particles such as protons, neutrons, and electrons from the radioactive parent atom that are slowed and absorbed by surrounding matter (in this case rocks). The energy of motion (kinetic energy) of these emitted subatomic particles is converted to heat by their collision with the surrounding matter.

Gravitational energy and **impact energy** produced a tremendous amount of heat during the initial formation of the Earth. Gravitational energy was released as the Earth grew in size by accretion in the original solar nebula. The deepening burial of material caused an increasing gravitational pull that continued to compact the interior and convert gravitational energy to heat. At the same time in Earth's early history, tremendous numbers of assorted-size space objects (asteroids, meteorites, and comets) were hitting the Earth. The impacts of these objects converted their energy of motion (kinetic energy) to heat. The large quantity of heat produced by the combination of these two sources did not easily escape to the surface due to the poor conductivity (see following heat transfer discussion) of rock material. The sum of heat energy generated from radioactive decay, impacts, and gravity is slowly rising to the surface of our planet, and along its journey is causing the continents to drift and ocean basins to form.

Methods of Heat Transfer

Any body, whether a solid, liquid, or gas, has a temperature associated with it. This thermal energy can be transferred to another body or its surroundings by the process of **heat transfer**. The movement occurs by heat moving from a hot body to a cold one through the processes of conduction, convection, or radiation, or any combination of these (**Figure 2.9**). Heat transfer never stops; it can only be slowed down. Earth's internal and external processes are powered by heat energy moving from hotter areas (such as the Earth's interior) to cooler ones (the crust and ultimately into space) and so it is important to understand how heat can move through materials.

heat transfer
Heat moving from a hot body to a cold one through processes of conduction, convection, or radiation, or any combination of these

conduction
Heat transfer directly from atom to atom in solids.

Conduction

Conduction transfers heat directly from atom to atom in mostly solids. Heat energy causes atoms to vibrate against each other, and these vibrations move from high-temperature areas (rapid vibrations) to low-temperature areas (slower vibrations). Heat from the Earth's interior moves through the solid crust by this mode of heat transfer, but very slowly. Each substance has a different heat conductivity based on how easily the atoms are affected by the heat energy. Metals are highly conductive, so they heat and cool rapidly which is why we use metal pots for cooking—heat is transferred quickly to the contents inside. Rocks are much less conductive and so heat up and cool down more slowly, which helps heat build up inside the Earth.

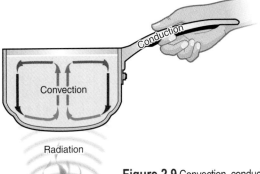

Figure 2.9 Convection, conduction, and radiation often occur together. These are three ways in which heat is transferred through objects and space.

Convection

Convection transfers heat by being carried along by the material that was heated. Since heat energy moves with the material, the materials that are able to move are mostly liquids and gases. We are familiar with convection of air in our houses when we heat them in the winter by a heat source (whether by a wood stove, radiator, electric heater, etc.). The warmed, less-dense air rises to the ceiling as a moving convection current, which then cools to a denser air mass that returns to the floor to be heated again. A complete circulation of hot and cold movement as just described is called a **convection cell**. Convection is an important method of moving Earth materials. Inside the Earth, the mantle appears to transfer heat by this method as well as the flow of magma to the surface. On the surface, heat is transferred in the atmosphere and oceans by this mode that influence climate and weather.

convection
(1) (Physics) Heat transfer in a gas or liquid by the circulation of currents from one region to another; also fluid motion caused by an external force such as gravity. (2) (Meteorology) The phenomenon occurring where large masses of warm air, heated by contact with a warm land surface and usually containing appreciable amounts of moisture, rise upward from the surface of the Earth.

convection cell
Within the geosphere it is the movement of the asthenosphere where heated material from close to the Earth's core becomes less dense and rises toward the solid lithosphere. At the lithosphere-asthenosphere boundary heated asthenosphere material begins to move horizontally until it cools and eventually sinks down lower into the mantle, where it is heated and rises up again, repeating the cycle.

Radiation

Radiation transfers heat using electromagnetic waves. Radiant electromagnetic energy from the Sun (produced by nuclear reactions) is transferred by this method. On Earth radiative heat transfer is responsible for warming the oceans and atmosphere, and for reradiating heat back into space to keep our planet at a constant average temperature.

radiation
Energy emitted in the form of electromagnetic waves. Radiation has differing characteristics depending upon the wavelength. Because the radiation from the Sun is relatively energetic, it has a short wavelength (ultra-violet, visible, and near infrared) while energy radiated from the Earth's surface and the atmosphere has a longer wavelength (e.g., infrared radiation) because the Earth is cooler than the Sun.

plate tectonics
The theory that the Earth's lithosphere consists of large, rigid plates that move horizontally in response to the flow of the asthenosphere beneath them, and that interactions among the plates at their borders (boundaries) cause most major geologic activity, including the creation of oceans, continents, mountains, volcanoes, and earthquakes.

lithospheric plate
A series of rigid slabs of lithosphere (12 major ones at present) that make up the Earth's outer shell. These plates float on top of a softer, more plastic layer in the Earth's mantle known as the asthenosphere.

Plate Tectonics and the Cycling of the Lithosphere

Plate tectonics theory provides us with a framework for interpreting the composition, structure, and internal processes of Earth on a global scale and how they contribute in producing natural hazards such as earthquakes and volcanoes. **Plate tectonics** (*tekton*, "to build") is a theory developed in the late 1960s to explain how the lithospheric outer layer of the Earth moves and deforms. The theory has caused a revolution in the way we think about the Earth and has proven useful in predicting geologic hazard events and explaining many aspects of the natural processes we see on Earth.

According to plate tectonics theory, the lithosphere is divided into segments called tectonic plates that move continuously over the asthenosphere (**Figure 2.10**). They are also referred to as **lithospheric plates,** or simply plates. The plates mostly include both continental and oceanic crust (note the North American plate), but some are composed largely of oceanic crust (e.g., the Pacific plate for example). The continuous movement of the plates is powered by internal heat forming convection currents within the mantle. The slow movements of different plates (usually a few centimeters per year, or about as fast as your fingernail grows) create natural hazard zones of volcanic activity and earthquakes where their margins meet.

Plate Boundaries

Along these margins, or **plate boundaries**, the plates diverge, converge, or slide horizontally past each other (**Figure 2.11**). Again, these plate boundaries are important because the plates interact where they meet and are zones where deformation of the Earth's lithosphere is taking place and generating potential hazard areas.

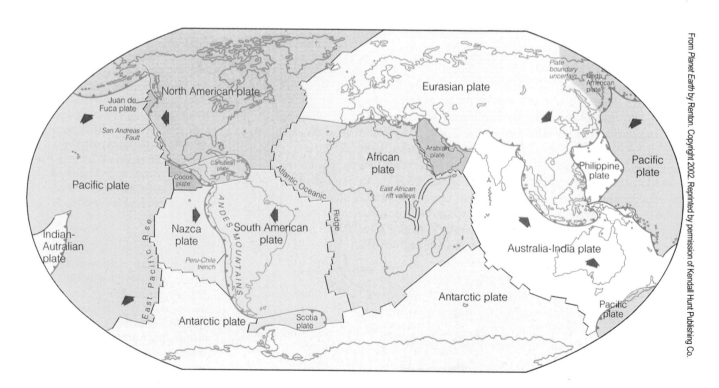

From *Planet Earth* by Renton. Copyright 2002. Reprinted by permission of Kendall Hunt Publishing Co.

Figure 2.10 The surface of the Earth is divided into major lithospheric plates, some of which are found in the oceans and some contain both continental material along with surrounding oceanic material.

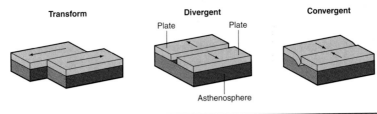

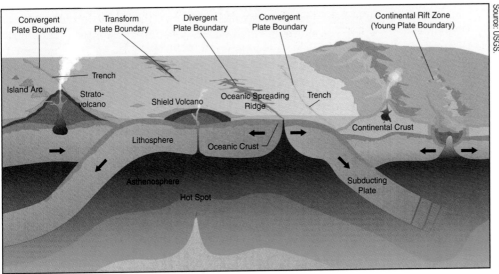

Source: USGS

Figure 2.11 Plate boundaries consist of three types: divergent, where plates move apart; convergent, where plates come together; and transform, where the plates slide past one another.

Divergent Plate Boundaries

At **divergent plate boundaries**, the plates move away from each other by tensional stresses. This is where new oceanic crust and lithosphere are created as magmas rises from the underlying asthenosphere (**Figure 2.12**). The magmas intrude, or can erupt,

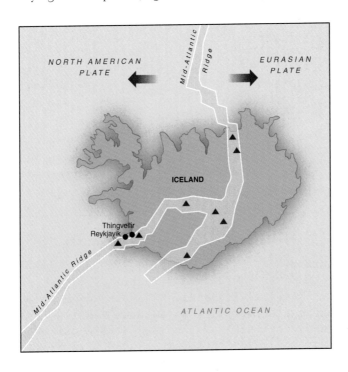

plate boundary
According to the theory of plate tectonics, the locations where the rigid plates that comprise the crust of the earth meet. As the plates meet, the boundaries can be classified as divergent (places where the plates are moving apart, as at the mid-ocean ridges of the Atlantic Ocean), convergent (places where the plates are colliding, as at the Himalayas Mountains), and transform (places where the plates are sliding past each other, as the San Andreas fault in California).

divergent plate boundary
A boundary in which two lithospheric (tectonic) plates move apart.

Figure 2.12 Iceland lies along a divergent plate boundary as two separate plates move away from each other. The island consists of volcanic material extruded upward from greater depths. Active volcanoes are shown by triangles.

oceanic ridge
An uplifting of the ocean floor that occurs when convection currents beneath the ocean bed force magma up where two tectonic plates meet at a divergent boundary. The ocean ridges of the world are connected and form a single global ridge system that is part of every ocean and form the longest mountain range on Earth.

convergent plate boundary
A boundary in which two plates collide. The collision can be between two continents (continental collision), a relatively dense oceanic plate and a more buoyant continental plate (subduction zone) or two oceanic plates (subduction zone).

subduction
Process of one crustal plate sliding down and below another crustal plate as the two converge.

subduction zone
Also called a convergent plate boundary. An area where two plates meet and one is pulled beneath the other.

oceanic trench
Deep, linear, steep-sided depression on the ocean floor caused by the subduction of oceanic crustal plate beneath either other oceanic or continental crustal plates.

volcanic arc
Arcuate chain of volcanoes formed above a subducting plate. The arc forms where the downgoing descending plate becomes hot enough to release water and gases that rise into the overlying mantle and cause it to melt.

island arc
An arc-shaped chain of volcanic islands produced where an oceanic plate is sinking (subducting) beneath another.

on the surface and solidify to form basalt-type rock attached to the moving plate. The margins of divergent plate boundaries occur mostly in the oceanic plates and they are marked by **oceanic ridges** (e.g., the Mid-Atlantic Ridge) that are essentially linear underwater mountain ranges uplifted by the hot nature of the rocks. These boundaries are also known as *spreading centers*, because oceanic lithosphere spreads away on each side of the margin at the ridge as a result of underlying *convection cells* in the mantle. While most diverging plate boundaries occur at the ocean ridges, sometimes continents are split apart along zones called *rift zones*, such as at the East African Rift in eastern Africa. Volcanism and earthquakes are common along diverging plate boundaries but are of relatively low intensity.

Convergent Plate Boundaries

Convergent plate boundaries are where two plates move toward each other by compressional stresses (**Figure 2.13**) and where most lithosphere is destroyed. In this manner, the Earth maintains a global balance between the creation of new lithosphere and the destruction of old lithosphere. At such boundaries, one of the plates sinks below the other in a process called **subduction**. Convergent boundaries are more complex because the margins can involve convergence between two plates carrying oceanic crust, between a plate carrying oceanic crust and another carrying continental crust, or between two plates carrying continental crust.

At **subduction zones**, the oceanic plate subducts (sinks) beneath another oceanic plate, *ocean-ocean convergence*, or the oceanic plate subducts beneath a continental plate, *ocean-continent convergence*. Where subduction occurs, an **oceanic trench** is formed on the seafloor that marks the plate boundary. The subducting plate is usually partially melted as it descends into the hotter asthenosphere and generates magma that rises to the surface through the nonsubducting overriding plate, and if it reaches the surface it forms a chain of volcanic eruptions (known as a **volcanic arc**). The oldest oceanic crust on Earth is about 200 million years old because the denser oceanic lithosphere continuously recycles back into the mantle at subduction zones. Rocks found in continents are as old as 3.96 billion years because subduction consumes very little of the less-dense continental crust.

In ocean-ocean convergence (**Figure 2.14**), the magmas rise to form a chain of underwater volcanoes, some growing to form chains of islands known as **island arcs**. Well-known examples of island arcs include the islands of the Caribbean, the Aleutian Islands, and the island chains of Japan, Indonesia, and the Philippines. In ocean-continent convergence (**Figure 2.15**), the magma rises through the continent plate to the surface, where it erupts to form a volcanic mountain chain along the edge of the continent. Well-known examples of this type of volcanic arc are the Cascade

Figure 2.13 Convergent plate boundaries display compressive stresses that often result in the destruction of lithospheric material.

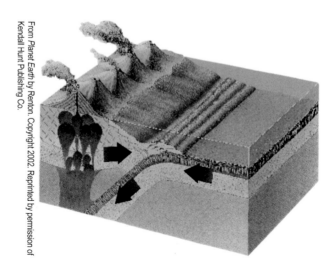

From *Planet Earth* by Renton. Copyright 2002. Reprinted by permission of Kendall Hunt Publishing Co.

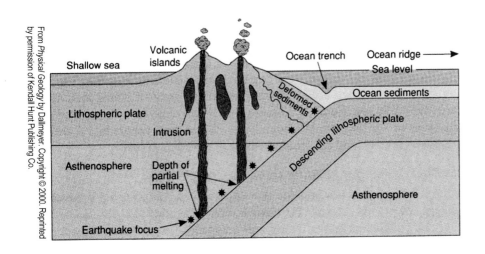

Figure 2.14 The convergence of two oceanic plates results in the subduction of one plate under the other. The molten material created by the subducting plate rises through the lithosphere and produces a string of volcanoes seated on the ocean floor.

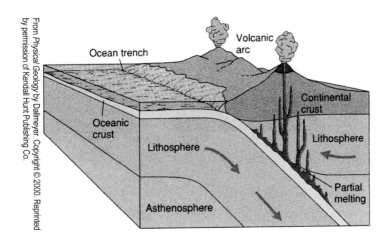

Figure 2.15 The convergence of an oceanic plate with a continental plate results in the subduction of the denser ocean plate under the less dense continental plate. Melting of the subducted plate produces magma that rise up through the continental plate, producing a string of volcanoes on the continent.

Mountains (includes Mount Saint Helens volcano) of the northwestern United States and the Andes Mountains of South America. The hazards associated with these two boundaries include frequent volcanic eruptions and earthquakes of high intensity.

When two continental plates converge, neither plate is subducted. Instead, they collide together to form a collision boundary (**Figure 2.16**). The force of collision forms a mountain range from the squeezing of the two continents at their juncture. The Himalayan Mountains between India and China are being formed in this way today, as were the ancient Appalachian Mountains about 450 million years ago (**Figure 2.17**). This boundary also produces frequent and powerful earthquakes.

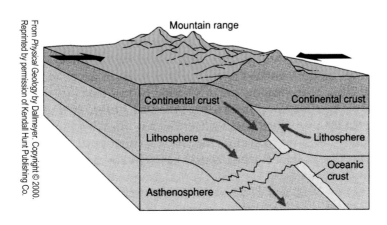

Figure 2.16 The collision of two continental masses does not produce subduction of the continental crust. This is because both masses are less dense than the material in the lithosphere and thus do not sink.

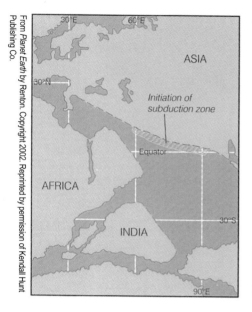
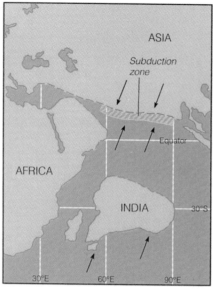
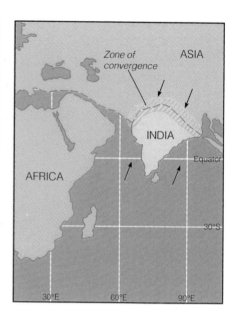

Figure 2.17 Over the past 70 million years the Indian subcontinent has moved northward and collided with the southern edge of the Eurasian Plate. This collision has pushed up the rocks on the surface and formed the Himalayan Mountains, the highest mountain range on Earth.

Transform Plate Boundaries

Transform plate boundaries occur when two plates slide past one another horizontally (**Figure 2.18**) along faults known as **transform faults**. These are also zones of frequent and powerful earthquakes, but generally not zones of volcanism since material is not transferred between the asthenosphere and lithosphere. A well-known example of a continental transform fault is the San Andreas fault of California, which forms one part of the boundary between the Pacific plate and the North American plate (**Figure 2.19**).

transform plate boundary
An area where two plates meet and are moving side to side past each other.

transform fault
A strike-slip fault with side to side horizontal movement that offsets segments of an oceanic ridge.

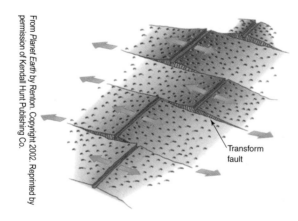

Figure 2.18 Transform faults occur perpendicular to the oceanic ridges as the material formed at the ridge moves outward.

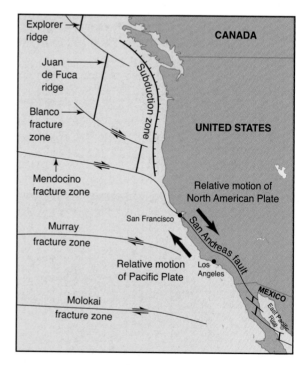

Figure 2.19 The Pacific Plate is sliding along against the western edge of North America. This contact has formed the San Andreas fault, a right-lateral strike slip fault. Notice that California actually lies in two different plates.

The Rock Cycle

A **rock** is an aggregate of one or more **minerals** which are naturally occurring inorganic solids, each having a definite chemical composition and a crystalline structure. Chemical composition and crystalline structure are the two most important properties of a mineral and give them their unique physical properties, such as hardness. Geologists classify minerals into groups that share similar ion groups. The most important and common minerals are the silicates, which are composed of combinations of oxygen and silicon with (or without) metallic elements. Geologists group rocks into three categories based on how they were formed: *igneous*, *sedimentary*, and *metamorphic*. Each group contains a variety of different rock types that differ from one another in composition or texture (size, shape, and arrangement of minerals).

Rocks may seem permanent and unchanging over a human lifetime. However, over geologic time rocks are constantly exposed to different physical and chemical conditions (or environments) that change them. The processes that change rocks from one to the other are illustrated in the rock cycle (**Figure 2.20**). Several interactions of energy, earth materials, and geologic processes act to form and destroy rocks and minerals. The rock cycle is the slowest of Earth's cyclic processes with rocks recycled on a scale of millions of years. This slow process is responsible for concentrating the planet's nonrenewable resources that humans depend upon. Plate movements are responsible for recycling rock materials and drive the rock cycle, as well as determining to a large degree where and which one of the rock groups will form. The rock cycle provides a way of viewing the interrelationship of internal and external processes and how the three rock groups relate to each other.

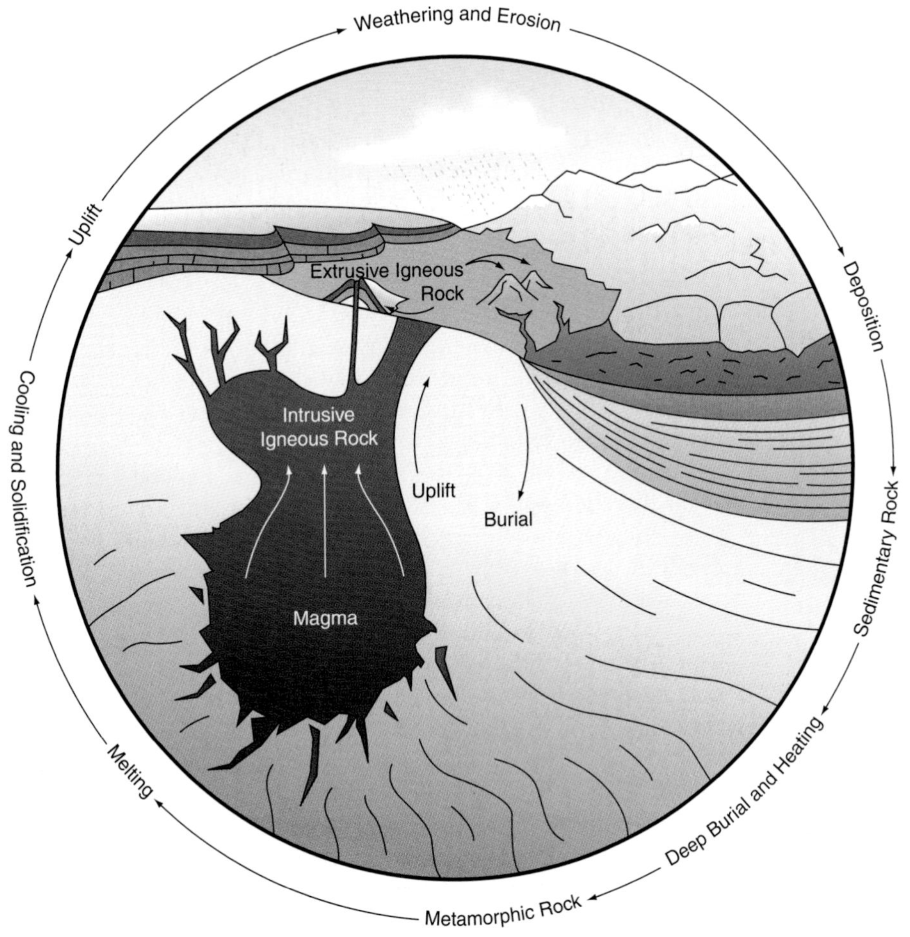

rock
A naturally occurring aggregate of minerals. Rocks are classified by mineral and chemical composition; the texture of the constituent particles; and also by the processes that formed them. Rocks are thus separated into igneous, sedimentary, and metamorphic rocks

mineral
Any naturally occurring inorganic substance found in the earth's crust as a crystalline solid

Figure 2.20 In the rock cycle material is continually reworked to form sedimentary, igneous, and metamorphic rocks.

Figure 2.21 Igneous rocks can form either from the cooling of magma in the subsurface or by solidifying after being erupted onto the surface. These rocks formed following a catastrophic eruption of a volcano in the southwestern United States more than 27 million years ago.

Figure 2.22 Sedimentary rocks are typically layered, having been formed by the deposition of sediments derived from weathering of the continents.

Figure 2.23 Metamorphic rocks form by the alteration of pre-existing rocks by being heated, subjected to pressure, or having chemically active fluids introduced into their composition.

Igneous Rocks

Igneous rocks result when magma cools and solidifies or volcanic ejecta such as ash accumulate and consolidate (**Figure 2.21**). Magma is generated in the asthenosphere and migrates upward into the crust. Magma that cools and solidifies (a process called crystallization) slowly beneath the surface produces *intrusive igneous rocks* whereas magma that erupts and cools on the surface produces *extrusive igneous rocks*. Igneous rocks make up approximately 95 percent of the Earth's crust, but their great abundance is hidden on the Earth's surface by a relatively thin but widespread layer of sedimentary and metamorphic rocks.

Rocks exposed at the Earth's surface break down physically and chemically by **weathering** caused by exposure to gases and water in the atmosphere and hydrosphere. Weathering processes form particles and dissolved materials that are eventually picked up and transported by erosional agents such as streams, glaciers, wind, or waves that are eventually deposited as *sediment*.

Sedimentary Rocks

Sedimentary rocks result from the lithification (the process of turning loose sediment into stone) of sediment (**Figure 2.22**). Sediments are usually lithified into sedimentary rock when compacted by the weight of overlying sediment layers, or when cemented by mineral matter as subsurface water moves through the pore spaces between sediment particles. Sedimentary rocks can form from the consolidation of rock fragments, precipitation from solution, or compaction of plant and animal remains. Since these rocks form from sediment at the Earth's surface, geologists can determine the type of environment in which sediments were deposited and the transporting agent. Sedimentary rocks also contain evidence of past life-forms preserved as fossils and thus are useful in interpreting Earth's history.

Metamorphic Rocks

Metamorphic rocks result from the alteration of other rocks when deeply buried and subjected to high heat and pressure as well as chemical activity with fluids (**Figure 2.23**). Rocks become foliated or nonfoliated. Foliation is the parallel alignment of minerals due to pressure, which gives the rock a layered or banded appearance. With the addition of additional pressures or still higher temperatures, the metamorphic rocks will melt, creating magma, beginning the cycle again.

The rock cycle can follow many different paths. For example, weathering may turn uplifted metamorphic rocks into sediment, which becomes lithified into a sedimentary rock. An igneous rock may become metamorphosed into a metamorphic rock. The rock cycle simply shows that rocks are not permanent but change over geologic time through internal and external processes. The rock cycle illustrates several types of interactions between the Earth system's spheres. Interactions occur among rocks, the atmosphere, the hydrosphere, and the biosphere. Water and air, aided by natural acids and other chemicals secreted by plants, weather solid rocks to form large amounts of *clay* (an important silicate mineral) and other sediments.

During these processes, water and atmospheric gases react chemically to become incorporated into the clay minerals and thus transfer matter from the atmosphere and hydrosphere to the solid material of the geosphere. Rain and gravity then wash the sediment into streams, which is later deposited (mostly in the oceans at the edge of continents). Energy from the Sun powers the hydrologic cycle to evaporate water to form rain, which in turn feeds flowing streams. The interaction of the hydrologic cycle is also part of the rock cycle, illustrating again that all of the Earth's processes are interrelated.

The Hydrologic Cycle

Radiant energy from the Sun heats the upper portion of water bodies, causing evaporation of water molecules, which then rise into the atmosphere as vapor. The heated air and water vapor are less dense than surrounding cooler air, and so they rise and carry moisture upward into the atmosphere, where temperatures become progressively colder with altitude. As the rising air cools, the water vapor molecules combine, or condense, back into microdroplets of water through the process of condensation. Depending on the air temperature, the condensed moisture forms precipitation, where liquid rain or solid snow falls under the influence of gravity. Precipitation that falls to Earth's surface replenishes surface water (oceans, lakes, and streams) and also infiltrates into the ground which replenishes streams and lakes from below or, where snow falls in constantly frigid regions, accumulates to form glacial ice. Gravity pulls on water and ice, converting their potential energy into kinetic energy as flowing streams and glaciers, which enables streams and glaciers to erode the surface. The rising and falling of water vapor in the atmosphere along with heating of atmospheric gases drives the Earth's wind systems that drive the surface currents in the oceans and create waves on the ocean surface, which causes erosion of continental coasts.

From the processes described we can see that water exists in three states on Earth: liquid, vapor, and solid. Water constantly changes from one state to another, depending on temperature, and is stored in a number of dynamic reservoirs (for example, oceans, glaciers, lakes, streams, groundwater, clouds, plants, animals). The continual circulation of water in its three states through the atmosphere, hydrosphere, biosphere, and geosphere is called the hydrologic cycle, or simply, the water cycle (**Figure 2.24**).

igneous rock
A rock formed when molten rock (magma) has cooled and solidified (crystallized). Igneous rocks can be intrusive (plutonic) and extrusive (volcanic).

weathering
A process that includes two surface or near-surface processes that work in concert to decompose rocks. Chemical weathering involves a chemical change in at least some of the minerals within a rock. Mechanical weathering involves physically breaking rocks into fragments without changing the chemical make-up of the minerals within it.

sedimentary rock
A sedimentary rock is formed from pre-existing rocks or pieces of once-living organisms. They form from deposits that accumulate on the Earth's surface. Sedimentary rocks often have distinctive layering or bedding.

metamorphic rock
A rock that has been altered physically, chemically, and mineralogically in response to strong changes in temperature, pressure, shearing stress, or by chemical action of fluids.

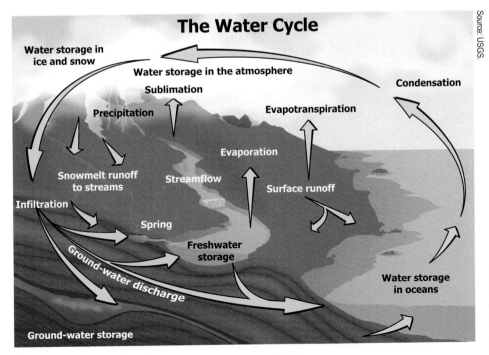

Source: USGS

Figure 2.24 The hydrologic cycle describes the movement of water through the atmosphere, as well as along the surface of the continents and the oceans.

As can be imagined, throughout the cycle various processes are operating. Energy from the Sun provides the heat for evaporation and atmospheric circulation and gravitational energy causes water to flow back to the oceans. Overall, the hydrologic cycle operates on a much shorter time scale than the rock cycle because the exchange of matter (water in this case) between reservoirs varies. Known as residence time, water resides in the atmosphere for only a few days, which means severe weather can be generated more frequently than volcanic eruptions from the slow-moving plates of the lithosphere.

Courtesy of David M. Best

Figure 2.25 A forest ecosystem contains trees, soil, moisture and the atmosphere, all of which interact with one another to provide a dynamic, living environment.

The Biogeochemical Cycle

Besides water cycling through the biosphere in plants and animals, other materials (e.g., carbon and nitrogen) have a high concentration in the biosphere. Cycles that involve the interactions between the biosphere and other reservoirs involve biological processes such as respiration, photosynthesis, and decomposition (decay), which are referred to as biogeochemical cycles. A biogeochemical cycle is a pathway by which a chemical element or molecule moves through both biotic and abiotic components of an **ecosystem** (**Figure 2.25**). All chemical elements occurring in organisms are part of biogeochemical cycles, and, in addition to being a part of living organisms, these chemical elements also cycle through the hydrosphere, atmosphere, and lithoshere. All the chemicals, nutrients, and elements—such as carbon, nitrogen, oxygen, and phosphorus—used by living organisms are recycled. An important example is the carbon cycle, which is important in regulating our global climate by affecting levels of greenhouse gases within the atmosphere.

ecosystem
A community of plants, animals and other organisms that interact together within their given setting.

carbon cycle
All carbon reservoirs and exchanges of carbon from reservoir to reservoir by various chemical, physical, geological, and biological processes. Usually thought of as a series of the four main reservoirs of carbon interconnected by pathways of exchange. The four reservoirs, regions of the Earth in which carbon behaves in a systematic manner, are the atmosphere, terrestrial biosphere (usually includes freshwater systems), oceans, and sediments (includes fossil fuels). Each of these global reservoirs may be subdivided into smaller pools, ranging in size from individual communities or ecosystems to the total of all living organisms (biota).

Carbon Cycle

Carbon (C) is the basic building block of life. As the fourth most abundant element in the universe (after hydrogen, helium, and oxygen), carbon occurs in all organic substances, including DNA, bones, coal, and oil. Carbon moves through the hydrosphere, geosphere, atmosphere, and biosphere, each of which serves as a reservoir of carbon and carbon dioxide. The **carbon cycle** (**Figure 2.26**) involves the biogeochemical movement of carbon as it shifts between living organisms, the atmosphere, water environments, and even solid rock. The least amount of carbon is contained in the atmosphere while the largest amount is found in the lithosphere.

Carbon dioxide occurs in all spheres on Earth and as a gas is readily mobile. Our increased use of fossil fuels has increased the amount of carbon dioxide in the atmosphere, with some of gas being taken up in the hydrosphere and biosphere. Marine and freshwater organisms use carbon dioxide in their life cycles and can increase the amount of carbon dioxide in the geosphere. An example is the coral reefs that are formed by corals extracting dissolved carbon dioxide from sea water.

Generally there is a dynamic equilibrium between and among the four reservoirs of carbon. However, these are often changed by natural processes that contribute to the production and hence increased abundance of carbon dioxide. Volcanic eruptions and wildfires are two natural hazards that upset the carbon equilibrium. These two examples are discussed elsewhere in the text.

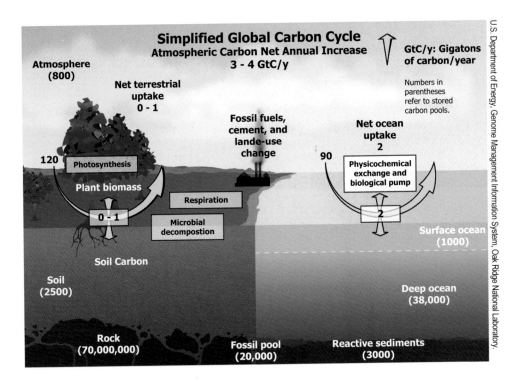

Figure 2.26 The carbon cycle shows where carbon is contained in the geosphere, hydrosphere, biosphere, and atmosphere. Each year more than 3 gigatons of carbon are added to the cycle.

Summary

The Earth consists of four spheres: (1) the geosphere, which is divided into three layers consisting of the core, mantle, and crust; (2) the hydrosphere, which consists of Earth's water in different reservoirs, with the oceans being the largest; (3) the atmosphere, which is a mixture of gases, mostly nitrogen and oxygen; and (4) the biosphere, which is the thin zone that life inhabits. Each of these spheres is continuously interacting with each other as a whole called the Earth system.

Interactions between the different spheres are accomplished through the exchange of matter and energy. Earth's geologic processes are powered by two energy systems: (1) the internal system powered by heat generated within the Earth and from gravity, driving mantle convection and tectonic plate movement; and (2) the external system powered by solar energy and from gravity, driving surface processes such as rivers, wind, and glaciers.

Rocks that make up the geosphere are classified into three groups: (1) igneous rocks, formed from the cooling of magma; (2) sedimentary rocks, formed from lithification of sediment composed of loose solid particle fragments or of precipitates from solution, and (3) metamorphic rocks, formed from preexisting rocks through application of heat and pressure. Over geologic time, each rock type is transformed into others in the rock cycle by internal and external processes.

The lithosphere includes the crust and solid upper mantle above the asthenosphere. Slabs of lithosphere called plates slowly move over the asthenosphere by convection in the mantle below. Plates move away from each other at divergent margins, move toward each other at convergent margins, and slide past each other along transform margins.

Solar energy heats the Earth's atmosphere, hydrosphere, and lithosphere, driving the interrelated surface processes of air and water circulation, erosion, and transportation and deposition of sediment. The hydrologic cycle describes water movement in the three states (solid, liquid, gas) between the atmosphere, hydrosphere, and lithosphere.

References and Suggested Readings

Coch, N. K. 1995. *Geohazards: Natural and Human.* Englewood Cliffs, NJ: Prentice Hall.

Kious, W.J., and R. I. Tilling, R. 1996. *This dynamic earth: The story of plate tectonics.* U.S. Geological Survey. http://geology.usgs.gov/publications/text/dynamic .html.

Lutgens, F. K. and E. J. Tarbuck. 2006. *Essentials of Geology*, 9th edition. Upper Saddle River, NJ: Pearson Prentice Hall.

McCall, G. J. H., D. J. C. Laming, and S. C. Scott. 1992. *Geohazards: Natural and man-made.* London: Chapman and Hall.

Murck, B. W., B. J. Skinner, and S. C. Porter. 1997. *Dangerous Earth: An Introduction to Geologic Hazards.* New York, NY: John Wiley and Sons.

Web Sites for Further Reference

http://biology.usgs.gov
http://geology.usgs.gov/publications/text/dynamic.html
http://www.ngdc.noaa.gov/mgg/global
http://www.platetectonics.com
http://www.usgs.gov
http://science.nationalgeographic.com/science/earth/the-dynamic-earth
http://ga.water.usgs.gov/edu/watercycle.html

Questions for Thought

1. Distinguish between the process of conduction and convection as ways to transfer heat energy within the Earth.
2. Describe the differences in formation of igneous, sedimentary, and metamorphic rocks.
3. What is the rock cycle, and what is the energy source driving it?
4. List the major internal layers of the geosphere on the bases of chemical composition and physical properties.
5. List and explain the types of plate boundaries.
6. List and briefly describe the four spheres of the Earth system.
7. What are the energy sources for the Earth system?
8. How did the Earth system form?
9. Describe the effects on the Earth's surface as a result of internal and external processes.
10. Explain how the hydrologic cycle operates and its role in external processes.

Volcanoes

3

Photo by K. Jackson, U.S. Air Force.

The eruption of Mount Pinatubo, Philippine Islands, June 1991.

Key Terms

active
ash
asthenosphere
basalt
block
bomb
caldera
cinder cone
composite volcano
decompressional melting
dormant
dust
extinct
isotope
lahar
lapilli
lava
lava dome
lithosphere

(Continued)

magma	pyroclastic	tephra
mitigation	rift valley	viscosity
phreatomagmatic eruption	shield volcano	volcanic explosivity index (VEI)

Regions at Risk: The Plate Tectonic Connection

magma
Molten rock that lies below the surface.

lava
Molten rock that flows onto the surface and cools.

basalt
An extrusive, fine-grained, dark volcanic rock that contains less than 50 percent silica by weight and a relatively high amount of iron and magnesium.

Earlier discussions of plate tectonics showed us that volcanic activity has played a major role in the development of Earth's outer crust. Volcanoes are the result of molten rock (called **magma**) that has worked its way to the surface and has been extruded onto either the continents or the ocean floor as **lava**. Lava can form volcanic mountains (**Figure 3.1**) and flows on land (**Figure 3.2**).

Magma rises under the mid-oceanic ridges as plates move away from each other in a divergent manner (**Figure 3.3**). The extruded lava is cooled relatively rapidly and forms lava flows and pillow **basalts**, pieces of hot material whose outer shell is cooled to produce a pillow shape (**Figure 3.4**).

As tectonic plates continue to move away from the ridge, more magma flows out onto the ocean floor and hardens, forming the basaltic oceanic lithosphere. This activity is not readily observed as it occurs under several kilometers of water.

Jennifer Adleman, Alaska Volcano Observatory/ U.S. Geological Survey (USGS) photo.

Figure 3.1 Augustine Volcano, Alaska, the most active volcano in the eastern Aleutian arc, rises to an elevation of 1,260 meters.

Courtesy of David M. Best.

Figure 3.2 Lava flow located in west central New Mexico.

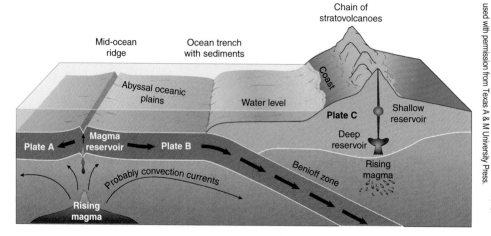

Figure 3.3 Cross section showing the divergent movement of oceanic lithospheric Plates A and B away from the mid-ocean ridge; Plate B is subducted underneath continental Plate C at a convergent plate boundary.

From *Volcanoes: An Introduction* by Alwyn Scarth, 1994. Copyrighted material used with permission from Texas A & M University Press.

Chain of stratovolcanoes

Mid-ocean ridge

Ocean trench with sediments

Abyssal oceanic plains

Water level

Coast

Plate C

Shallow reservoir

Deep reservoir

Plate A

Magma reservoir

Plate B

Rising magma

Probably convection currents

Benioff zone

Rising magma

Rising magma

Asthenosphere   Magma   Oceanic lithosphere   Continental lithosphere

Oceanic plates that move away from the mid-oceanic ridge in one area of the Earth eventually collide with other plates. When this collision occurs between two oceanic plates or between an oceanic plate and a continental plate, basaltic oceanic lithosphere and sediments that have accumulated on the oceanic plate are driven downward where they encounter increased heat. The trapped water in these subducted rocks is driven out of the sediments by the increased heat at depth. This dehydration process triggers melting of the subducted slab (**Figure 3.5**). At depths of roughly 100 kilometers there is sufficient heat to begin melting the subducted slab and produce new magma, which begins to rise toward the surface, forcing its way through the overlying rock.

Figure 3.4 Underwater lava flows add material to the ocean floor. Notice the spherical shape of the pillow basalts.

Lithospheres

Near the continental margin the subducted oceanic **lithosphere** is relatively cold as it is far removed from its heat source at the mid-oceanic ridge. The descending oceanic lithosphere is being pulled down by the effect of gravity tugging on the sinking, colder slab. In Figure 3.3 we see the Benioff zone, a region where earthquakes occur along the upper edge of the more brittle, subducted slab. As the subducted oceanic plate descends, sometimes it snaps into two pieces. Seismic evidence shows that there are regions in the subsurface where no earthquakes are occurring. This produces a seismic gap on the surface, a situation we will look at when we discuss earthquakes in detail.

Earth's surface consists of about a dozen lithospheric plates (**Figure 3.6**). Notice that continental plates such as the African plate and the North American plate also include oceanic areas as the plates extend to the mid-oceanic ridges and grow out from that boundary. The continents themselves consist of less dense, granitic material that "floats" atop the denser, underlying **asthenosphere** (Figure 3.5).

During recorded history, more than 500 volcanoes have erupted worldwide. More than 90 percent of these are associated with plate boundaries and approximately two-thirds of all active volcanoes are located around the Ring of Fire that circumscribes

lithosphere
The solid outer layer of the Earth that rests on the mobile, nonsolid asthenosphere. Continental lithosphere, having a granitic or granodioritic composition, ranges in thickness from about 30 to 60 km; basaltic oceanic lithosphere ranges between 2 and 8 km thick.

asthenosphere
The semi-plastic to molten zone beneath the rigid lithosphere.

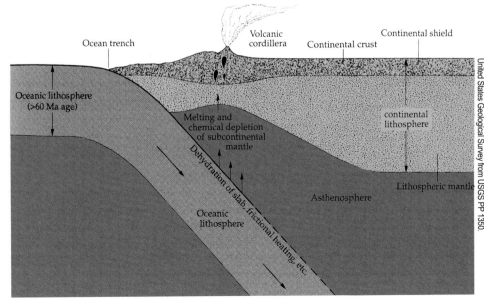

Figure 3.5 Subduction of oceanic lithosphere at a continental margin. Melting of the oceanic lithosphere, subcontinental mantle, and continental lithosphere and crust intermix to produce a magma that forms a chain of volcanoes near the edge of the continent.

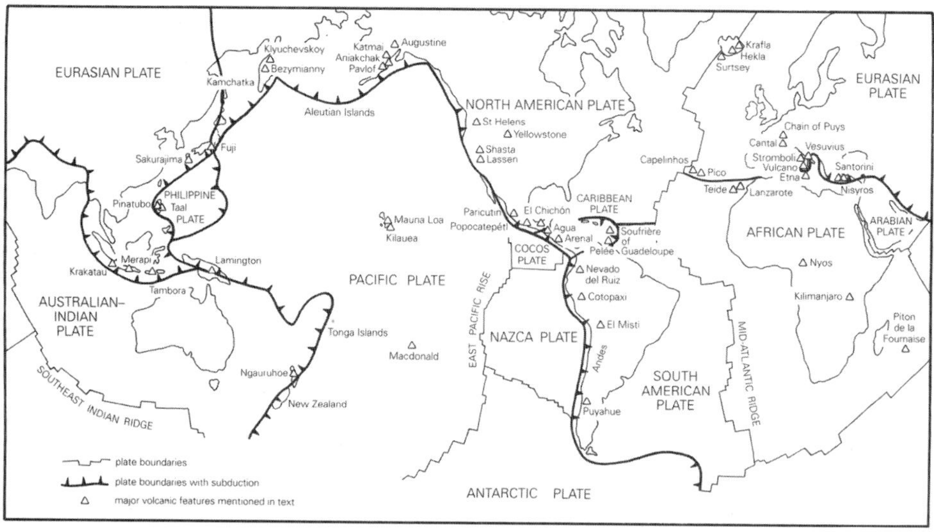

Figure 3.6 The major oceanic and continental plates on Earth's surface. Major volcanoes are shown by the triangles. Notice that many of the volcanoes are located along convergent plate boundaries where subduction is occurring.

the Pacific Ocean (**Figure 3.7**). Those areas on the Ring of Fire that display an almost continuous line of active volcanoes, such as in the Aleutian Islands southwest of Alaska or the area around Japan and the Kurile Islands, are regions where very active subduction is taking place. Both volcanic eruptions and significant earthquake activity occur in these areas as the Pacific Plate is subducted under the adjacent continental and other oceanic plates.

Of the world's 15 most populous countries, portions or all of five of the countries (the United States, Indonesia, Japan, Mexico, and the Philippines) lie on the Ring of Fire and have areas that could experience the direct effects of major volcanic eruptions. The hazard potential for large portions of Indonesia, Japan, and the Philippines is high and we hope that people in these areas are always on alert. Volcanic eruptions during the past two decades have been common in these regions.

Volcanoes also occur in **rift valleys**—regions on the Earth that are undergoing stretching due to large scale, extensional tectonic forces. There are also areas on the continents and in the oceans above sea level, such as across Iceland, where the land

rift valley
A depression formed on the surface caused by the extension of two adjacent blocks or masses of rock.

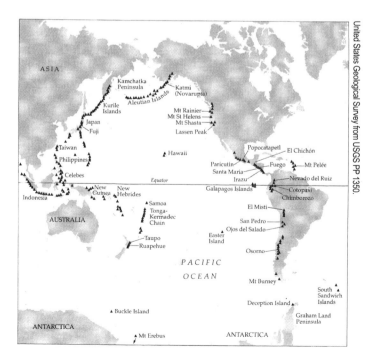

Figure 3.7 Major volcanoes found in the circum-Pacific Ring of Fire and adjoining regions.

is being pulled apart. The Rio Grande rift, which is situated in the southwestern United States, has volcanoes and lava flows. The East African Rift System, which passes through Uganda, Kenya, Tanzania, and Ethiopia, is home to many volcanic peaks, including Mount Kilimanjaro (**Figure 3.8**). At an elevation of 5,895 m (19,340 ft), it is the world's highest free-standing mountain.

Figure 3.8 Mt. Kilimanjaro, the world's highest free-standing mountain, is the result of volcanism related to the East African Rift System.

Magma Generation and Eruption Products

Volcanoes form from magma, which is termed lava when it is extruded onto either a continental or ocean floor surface. How does melting of rock occur in the subsurface? Three contributing factors come into play: heat, pressure, and fluids. Obviously we need a lot of heat to melt rock. Depending on the conditions and composition of the rocks, they can usually melt between 750°C and 1200°C.

Heat Sources

One source of heat is thermal energy given off when naturally occurring radioactive elements decay in the crust and upper mantle. Many elements have radioactive **isotopes**, which are unstable variations of a chemical element. However, three elements have radioactive isotopes that produce the vast majority of the heat in the outer portions of the Earth. Uranium, thorium, and potassium have relatively abundant isotopes (U-235 and U-238, Th-232, and K-40, respectively) that are unstable, causing them to decay. This natural disintegration process generates a great deal of heat. Although uranium and thorium are relatively rare in the overall makeup of the Earth, they are present in sufficient quantities to contribute significantly to the melting process.

In subduction zones some heat is generated by the friction of the subducted slab with the surrounding rock. There are debates about the role that frictional melting plays in the subduction of an oceanic plate, but obviously there is friction present; however, it is lessened by water that is driven down along the upper edge of the plate. This water plays a large role in reducing the temperature needed to melt the rock.

Rocks are poor conductors of heat. You know this is true if you have ever gone into a cave on a warm day. Air in the cave is much cooler than the outside air because surface heat has not penetrated the ground overlying the cave. This property of poor conduction allows heat generated in the subsurface from radioactive decay to remain in the vicinity of the heat-producing elements. Over time, sufficient heat builds up to help produce melting of the surrounding rock.

Remember that major melting takes place at a depth of approximately 100 km. At that depth there are intense pressures compacting rocks in the upper mantle. This compression on the minerals increases the temperature required to bring about their melting. As the magma increases in temperature, it begins to expand and rise to the surface. This decrease in depth of burial in turn decreases the pressure exerted on the rocks and minerals. Accordingly, less heat is now required to maintain the molten state. This process is termed **decompressional melting**.

Water can be added to magmas that originate near mid-oceanic ridges or subduction zones. Continental sediments that have been carried into ocean trenches can be dragged down as subduction takes place. These sediments are subsequently melted and become incorporated into the magma. Water molecules could also be present in some minerals and rocks and these become assimilated by the magma as it rises to the surface (**Figure 3.9**).

Another source of heat for localized melting is associated with hot spots, plumes of molten rock rising from deep within the lower crust and mantle. One of these features can be thought of as a stationary mantle blow torch aimed at the surface. As a tectonic plate moves over this heat source, the plate is partially melted, thereby

isotope
A different form of an element that has the same atomic number but a different atomic weight due to a change in the number of neutrons present.

decompressional melting
Melting of hot rocks in the subsurface caused by a reduction in overlying pressure as the material rises toward the surface.

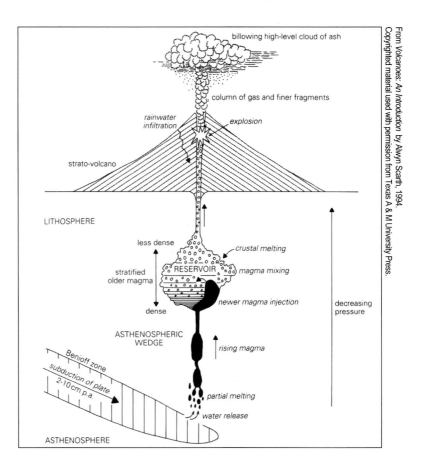

Figure 3.9 Geologic conditions associated with volcanic eruptions at a subduction zone.

producing magma for a volcano. As the plate's position shifts, a new volcano is formed while the previously formed volcanoes have moved off the hot spot. The string of islands that make up the Hawaiian Islands and the Emperor Seamount chain are excellent examples of this process (**Figure 3.10**). We know the hot spot stretches more than 120 km since significant activity is occurring at Mauna Loa, Kilauea, and under the Loihi, which is a submarine volcano situated off the southern coast of the Island of Hawaii.

We can analyze the ages of the volcanoes and their distances from the hot spot to determine rates of motion of the Pacific plate, whose motion has averaged about 8.6 cm per year (or 860 km per 10 million years; **Figure 3.11**). Notice that a change in the rate of motion of the Pacific plate occurred between 45 and 50 million years ago. Prior to that time the bend in the trace of the sea floor features near Abbott and Yuryaku Seamounts indicates a change in direction of plate motion in the Pacific plate (Figure 3.10).

In the northwestern United States, extensive basalt flows indicate a possible hot spot positioned under Yellowstone National Park in northwestern Wyoming signifying a present-day heat source underneath a continent. Progressively older volcanic eruptions to the west demonstrate this western movement of the North American continent (**Figure 3.12**). The "yellow stones" seen in the national park are more silicic-rich volcanic tuffs. Another possible explanation of the volcanic activity in this region is that extensional forces are pulling apart, or extending, the continent and hence magmas are rising from the upper mantle and being extruded on the surface.

Other influences contribute to heat production, including residual heat from Earth's early history. Earth's outer core is molten and supplies heat to the mantle. This heat produces large convection cells that move the continental and oceanic plates. The internal temperature of the Earth is estimated to be between 4,000°C and 5,000°C. Also, tidal friction related to the slowing down of Earth's rotational velocity generates heat as the deep layers inside the Earth rub past one another.

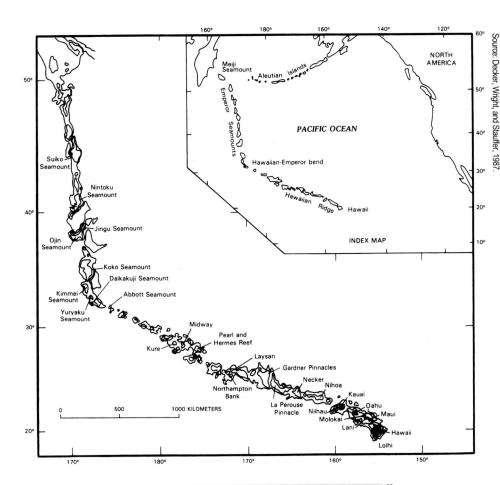

Source: Decker, Wright, and Stauffer, 1987.

Figure 3.10 The Hawaiian islands and the Emperor Seamount chain extend out from the hot spot lying under the island of Hawaii. Ages of the rocks are progressively older to the northwest—evidence that the Pacific plate has moved across the hot spot. Exposed islands lie to the southeast of Midway Island; the remaining islands to the northwest are underwater.

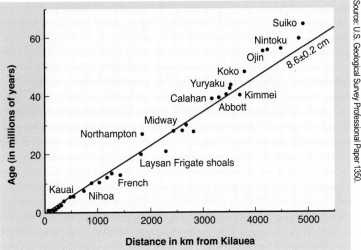

Source: U.S. Geological Survey Professional Paper 1350.

Figure 3.11 A plot of isotopic ages of samples from islands shown vs. distance from Kilauea. Vertical axis is age in million of years.

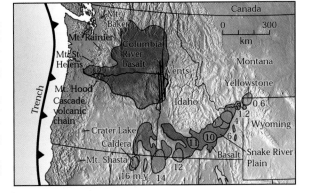

Figure 3.12 Episodes of volcanic eruptions in the Pacific Northwest. Note the progressive increase in ages of flows lying to the southwest of Yellowstone National Park (ages in millions of years from the present). Movement of the North American plate has been to the southwest over the time period shown.

Volcanic Eruptions

The type of ejecta and the way in which it is erupted determine the shape and structural form a volcano takes. Magma rising from the magma chamber or reservoir ascends through dikes and a central conduit. Several factors control the way in which lava and other volcanic debris are spread onto the surface and thrown into the atmosphere.

Role of Magmatic Gases

In addition to the molten rock that is being produced, gases are often trapped in the magma as it rises to the surface. This combination of molten rock and gas is under pressure at depth, so when the gases reach the surface, they expand rapidly. This concept is illustrated when we shake a bottle of carbonated soda water and open it. Carbon dioxide dissolved in the soda escapes rapidly and explosively when the cap is removed.

Volcanic gases undergo the same expansion as they work their way to the surface, and the resulting expansion throws gas, lava, and surface rocks into the atmosphere and across the landscape. Fragmented material can include volcanic **dust**, **ash** (less than 2 mm in average particle diameter), **lapilli** (2 to 64 mm), and angular **blocks** and smooth **bombs** (greater than 64 mm). In some cases very large blocks and bombs ranging up to several meters across are located near the vent. Many different gases are produced by volcanic eruptions. The primary gas is water vapor, which comprises from 50 percent in basaltic and less silica-rich lavas to as much as 90 percent or more in some silica-rich magmas. Primary volatiles dissolved in magma come mostly from the mantle source area of the magma.

Water that is picked up in the shallow subsurface contributes very little to the overall magma composition. Interaction between hot magma and water on or near the surface produces steam explosions that are termed **phreatomagmatic eruptions**.

Scientists think that over several billions of years of eruptions, enough water has been ejected to have created Earth's oceans. In addition to all the expelled water vapor, other superheated gases have been thrown into the atmosphere. These include carbon dioxide (CO_2), sulfur dioxide (SO_2), hydrogen sulfide (H_2S), carbon monoxide (CO), methane (CH_4), chlorine (Cl_2), nitrogen (N_2), and gaseous acids such as HCl and HF. Several of these gases, when combined with water, produce strong acidic conditions that naturally generate acid rain. This process occurs today when large volcanic eruptions propel gases into the air.

Carbon dioxide is a common gas that is both colorless and odorless. Because its molecule contains an atom of carbon, carbon dioxide (CO_2) is heavier than oxygen (O_2). When carbon dioxide is expelled from a volcano, it can settle along low-lying areas and displace the oxygen. Numerous examples exist of animals and humans being affected by carbon dioxide. In 1986 an eruption of carbon dioxide killed many people and animals in Cameroon (see later in the chapter). In Iceland, sheep and small birds have died just before major volcanic events (Sigurdsson, 1999) and field geologists have noticed buildups of carbon dioxide in low-lying areas on the Island of Hawaii (Duffield, 2003) during eruptions.

Mineral Composition of the Magma

Another factor controlling eruptions is the mineral composition of the magma itself. Magmas rich in silicate minerals such as orthoclase, feldspars, and quartz have complex bonds that increase the **viscosity** of the lava and hence prevent it from flowing easily to the surface (see **Box 3.1**). Magma viscosity is controlled by the amount of silica present. These magmas sometimes cool and solidify in the subsurface to produce plutons, which are typically granitic in composition.

However, if magma has a higher percentage of silicate minerals with compositions containing simpler silicate structures, the magma moves readily because of its lower viscosity (Table 3.1). Once on the surface, lava spreads out forming basaltic flows that can cover hundreds of square kilometers. Note that a given volcano can display both an explosive and a quieter, effusive style throughout its lifetime depending on the amount of gas present in the magma and the composition of the magma, which can change over time.

---

dust
Pyroclastic material that is silt- or clay-size (less than 1/256 mm).

ash
Pyroclastic material that has an average particle size less than 2 mm.

lapilli
Pyroclastic material that ranges in size from 2 mm to 64 mm; sometimes referred to as cinders.

block
A large angular fragment of lava measuring more than 64 mm in diameter.

bomb
A large lava fragment larger than 64 mm in diameter that becomes rounded as it is thrown into the air.

phreatomagmatic eruption
Term used to describe volcanic eruptions that involve magma and water coming in contact; the resulting steam caused extremely explosive activity.

---

viscosity
A measure of the internal resistance of a substance to flow; a lower viscosity means the material flows easily.

| BOX 3.1 | **Influence of the Silicon-Oxygen Tetrahedron on Viscosity** |

Many of the minerals that make up common volcanic rocks contain the silicon-oxygen tetrahedron, the basic building block for more than one-third of the more than 4,000 minerals found on Earth. Oxygen and silicon are the two most common elements in Earth's oceanic and continental lithospheres. The silicon-oxygen tetrahedron consists of one silicon cation surrounded by four oxygen anions, thereby producing a negatively charged anionic complex having a charge of -4. To offset this negative charge, cations need to be present.

Structures of Silicate Minerals Based on Combinations of the Silica Tetrahedon

	Structure	Formula of *Silicon-Oxygen Unit*	Examples
	Single Tetrahedron	(SiO_4)	Olivine
	Chains	(SiO_3)	Pyroxene: augite
	Double Chains	(Si_4O_{11})	Amphibole: hornblende
	Sheets	(Si_2O_5)	Mica: muscovite biotite
	Three Dimensional Framework	(SiO_2)	Quartz Feldspar

Measuring Volcanic Explosivity

Scientists have devised a measure of the degree of explosivity observed in volcanic eruptions, the **volcanic explosivity index** (VEI). This index assigns a single value that describes the severity of an eruption (Table 3.2). This scale is roughly analogous to one of the magnitude scales used to measure the severity of an earthquake, a tornado, or a hurricane. Through a great deal of research, the developers of the VEI were able to assign values to ancient eruptions. Observations of the duration of the

volcanic explosivity index (VEI)
A measure of the intensity of a volcanic eruption, with a value of 0 representing the quietest and 8 being the most explosive.

TABLE 3.1	Descriptive Terms for Extrusive Igneous Rocks and Their Relationships Based on Chemical Content and Color		
Characteristic	**Rhyolite**	**Andesite**	**Basalt**
	Felsic	Intermediate	Mafic
Silica content	>65%	55–65%	50–55%
Viscosity	High	Intermediate	Low
Color	Light	Gray	Dark
	High amounts ⟷ Low amounts Sodium (Na), potassium (K), and aluminum (Al)		
	Low amounts ⟷ High amounts Calcium (Ca), magnesium (Mg), and iron (Fe)		

Silica content refers to the amount of silicon dioxide present.

TABLE 3.2	Volcanic Explosivity Index (VEI)						
VEI	**Volume of Ejecta (m^3)**	**Description**	**Height of Eruption Column (km)**	**Classification**	**Duration (hrs)**	**Troposphere**	**Stratosphere**
0	$<10^4$	nonexplosive	<0.1	Hawaiian		negligible	none
1	10^4–10^6	small	0.1–1	Strombolian		minor	none
2	10^6–10^7	moderate	1–5			moderate	none
3	10^7–10^8	moderate-large	3–15	Vulcanian	1–6		possible
4	10^8–10^9	large	10–25		6–12		definite
5	10^9–10^{10}	very large	>25			substantial	
6	10^{10}–10^{11}			Plinian	>12		significant
7	10^{11}–10^{12}			Ultra-Plinian			
8	$>10^{12}$						

Note that 1 cu km = 10^9 cu m. Modified from Newhall and Self (1982).

main eruption, height of the ash and debris column, volume of material erupted, and volume of ash cast into the upper atmosphere are used to determine the VEI of a volcanic event. Values range from 0 to 8, with the vast majority of events receiving values of 1 or 2. This range in eruptive styles generates a variety of surface volcanic features (**Figure 3.13**).

The degree of explosivity of a volcano is related to the viscosity or thickness of the magma. Highly silicic magmas have a high viscosity and do not flow very well, if at all. Hence magmas of this type tend to plug up the vent and produce an explosive event.

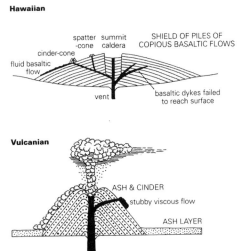

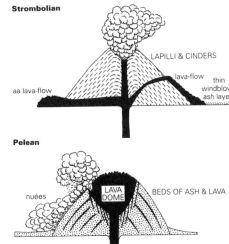

Figure 3.13 Examples of different eruptive styles. Refer to the discussion on the VEI to compare the degree of explosiveness with the resulting eruptions.
From *Volcanoes: An Introduction* by Alwyn Scarth, 1994. Copyrighted material used with permission from Texas A & M University Press.

Magmas that are low in silica, such as basaltic magmas, flow very easily and thus the lavas move away quickly once the molten material works its way to the surface. These are called Hawaiian eruptions (see Figure 3.13). If a large amount of gas is present in the magmas, Strombolian eruptions can form. These are mainly cinder cone types of features. Magmas that are intermediate in composition between the highly silicic and the low silica magmas are in the majority on the continents. These eruptions consist of combinations of flows and explosions, thus forming the common composite volcanoes. Vulcanian, Pelean, and Plinian types of eruptions form when there is a combination of lava flows and high amounts of gases in the magma. The structural forms that these types of eruptions create is discussed in the section on types of volcanoes.

Figure 3.14 A series of basaltic lava flows along the Columbia River on the Washington-Oregon border.

Low-Level Volcanism

Quiet eruptions are generally characterized by basaltic magmas that have small amounts of gases and a low silica content and thus flow easily once they are extruded onto the surface (**Figure 3.14**). Any gases contained in the magma are able to escape fairly quickly due to the low viscosity of the lava. Most eruptions on the island of Hawaii are basaltic lava flows that gently move along the surface. These fissure eruptions, which commonly evolve quickly (within a few hours), occur along cracks in the surface that display curtains of fire and ejected material that sometimes rises several hundred meters into the air. However, the **pyroclastic** material falls back to Earth a short distance from the source area. Such eruptions are termed Hawaiian after the style of eruption that occurs in the Hawaiian Islands and produces low relief volcanic structures, termed shield volcanoes.

pyroclastic
Related to material that is thrown out by a volcanic eruption.

Types of Volcanoes
Shield Volcanoes

Shield volcanoes are the largest volcanoes in terms of the area they cover and are so named because they have the broad appearance of an inverted shield. They are formed when low viscosity magmas are slowly extruded onto the surface and spread out over large areas. These flows build up over time and produce very large structures such as those found in the Hawaiian Islands and Iceland. Mauna Loa, probably the

shield volcano
A broad volcano that has gentle slopes consisting of low viscosity basaltic lava flows.

largest volcano on Earth, has an elevation of 4,175 m (13,697 ft) above sea level and rests on the ocean floor in 5,800 m (19,000 ft) of water (**Figure 3.15**). Generally eruptions associated with shield volcanoes are quiet and do not move across the surface very fast. A car could certainly move faster than one of these eruptions; it would be possible to outrun one on foot.

Figure 3.15 Mauna Loa, on the island of Hawaii, appears along the horizon.

Cinder Cones

Cinder cones build up when small blobs of basaltic lava are thrown through the air, solidify during flight, and fall to the ground to construct a pile of layered pyroclastics near the vent (**Figure 3.16a**). These particles include ash, lapilli, and some bombs and blocks (collectively referred to as **tephra**). As they are cast out of the volcano, they pile up and create a conical structure called a cinder cone. Because little or no lava flows out the top of the volcano, the cinders remain as loose grains that are susceptible to erosion. Occasionally lava flows are forced out the base of a cinder cone as the central conduit begins to fill with debris (**Figure 3.16b**).

Cones are generally no more than 300 m (1,000 ft) in height and pose a threat only to areas in the immediate vicinity of the eruption. Wind can transport some of the smaller ejecta over considerable distances. As the ash and lapilli break down by chemical weathering, they produce very rich soils that allowed civilizations to develop their agriculture-based societies. The problem for these people was that once they established an area as a village or town, they were often directly affected by future eruptions and either had to move or were wiped out by the eruptions.

As the level of eruptive activity in the volcano increases, lava and pyroclastics are ejected onto the surface and into the atmosphere and can build a cone-shaped structure that increases in size and height. If the majority of material thrown out is cinders, the resulting form is a cinder cone (**Figure 3.17**).

Composite Volcanoes

Perhaps the most widely recognized is a **composite volcano**, which has a symmetrical, generally conical shape. The volcanic peaks of Japan, the Philippines, Alaska, and the Pacific Northwest in the United States are excellent examples of composite volcanoes. They form from deposits of pyroclastic debris that build up near the vent

cinder cone
A conical-shaped hill created by the build up of cinders (lapilli) and other pyroclastic material around a vent.

tephra
A general term for all types of pyroclastic material produced by a volcano.

composite volcano
A volcano that forms from alternating layers of lava and pyroclastic debris; also known as a stratovolcano.

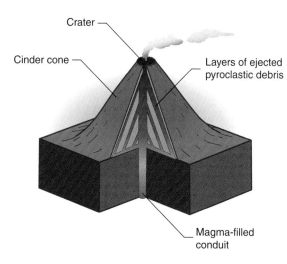

Figure 3.16a Block diagram showing how pyroclastic debris builds up to form a cinder cone.

Figure 3.16b SP crater in northern Arizona consists of unconsolidated material.

Figure 3.17 Volcanic cinder cones consist of unconsolidated material.

Figure 3.18 Mayon volcano, rising to an elevation of 2462 m in southeast Luzon, is the most active volcano in the Philippines. It has steep upper slopes averaging 35–40 degrees that are capped by a small summit crater. A single central conduit system has allowed the volcano to build up into a classic, symmetrical profile.

and then are covered by occasional lava flows that move down the flanks to "glue" the looser debris in place. This repeated combination of loose material and lava produces stratified layers that create the awe-inspiring, cone-shaped mountains we associate with volcanoes (**Figure 3.18**). These are made mostly of andesite but usually include some rock types that are both higher and lower in silica.

Magma with higher gas and silica content throws material much farther into the atmosphere. Vulcanian and sub-Plinian eruptions are characterized by profuse columns of ash and gas that rise to heights of 1 km to almost 25 km above sea level. A very gas-rich explosion that carries material in a fairly well-defined vertical column to heights ranging from 10 to more than 25 km into the atmosphere is termed Plinian, named for the Greek writer Pliny the Younger who described the eruption of Mt. Vesuvius that destroyed Pompeii and Herculaneum and killed his uncle, Pliny the Elder, in A.D. 79.

A Pelean-style eruption is characterized by highly viscous, gas-rich magmas that contribute to extreme explosions. Glowing, incandescent pyroclastic flow (sometimes called nuée ardentes) rolls down the volcanic slopes. These superheated dense mixtures of solids and gas hug the terrane and destroy everything in their path.

Figure 3.19 View of the dome inside the crater of Mount St. Helens, Washington.

Lava Domes

During the extrusive phase of many volcanoes, the more fluid portions of the magma chamber move out onto the surface or are thrown out by dissolved gases. As an eruption slows down, the remaining magma becomes thicker as the composition of the magma becomes more silica rich and develops a higher viscosity. Magma extruded at the vent thickens and creates a **lava dome** either just under or on the surface. Within a year or two of the major eruption of Mount St. Helens in May 1980, a dome formed in the inner crater and activity slowed down (**Figure 3.19**). A dome located at the top of a volcano can act like a cork in a bottle and hold in gases and magma until there is sufficient force to cause a later eruption that will blow the dome out of the way. Following the main eruption of Mount St. Helens in May 1980 there have been additional eruptions that have added almost 150 million cubic yards of material to the dome and surrounding area (see **Box 3.2**).

lava dome
A dome-shaped mountain formed by very viscous lava flows.

BOX 3.2	The Eruption of Mount St. Helens May 18, 1980

Facts about the Volcano

Summit elevation
    Before   9,677 ft (2950 m)   After 8,363 ft (2549 m)
    Removed   1,314 ft (401 m)
Crater dimensions
    East-west   1.2 miles (1.93 km)
    North-south   1.8 miles (2.9 km)   Depth 2,084 ft (635 m)
    Volume of material removed   3.7 billion cu yds (2.8 billion cu m)

Facts about Fatalities

Human loss of life   57
Wildlife   At least 7,000 big game animals and 12 million salmon fingerlings in hatcheries;
        countless nonburrowing wildlife in blast zone

Facts about the Eruption Column and Cloud

Height	Reached 80,000 ft (24,384 m) in 15 minutes
Downwind extent	Spread across the United States in 3 days; circled the globe in 15 days
Volume of ash	1.4 billion cu yds (1.07 billion cu m)
Ash fall area	Detectable amounts covered 22,000 square miles (56,980 sq km)
Ash fall depth	10 in (25.4 cm) at 10 miles (16 km) downwind (ash and pumice); 1 in (2.5 cm) at 60 miles (100 km) downwind; fractions at 300 miles (500 km) downwind

Facts about Pyroclastic Flows

Area covered	6 sq mi (15.5 sq km); reached 5 mi (8 km) north of crater
Volume and depth	155 million cu yds (199 million cu m); many flows 3 to 30 ft thick; up to 120 ft thick (37 m) in some locations
Velocity	Estimated at 50 to 80 mph (80 to 130 kph)
Temperature	At least 1,300°F (700°C)

Facts about the Landslide

Area and volume removed	23 sq miles; 3.7 billion cu yds (60 sq km; 2.83 billion cu m)
Deposit depths	Buried 14 miles (22.5 km) of North Fork of Toutle River Valley to an average depth of 150 ft (46 m) (deepest deposits 600 ft or 183 m)
Velocity	70 to 150 mph (113 to 241 kph)

Source: http://pubs.usgs.gov/fs/2000/fs036-00/-United States Geological Survey.

Calderas

In terms of the destructive potential of various types of volcanoes, **calderas** are by far the biggest threat. However, they are also the least likely to occur, so we are relatively safe from the mega-hazard their explosions would produce. Mega-eruptions are termed ultra-Plinian and are characterized by extreme volumes of ejected material exceeding 10^9 or more cubic meters (1 cubic km) of material. The largest of these result in subsidence bowls (calderas) over the vent, which form toward the end of the eruption process. Yellowstone National Park, located in the northwestern corner of Wyoming and stretching into portions of Idaho and Montana, is centered

BOX 3.2	**The Eruption of Mount St. Helens May 18, 1980** (*continued*)

Facts about the Lateral Blast

Area covered	230 sq mi (596 sq km); extended to 17 mi (27 km) NW of crater
Volume of deposit	250 million cu yds (190 million cu m)
Velocity	At least 300 mph (480 kph)
Temperature	Up to 660°F (350°C)
Energy released	24 megatons (MT) of thermal energy (blast generated 7 MT; remainder was released heat)
Trees flattened	4 billion board ft (enough to build 300,000 2BR homes)

Facts about Lahars

Velocity	Between 10 and 25 mph (16 to 40 kph) (50 mph or 80 kph) on steep slopes on side of volcano)
Destruction caused	27 bridges; almost 200 homes
Effects on Cowlitz River	Reduced flood stage capacity at Castle Rock from 76,000 cfs (2150 cms) to less than 15,000 cfs (425 cms)
Effects on Columbia River	Reduced channel depth from 40 ft (12 m) to 14 ft (4.3 m); 31 ships stranded in ports upstream

Source: http://pubs.usgs.gov/fs/2000/fs036-00/

The U.S. Geological Survey (USGS) reported that the bulge and surrounding area slid away in a gigantic rockslide and debris avalanche, releasing pressure and triggering a major pumice and ash eruption of the volcano. Thirteen-hundred feet (400 meters) of the peak collapsed or blew outward. As a result, 24 square miles (62 square kilometers) of valley were filled by a debris avalanche; 250 square miles (650 square kilometers) of recreation, timber, and private lands were damaged by a lateral blast; and an estimated 200 million cubic yards (150 million cubic meters) of material were deposited directly by lahars (volcanic mudflows) into the river channels. Fifty-seven people were killed.

Economic losses resulting from the eruption of Mount St. Helens included almost $900 million in revenues lost for timber, agriculture, and fisheries. Damage to the surrounding infrastructure, including roads and bridges, along with dredging costs of rivers, amounted to almost $460 million (Blong, 1984).

United States Geological Survey. Photo by Austin Post.

Mount St. Helens, May 18, 1980.

on a massive caldera covering almost 3,500 sq km (1,350 sq mi). This area has experienced three major eruptions. The first occurred approximately 2 million years ago, followed by another about 700,000 years later. The most recent major eruption took place about 640,000 years ago. The result of these eruptions was the ejection of almost 3,800 cu km (900 cu miles) of highly silicic volcanic debris. The area continues to be geologically active today with a well-developed geyser system emitting steam into the atmosphere (**Figure 3.20**). Earthquake activity is ongoing in the park and adjoining regions.

Courtesy of David M. Best

Figure 3.20 Fuming geysers at Yellowstone National Park.

Types of Volcanic Hazards

Emission of Gases

Figure 3.21 Scientists sample noxious gases following a minor eruption of Mount St. Helens.

Gases are present in magma and represent a significant component of volcanic eruptions. The hazardous nature of the lethal emissions is apparent when we consider that large amounts of carbon dioxide, sulfur dioxide, hydrogen sulfide, carbon monoxide, and nitrogen are emitted (**Figure 3.21**). All these are highly poisonous to breathing organisms and plant life. Many of these gases combine with rain and surface water to produce strong acids that increase chemical weathering of rocks.

On rare occasions, an eruption of only gas can occur. A documented case happened in August 1986 in Cameroon, West Africa. Lake Nyos, situated in the crater of a volcano at an elevation of 1,091 m, is relatively deep given its small size (about 200 m deep, 1 to 1.5 km wide). Little intermixing of its cold, CO_2-saturated bottom waters with shallow, warm water normally takes place. However, as the result of the rapid expulsion of gas at velocities of up to 100 km per hour, more than 1,700 people were suffocated along with several thousand head of cattle when a dense cloud of carbon dioxide was expelled. The denser carbon dioxide cloud moved down several stream valleys inflicting its wrath over 14 km from the source. Most researchers felt the cause of this tragedy was a convective overturn of carbon dioxide dissolved in water at the bottom of the lake. Some scientists thought the carbon dioxide was volcanic in origin. Thus two diametrically opposed groups of scientists, based mainly on differing political views and hindered by language problems, continue to argue about the actual cause of the event (Scarth, 1999).

Lava Flows

Figure 3.22 Basalt flows at Kilauea, December 1967.

Lava flows are produced when molten rock extrudes out onto Earth's surface. The ease with which it flows is controlled by the composition and temperature of the magma. Magmas that are rich in silica, such as those with a rhyolitic composition, have a high viscosity and barely flow at all. Basaltic magmas, those containing less silica, flow readily onto and across the surface. The island of Hawaii has been constructed by many episodes of eruption of basaltic lavas that continue today (**Figure 3.22**). As the lava cools and hardens, it forms new land on which future flows will travel and subsequently harden. Given the relatively slow movement of lava flows, it is easy to avoid them. However, many homes have been destroyed by these flows on the island of Hawaii in the past 25 years while Kilauea has been erupting.

Lahars

lahar
A volcanic mudflow or landslide that contains unconsolidated pyroclastic material.

Superheated material coming out of a volcano, especially any pyroclastic clouds, can transform water, snow, and ice on the surface into a roiling mass of hot slurry that rapidly moves downslope across the landscape and through preexisting stream channels. This moving mass is termed a **lahar**, an Indonesian term that describes this common occurrence in that country. Lahars can move at speeds of more than 80 km per hour.

Approximately two-thirds of all known volcanoes are found in the Northern Hemisphere, which is where a similar proportion of Earth's landmass is also situated. Roughly 38 percent of known volcanoes are located in temperate latitudes or higher (**Figure 3.23**). Volcanoes located in equatorial regions often have high elevations that allow them to be snow covered. Numerous volcanoes in wet climates have ash falls that will produce lahars during the wet seasons.

These regions experience significant snowfalls that can build up massive amounts of water and become lahars during volcanic eruptions. The eruption of Mount St.

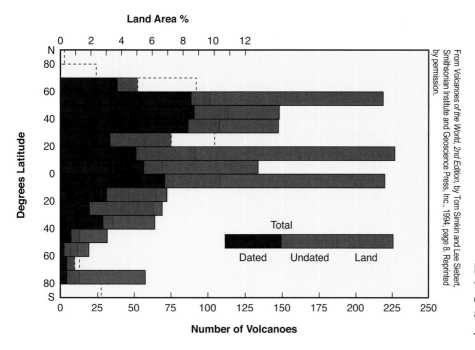

From *Volcanoes of the World, 2nd Edition,* by Tom Simkin and Lee Siebert, Smithsonian Institute and Geoscience Press, Inc. 1994, page 8. Reprinted by permission.

Figure 3.23 Distribution of known volcanoes per 10 degrees of latitude. The dashed line shows the percentage of land area per 10 degrees.

Helens melted ice and snow on the flanks of the peak and sent debris through the Toutle River, destroying roads, bridges, and homes along the river (**Figure 3.24**).

The far-reaching effects of lahars became evident during the eruption of Nevado del Ruiz in Colombia in November 1985, when the resulting lahar traveled more than 75 km from the mountain, covering the city of Armero and surrounding villages (**Figure 3.25**). More than 23,000 people died. The eruption of Mt. Pinatubo in the Philippines in 1991 was followed by more than 200 lahars produced by the rainy season in the spring and summer. These have continued for years, and some current-day lahars are undoubtedly related to the 1991 explosion.

Pyroclastic Flows

A pyroclastic flow, which has been termed by some researchers a volcanic hurricane, is a very hot, ash-rich flow of dust, rocks, and surface debris that blasts out of an eruption at velocities reaching 200 km per hour (**Figure 3.26**). The flow can cover hundreds or even thousands of square kilometers and produce deposits more than 100 meters thick. The larger grained, denser material rolls along the Earth's surface, following the surface topography and destroying everything in its path. The term *nuee ardente* was used to describe the fiery cloud associated with the 1902 eruption of Mount Pelée on the island of Martinique in the Lesser Antilles.

Directed Blasts

Major eruptions are accompanied by explosions that are produced by expanding gas as it rises to the surface and blasts out into the atmosphere. If an explosion is directed upward, pyroclastic material can rise to more than 30 km into the atmosphere. Finer particles are carried airborne for hundreds of kilometers or more, coating the landscape with a fine ash layer.

In most instances venting is directed vertically, thus throwing a large amount of ash and small debris into the atmosphere. Large eruptions can eject massive amounts of ash that can remain in the atmosphere for months or longer. Evidence

Photo by R. L. Schuster, United States Geological Survey.

Figure 3.24 Debris in North Fork of the Toutle River, northwest of Mount St. Helens, July 1980.

Photo by R. J. Janda, United States Geological Survey.

Figure 3.25 Aftermath of the eruption of Nevado del Ruiz, Colombia in November 1985.

of this is seen in spectacular sunsets caused by this ash. The eruption of Tambora in Indonesia in April 1815 had a worldwide effect the following year on weather to the degree that in the northeastern United States there was a significant decrease in solar radiation and temperature. The year 1816 became known as the "year without a summer" (see **Box 3.3**). Lower temperatures had a direct effect on crops. This climatic effect was caused by dust particles and sulfur dioxide reflecting and absorbing solar radiation. Atmospheric gases, particularly sulfur derivatives such as sulfur dioxide and sulfur trioxide, are capable of absorbing large amounts of infrared radiation energy, thus producing a longer term effect.

If the blast is directed laterally out the side of the volcano, the energy is oriented along the ground and can have a devastating effect on everything in its path (**Figure 3.27**). Such was the case with the May 1980 eruption of Mount St. Helens in southwestern Washington. The initial blast was directed to the northeast and thousands of acres of trees were flattened and large amounts of debris clogged the rivers and streams. The volcano sent ash and debris vertically to a height of 25 km (80,000 ft) in 15 minutes (see Box 3.2). The U.S. Geological Survey reported that ash clouds stretched across the United States in a period of three days and ash circled the Earth in 15 days. Ash remained aloft for several years.

United States Geological Survey.

Figure 3.26 On August 7, 1980, an eruption of Mount St. Helens produced this fully developed ash flow that moved down the north flank at speeds in excess of 100 km/hr. An ash cloud rises above the basal pyroclastic flow. The main Plinian column is the higher, more diffuse cloud in background.

BOX 3.3	One Volcano's Influence on People's Lives

In the spring of 1816, Percy Bysshe Shelley, the English poet, planned to spend some time in Switzerland with his new wife, Mary. Their plans were drastically upset by a major change in the anticipated summer weather by the eruption of Tambora the previous year. Because of a prolonged period of very inclement weather, they along with the poet Lord Byron spent a lot of time indoors. They entertained each other by writing ghost stores—the end result of which was Mary Shelley's opus *Frankenstein*.

The effects of Tambora on New England and adjoining areas were also considerable. Early plantings of corn, the staple crop in the early 1800s, were devastated by severe frosts. These were followed by a drought and several more frosts. What was normally a growing season ranging between 120 and 160 days was reduced by up to 50 percent. The cause for these crop failures and weird weather in North America and Europe was not determined for more than 80 years. The eruption of Tambora almost halfway around the globe indeed had a far-reaching effect on Earth's climate.

United States Geological Survey.

Figure 3.27 Notice the distinct line separating downed and standing trees from the blast from Mount St. Helens. Standing trees were killed as a result of the blast.

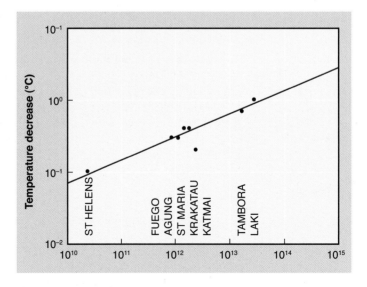

Figure 3.28 Global temperature decreases due to increases in sulfur emissions from selected volcanic eruptions.

Landslides and Tsunami

The steep slopes of a volcano become very unstable during an eruption. In addition to lava and other pyroclastic debris, the slope itself can disintegrate and race down the side of the volcano at velocities approaching more than 200 km per hour, producing larger landslides. Researchers believe that a catastrophic landslide could result if there is an eruption of Cumbre Vieja Volcano on the island of La Palma in the Canary Islands (**Figure 3.29**).

A review of geological evidence on the island and offshore indicates that the western edge of the volcano could slide into the Atlantic Ocean. When this happens, an estimated 150 to 500 cu km of material entering the ocean will generate a tsunami that could inundate the east coast of Florida with waves exceeding 10 m in height. If this did occur, coastal residents would only have a few hours warning. This would be an insufficient amount of time to allow any significant evacuation of the millions of people who would be affected.

In addition to major earthquakes that occur near or under the ocean floor, submarine and near-shore volcanic eruptions can generate tsunamis or seismic sea waves that travel at very high speeds across the open ocean. Reaching heights of several tens or even hundreds of meters, these waves wreak havoc on coastal regions. Large landslides are associated with volcanic activity near the shore on the island of Hawaii. Massive blocks measuring hundreds of cubic kilometers are moving into the sea at a relatively slow rate (several cms per year). However, the recurrence rate is about every 100,000 years—we shouldn't experience that any time soon (Decker and Decker, 2006). In 1975 a magnitude 7.5 earthquake on the island of Hawaii created a landslide that moved large blocks of the island into the sea, which resulted in a tsunami that killed two people.

In addition, landslides, heavy rainfall, or earthquakes can produce a downslope movement of volcanic deposits, which usually contain a large amount of ash and other small debris.

Figure 3.29 Cumbre Vieja Volcano in La Palma Island in the Canary Islands.

Earthquakes

Magma moving toward the surface produces forces that trigger earthquakes in the vicinity of a volcano. Although relatively small in magnitude, these earthquakes can destroy buildings in the near vicinity of the volcanic eruptions before any direct effects of the volcano are felt. Earthquakes are commonly associated with volcanic activity in Hawaii, and their subsurface locations have been useful in outlining the magma chambers (**Figure 3.30**).

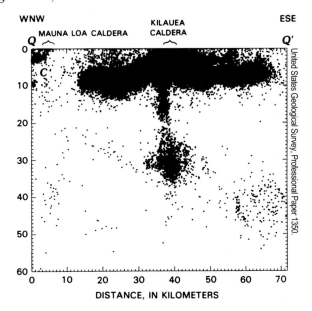

Figure 3.30 Plot of earthquake epicenters beneath the island of Hawaii. Notice the concentration of activity at shallow depth, mostly less than 12 km. This underlies the major area of volcanic activity at Kilauea and stretches along the major fissure zone.

Lessons from the Geologic Record

Volcanoes have occurred on Earth for billions of years. The principle of uniformitarianism states that "the present is the key to the past." In other words, geologic events we observe taking place today have occurred in the past.

When we examine extrusive igneous rocks that formed millions of years ago, we see the preserved explosions and fallout that covered areas in the vicinity of the central eruption. It is not easy to study these geologic deposits because they are often buried under many meters or kilometers of newer rock material. When we do extract samples, either through drilling into the continental crust or ocean floor or by examining older deposits now exposed on the surface, we can determine the type of igneous rock in the sample and perhaps gain some understanding of its mode of emplacement.

Caldera-style eruptions, such as those associated with the Yellowstone hot spot, would be utterly devastating if they occurred today. In addition to literally wiping out millions of people in the initial stages of a blast, the immediate and long-term atmospheric effects would be felt around the globe. Air quality would be adversely affected, leading to countless deaths in all countries. For the survivors, the economy would be totally disrupted, and the effects on agriculture and water supplies would be catastrophic.

Lessons from the Historic Record and the Human Toll

Throughout recorded history there have been many well-documented descriptions of volcanic eruptions (Table 3.3). Among the earliest are those written by Pliny the Younger, who described the eruption of Mount Vesuvius in southern Italy in A.D. 79. During that eruption, which covered the cities of Pompeii and Herculaneum, his uncle Pliny the Elder died trying to rescue people in the path of the volcano's wrath.

TABLE 3.3	Some Significant Volcanic Eruptions and Their Associated Loss of Life			
Year	Country	Volcanic Structure	Estimated Deaths	Notes
79	Italy	Mount Vesuvius	>3,000	Pompeii and Herculaneum destroyed by pyroclastic flow
1169	Italy	Mount Etna	16,000	
1631	Italy	Mount Vesuvius	4,000–6,000	Torre del Greco destroyed
1669	Sicily	Mount Etna	20,000	Catania destroyed
1672	Indonesia	Mount Merapi	3,000	
1683	Sicily	Mount Etna	60,000	Earthquakes occurred
1782	Japan	Mount Unzen	15,000	Tsunami produced
1783–1784	Iceland	Laki fissure on Mount Skaptar	9,800–10,500	Poisonous gases killed 280,000 cattle; major famine; 15 cubic km of lava
1792	Japan	Mount Unzen	15,000	Avalanche produced tsunami

TABLE 3.3	Some Significant Volcanic Eruptions and Their Associated Loss of Life (*continued*)			
Year	Country	Volcanic Structure	Estimated Deaths	Notes
1815	East Indies	Mount Tambora	66,000–162,000	10,000 died in initial eruptions; famine killed 50,000–80,000; produced "Year without a summer"
1845	Colombia	Nevado del Ruiz	1,000	mudslides from melting snow
1883	East Indies	Krakatua	36,000	90 percent killed by tsunami that flooded islands; global climate change resulted
1902	West Indies	La Soufriére, St. Vincent	>1,500	Occurred one day before Mount Pelée eruption
1902	West Indies	Mount Pelée, Martinique	29,000	Town of St. Pierre covered by nuee ardente
1902	Guatemala	Santa Maria	Thousands	28 km-high column; ash covered >1 million sq km
1911	Philippine Islands	Taal, near Manila	1,300	Lateral blast destroyed nearby villages
1912	Alaska	Novarupta	0	Uninhabited island saw largest eruption of twentieth century
1919	Indonesia	Mount Kelud	>5,000	Boiling waters of crater lake on Mount Kelud broke through side of mountain
1931?	Indonesia	Mount Merapi	>1,300	Major ash flow
1943	Mexico	Parícutin	0	Towns buried by lava and cinders
1951	New Guinea	Mount Lamington	>3,000	Eruption similar to style of Mount Pelée; 180 sq km destroyed
1963	Indonesia	Mount Agung	>1,100	Ash and mud flows
1980	United States	Mount St. Helens	57	Earthquake triggered landslide that resulted in eruption
1982	Mexico	El Chichón, Chiapas	2,000	25 km high dust cloud circled the globe; climate impact
1985	Colombia	Nevado del Ruiz	25,000	Mudslides from melting snow covered town of Armero
1986	Cameroon	Lake Nyos	1,700	Expulsion of carbon dioxide also killed >3,000 cattle
1991	Japan	Mount Unzen	43	
1991	Italy	Mount Etna	0	Lava flowed out for 473 days, longest period on record in 300 years
1991	Philippine Islands	Mount Pinatubo	754	Eruption predicted; 200,000 people evacuated; 30 km high dust circled the earth; global climate impact
1993	Philippine Islands	Mayon Volcano	70	60,000 people evacuated
2002	Congo	Nyirangongo	45	Lava flows

The term *plinian* is used to describe extremely explosive eruptions. Other volcanic eruptions in Italy and Greece were documented by early scholars.

Because volcanoes produce either rich soils or spectacular scenery, or both, many cities have grown up in the vicinity of these features and hence there is a large population that lives within the direct path of a major eruption. Areas that fall within this grouping include Japan, Indonesia, the Philippine Islands, southern Italy, Central America, the eastern Caribbean, and the Pacific Northwest in the United States.

In the past four centuries, approximately two-thirds of all deaths caused by volcanic eruptions have occurred in Indonesia; the Caribbean and Japan have experienced 21 percent of all deaths. Until Mount St. Helens, no deaths from eruptions had been recorded in the continental United States (Blong, 1984).

Predicting Volcanic Eruptions

active
A term applied to a volcano that has erupted in recorded history or is currently erupting.

dormant
A term applied to a volcano that is not currently erupting but has the likelihood to do so in the future.

extinct
A term applied to a volcano that no longer is expected to erupt.

The terms active, dormant, and extinct have been used to indicate the eruptive potential of volcanoes. Different definitions exist for these terms, but generally they include the following guidelines. An active volcano is one that has shown some activity in the past several thousand years (historic time), but it could be currently inactive. A dormant volcano is one we think of as sleeping or inactive but could awaken at any time. An extinct volcano has shown no activity in historic time and is not considered likely to do so.

Blong (1984) points out that the terms active, dormant, and extinct are unsatisfactory because some extinct volcanoes become active. A review of past eruptions shows that many volcanoes have periods of no activity much longer than their historic record. Also the historic record differs for various parts of the world. The Mediterranean Sea region has a longer historic record than does the western United States. Hence, more volcanism has been documented in Italy and Greece than in the Pacific Northwest of the United States.

Precursors to Volcanoes

Scientists can make several observations to forecast the possible eruption of a volcano. Among these are changes in seismic activity near the volcano, an increase in measured heat on the surface, and the bulging of the volcano's surface.

Seismic Activity

The rise of magma is a relatively slow process that can be monitored by scientists who can detect precursors to an eruption. The convective overturn of molten material within the magma chamber produces a series of small earthquakes (usually with magnitudes <4) that have a rather specific pattern in terms of Earth movement. This is caused by the reverberation of energy within the closed magma chamber as energy bounces off the walls of the chamber. Seismic detection equipment records harmonic tremors that produce a unique pattern. The occurrence of these tremors is a sign that magma is rising and that an eruption could happen; an increase in seismic activity often signals that an eruption is on the verge of occurring.

Surface Heat

Ascending magma brings increased heat closer to the surface. This heat can be directly measured by heat flow instrumentation. Should the magma be rising beneath a volcano that is covered with snow and ice, the heat will melt the frozen water, which will then begin to flow downhill. Such was the case with the imminent eruption of Mount St. Helens in the spring of 1980. Usually the top of Mount St. Helens was covered by a solid snow pack, but scientists began to notice in March of that year the snow was becoming much thinner. This was an obvious signal to everyone that magma was moving closer to the surface. Monitoring heat flow and increased seismic activity provided evidence that magma was approaching the surface. The first eruption of Mount St. Helens in almost 125 years occurred in late March 1980 and continued with increased frequency until the main eruption on May 18 (see Box 3.2).

Surface Bulge

Ascending magma also causes the Earth's surface to bulge. Measurements of the bulging can be taken using tiltmeters capable of measuring one unit of elevation change over a distance 1 million times greater than the rise. This is equivalent to detecting the increase in elevation produced by a dime coin over a distance of 1 kilometer. The detected bulges show scientists where the upward forces are the greatest and give an indication of where magma could be forced out. Laser beams reflected off targets placed on a volcano will show changes in travel times of the beams and hence a change in the distance as the volcano expands or contracts.

Other Techniques

Other techniques such as recording gas compositions and small changes in the magnetic and electrical fields give an indication of stress and pressure variations in the ground. Changes in gas compositions appear to be effective indicators of changes in the magma chamber. Emissions of sulfur dioxide and hydrogen have been indicators of eruptions at Mount St. Helens and at several Hawaiian volcanoes.

Research into past events allows geologists to produce hazard maps for areas that are likely to experience volcanic eruptions in the future. Analysis of historical records along with detailed mapping of volcanic deposits permits scientists to generate maps that show the likelihood of volcanic deposits covering areas in the vicinity of potential eruptions (**Figure 3.31**). Many people living in the region south of Seattle and in the suburbs of Tacoma are located in areas that could be affected by an eruption of Mount Rainier.

Volcanic Hazard Mitigation

The term **mitigation** refers to the lessening in intensity or force. Obviously in volcanic eruptions we cannot reduce the amount of energy expended by the volcano but we can reduce the intensity of the damage caused by the event. Thus our goal would be to make the impact of an event less severe on the people and area surrounding a volcano.

mitigation
The process of making something less severe.

During the past 500 years more than 350,000 people have been killed by volcanic eruptions. Usually there is some indication that eruptions may be imminent—small-scale earthquakes, restless animals, and small gas emissions. People generally take these signals as a clue to leave the area. Because communications are much better than they were centuries ago (or even 50 years ago), inhabitants of an area can be better informed about pending eruptions. Our ability to monitor changes in the heat flow, surface height and tilt changes, and seismic activity provide information to warn people of eruptions. A forewarning would be essential to save tens of thousands of lives if a major eruption occurred near any of the large cities in the Pacific Northwest.

Many areas that are prone to volcanic eruptions have seen significant population growth over the past several decades. The likelihood of these areas experiencing a

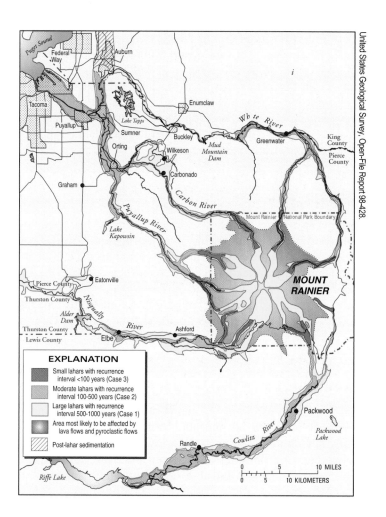

United States Geological Survey, Open-File Report 98-428.

Figure 3.31 Hazard zones for lahars, lava flows, and pyroclastic flows from Mount Rainier.

volcanic eruption increases each year. The Pacific Northwest ranks high on the list of potential eruptions, and we see that areas in the vicinity of volcanoes with a historic record of activity are lying in harm's way.

An example of an area in such danger is the small town of Orting, Washington, which has a population of about 3,500 people. The town lies in the shadow of Mount Rainier. Built on old lahar flows that occurred several thousand years ago, the town is considered to have the greatest risk of being affected by lahars from Mount Rainier (Figure 3.31). The web site for the town outlines explicit instructions for a lahar evacuation plan. The municipality is prepared in the event of an eruption, but the plan is good only if the citizens of Orting and surrounding areas are familiar with the plan and carry it out as instructed.

Heimaey Island

In the early morning of January 23, 1973, an area off the southwestern edge of Iceland experienced an unexpected volcanic event. On the island of Heimaey, which is part of the Vestmann Islands, there was an eruption of lava and small-scale tephra pyroclastics that came from a fissure along the northern edge of the island. The duration of the primary eruptions, which were Strombolian in nature and threw out ash, lapilli, and bombs as well as lava flows, lasted about three weeks. The complete episode extended for more than five months and ended in June 1973.

The entire population of about 5,300 inhabitants was evacuated using more than 75 fishing boats that had been ported in the local harbor as a result of an earlier impending storm (Scarth, 1994, p. 112). Given the geological setting of the island and its surrounding neighbors, a well-founded disaster plan had been in effect—and now it was time to implement it! In an orderly fashion the people, along with their cattle, cars, and even the money from the bank, were taken off the island. Because of this plan, the only items of value lost in the ensuing eruptions were more than 300 homes that were burned, inundated by lava, or crushed by the fallout. In the end, lava and other material added about 2.5 sq km (1 sq mi) of new land to the island.

A unique aspect of this particular event was that the U.S. Navy came in and used powerful pumps to spray sea water on the lava and lower its temperature from about 1,000°C to roughly 100°C. This stopped the forward progress of the lava and spared much of the island from burial. This is an instance when inhabitants rose to the challenge of nature and succeeded in saving the land and later reclaiming their homesteads. However, there are times when this is not so.

El Chichón

Fisher and others (1997) describe conditions that led up to the eruption of El Chichón, a moderate-size volcano in Chiapas, southern Mexico, in March 1982. A report submitted by two field geologists who had studied the area 18 months earlier indicated that an eruption could occur, but their report did not generate any action by authorities who could have taken steps to mitigate a disaster. After the volcano experienced a number of explosive eruptions over the course of a week, the intensity increased, terminating in a series of pyroclastic flows and surges that killed more than 2,000 people and erased more than nine villages in an 8-km radius around the central vent. More than 140 sq km were covered by the flows. Had local authorities heeded the warning, many fewer lives would have been lost.

Mount Etna

Mount Etna, located in southern Italy, is related to subduction of the African-plate underneath the Eurasian plate (**Figure 3.32**). During the eruption of Mount Etna that occurred nonstop from mid-December 1991 to late March 1993, numerous attempts were made to impede or direct lava flows headed toward nearby towns. Most efforts proved fruitless as the relentless inertia of the lava mass could not be stopped. In a few instances, building diversion berms with bulldozers was successful but only because of favorable circumstances and some luck. Overall, such techniques might buy some time, but if nature wants to overrun an area with lava, it will.

Figure 3.32 Mount Etna is one of several significant volcanoes in southern Italy.

Dissemination of Information

If we look back over several eruptions that occurred in modern times, we can see an unfortunate pattern in how information is disseminated to people in harm's way. Developed countries that have well-established communication and transportation systems are able to move people away from potential disaster areas. However, sometimes emergency agencies in economically stable countries experience problems, as when reports involving potential hazards contain clauses to control the sharing of findings. "This is not a problem confined to economically less developed countries [and] all contract research carries this danger. Poor countries do not have a monopoly on administrative secrecy. . . . [L]ong-term volcanic hazards are often given a lower priority than attempts to improve living standards" (Chester, 1993, pp. 210–211).

Fisher and others (1997) refer to the United Nations designated International Decade for Natural Disaster Reduction (IDNDR), which ran from 1990 to 2000.

TABLE 3.4	Fifteen Volcanoes Designated by United Nations International Decade for Natural Disaster Reduction Report as Receiving Intense Study		
Volcano	Country	Type or Style	Recent Eruptions
Colima	Mexico	Stratovolcano	2008
Galeras	Colombia	Stratovolcano	2008
Mauna Loa	Hawaii	Shield	1984
Merapi	Indonesia	Stratovolcano	2007
Mount Etna	Italy	Stratovolcano	2008
Mount Rainier	United States	Stratovolcano	1894
Mount Sakurajima	Japan	Stratovolcano	2008
Mount Vesuvius	Italy	Complex	1944
Mount Unzen	Japan	Complex	1996
Nyiragongo	Democratic Republic of the Congo	Stratovolcano	2008
Pico del Teide	Canary Islands, Spain	Stratovolcano	1909
Santa Maria	Guatemala	Stratovolcano	2008
Taal	Philippine Islands	Stratovolcano	1977
Santorini (Thera)	Greece	Shield	1950
Ulawun	Papua, New Guinea	Stratovolcano	2007

Source: www.volcano.si.edu

Among its themes of study was the Decade of the Volcano, which addressed detailed studies of selected volcanoes. Fifteen volcanoes were the focus of intense investigations (Table 3.4).

The potential for volcanic disasters is obvious. These events occur without much warning and can have major impacts on communities, their inhabitants, and the surrounding region. Volcanoes are a threat to air travel as they throw particles into the air that can foul jet engines. Volcanoes have generated about 5 percent of all tsunami over the past 250 years and have killed tens of thousands of people.

The U.S. Geological Survey reports that the United States has 170 active and dormant volcanoes. Significantly, approximately half of the 55 volcanoes that represent the biggest threat to the country are monitored with too few instruments. Volcanoes in the Cascade Range of the Pacific Northwest are the most likely to erupt in explosive forms. When this does occur, they will generate pyroclastic flows and ash falls that could threaten millions of people. More discussion of the seismic potential of this region and the Cascadia Subduction Zone of the Pacific Northwest coast is in Chapter 4, Earthquakes.

Summary

Volcanoes result from molten rock rising through the Earth's crust and being extruded onto the continents or the ocean floor. Plate tectonics explains the presence of many volcanoes that form in the vicinity of plate boundaries. Collisions of oceanic and continental plates subduct one plate to a depth sufficient to produce melting. The circum-Pacific belt consists of numerous volcanoes, such as those in Japan, the Aleutian Islands of Alaska, and the Andes Mountains of South America. The Hawaiian Islands were produced by the Pacific plate overriding a hot spot that generated a string of several hundred islands and seamounts.

Volcanic eruptions range in severity from very quiet outflows of basaltic lava such as those of Kilauea in Hawaii to historic events such as Vesuvius and Krakatua and the recent catastrophic eruptions of Mount St. Helens and Mount Pinatubo. In addition to projecting massive amounts of ash and other pyroclastics into the atmosphere, volcanoes generate a wide range of gases and large amounts of water vapor. Massive landslides and lahars and debris flows are devastating events that claim many lives of those living at the foot of a volcano. Although governments can plan for an unspecified eruption of a volcano, little can be done in the event of a large eruption occurring in close proximity to a heavily populated area. To minimize the loss of life, people must heed warnings by evacuating areas in the hazard area.

References and Suggested Readings

Bardintzeff, Jacques-Marie and Alexander R. McBirney. 2000. *Volcanology*. 2d ed. Sudbury, MA: Jones and Bartlett.

Blong, Russell J. 1984. *Volcanic Hazards: A Sourcebook on the Effects of Eruptions*. Sydney: Academic Press.

Bullard, Fred M. 1984. *Volcanoes of the Earth*. 2d rev ed. Austin: University of Texas Press.

Chester, David. 1993. *Volcanoes and Society*. London: Edward Arnold.

Decker, Robert W., Thomas L. Wright, and Peter H. Stauffer, eds. 1987. *Volcanism in Hawaii*. Vol. 1. Professional Paper 1350. Washington: U.S. Geological Survey.

Decker, Robert and Barbara Decker. 2006. *Volcanoes*. 4th ed. New York: W. H. Freeman.

Duffield, Wendell A. 2003. *Chasing Lava—A Geologist's Adventures at the Hawaiian Volcano Observatory*. Missoula, MT: Mountain Press.

Fisher, Richard V., Grant Heiken, and Jeffrey B. Hulen. 1997. *Volcanoes—Crucibles of Change*. Princeton, NJ: Princeton University Press.

Francis, Peter and Clive Oppenheimer. 2004. *Volcanoes*. 2d ed. Oxford, England: Oxford University Press.

Klein, Fred W. and Robert Y. Koyanagi. 1989. The seismicity and tectonics of Hawaii. In *The Geology of North America, Volume N, The Eastern Pacific Ocean and Hawaii*, ed. E. L. Winterer, Donald M. Hussong, and Robert W. Decker, 238–252. Boulder, CO: Geological Society of America.

Lipman, Peter W. and Donal R Mullineaux, eds. 1981. *The 1980 Eruptions of Mount St. Helens*. Professional Paper 1250. Washington: U.S. Geological Survey.

MacDonald, Gordon A. 1972. *Volcanoes*: Englewood Cliffs, NJ: Prentice-Hall.

Marshak, S. 2004. *Essentials of Geology*. New York: W. W. Norton.

McCoy, Floyd W. and Grant Heiken. 1990. Anatomy of an Eruption. *Archeology* 43: 42–49.

McGuire, Bill. 2002. *Raging Planet—Earthquakes, Volcanoes, and the Tectonic Threat to Life on Earth*. Hauppauge, NY: Barron's Educational Series.

Nations, J. and E. Stump. 1996. *Geology of Arizona*. 2d ed. Dubuque: Kendall/Hunt.

Newhall, Christopher G. and Raymundo S. Pungongbayan, eds. 1996. *Fire and Mud—Eruptions and Lahars of Mount Pintatubo, Philippines*. Queen City: Philippine Institute of Volcanology and Seismology, and Seattle, WA: University of Washington Press.

Newhall, Christopher G. and Stephen Self. 1982. The Volcanic Explosivity Index (VEI): An Estimate of Explosive Magnitude for Historical Volcanism. *Journal of Geophysical Research* 87: 1231–1238.

Prager, Ellen J. 2000. *Furious Earth—The Science and Nature of Earthquakes, Volcanoes, and Tsunamis*. New York: McGraw-Hill.

Ritchie, D. and A. E. Gates. 2001. *Encyclopedia of Earthquakes and Volcanoes*. New York: Facts on File.

Scarpa, R., and R. I. Tilling, eds. 1996. *Monitoring and Mitigation of Volcanic Hazards*: Berlin: Springer-Verlag.

Scarth, Alwyn. 1994. *Volcanoes, An Introduction*. College Station: Texas A & M University Press.

Scarth, Alwyn. 1999. *Vulcan's Fury—Man Against the Volcano*. New Haven, CT: Yale University Press.

Sigurdsson, Haraldur. 1990. Assessment of the atmospheric impact of volcanic eruptions. In *Global Catastrophes in Earth History—An Interdisciplinary Conference on Impacts, Volcanism, and Mass Mortality*, ed. V. L. Sharpton and P. D. Ward. Geological Society of America Special Paper 247: 99–110.

Sigurdsson, Haraldur. 1999. *Melting the Earth—The History of Ideas on Volcanic Eruptions*. New York: Oxford University Press.

Sigurdsson, Haraldur, ed. 2000. *Encyclopedia of Volcanoes*. San Diego, CA: Academic Press.

Simkin, Tom and Lee Siebert. 1994. *Volcanoes of the World—A Regional Directory, Gazetteer, and Chronology of Volcanism During the Last 10,000 Years*. 2d ed. Tucson, AZ: Geoscience Press.

Scarpa, R. and R. I. Tilling, eds. 1996. *Monitoring and Mitigation of Volcanic Hazards*. Berlin: Springer-Verlag.

Sutherland, Lin. 1995. *The Volcanic Earth—Volcanoes and Plate Tectonics Past, Present and Future*. Sydney, Australia: University of New South Wales Press.

Tilling, Robert I. 1993. *Monitoring Active Volcanoes*. Pamphlet. Washington: U.S. Geological Survey Information.

Williams, Stanley and Fen Montaigne. 2001. *Surviving Galeras*: Boston, MA: Houghton Mifflin.

Web Sites for Further Reference

http://pubs.usgs.gov/fs/2005/3024/
http://volcano.oregonstate.edu/
http://volcanoes.usgs.gov/About/Where/WhereWeWork.html
http://water.usgs.gov/wid/index-hazards.html
http://www.cityoforting.org/emergency_management/evacuation_plan.html
http://www.cumbavac.org/Earthquakes_Volcanoes.htm
http://www.geo.mtu.edu/volcanoes/world.html
http://www.geology.sdsu.edu/how_volcanoes_work
http://www.swisseduc.ch/stromboli/
http://www.volcano.si.edu

Questions for Thought

1. Explain the difference between magma and lava.

2. What processes were involved with the formation of the Andes Mountains along the western edge of South America?

3. Why does a volcano such as Mayon have a steep slope while Mauna Loa in Hawaii has a very gentle slope?

4. What role do dissolved gases in magma play in the eruption of a volcano?

5. How does a lahar differ from a pyroclastic flow?

6. Explain the differences between a strombolian-type eruption and a flood basalt.

7. Why was the death toll from the eruption of Mount St. Helens relatively low?

8. Potential activity of a volcano can be described by its being either active, dormant, or extinct. What does each of these terms mean?

9. Describe how surface heat can be used as a precursor of a volcanic eruption.

10. What volcano in the continental United States has the greatest likelihood of erupting in the near future?

Earthquakes

4

The third story of an apartment building lies above a crushed automobile in the Marina District of San Francisco following the Loma Prieta earthquake of October 1989.

Key Terms

body wave
convergent boundary
deep-focus earthquake
divergent boundary
epicenter
focus
intermediate-focus earthquake
local magnitude
moment magnitude
normal fault
P-wave
recurrence interval
reverse fault
Richter scale
S-wave
seismic gap
seismogram
seismograph
seismometer
shallow-focus earthquake
stress
strike-slip fault
surface wave

Thousands of earthquakes occur every day, but fortunately very few of them are large enough to create problems. Most major earthquake activity is associated with regions where tectonic plates are colliding with one other to generate stresses that cause rocks to snap, sending seismic energy through the Earth. As we saw earlier in the discussion of global plate tectonics, earthquakes and volcanoes—which can cause two of nature's greatest types of catastrophes—are associated with one another and will often occur together. Most earthquakes occur in regions that are well defined because of plate tectonics.

Regions at Risk: The Plate Tectonic Connection

Major lithospheric plates move about on Earth's surface and eventually collide with one another, generating forces that result in earthquakes. Depending on the actual collision mechanism, which creates movement and fracturing in rocks, the earthquakes occur at varying depths, ranging from near the surface to depths approaching 700 km. The majority of earthquakes occur at depths less than 100 km in the lithosphere.

Types of Plate Boundaries

There are three types of plate boundaries that produce earthquakes (**Figure 4.1**). **Divergent boundaries** are associated with areas where plates are moving apart—for example, along mid-oceanic ridges where plates are separating because of tensional forces pulling on the plates. Rising magma from the upper mantle also contributes to the outwardly directed stresses. Because magma is hot, stresses are unable to build up and the rocks at depth do not normally break. Seismic activity along mid-oceanic ridges is caused by rising magma as it bubbles its way to the surface. As magma ascends, the overlying material is pushed up and out by these forces. Earthquake activity along mid-oceanic ridges is typically confined to relatively shallow depths, taking place in the upper 20 km of Earth's crust. Divergent motion can also occur in continental regions where the land surface is being split apart by extensional stresses, such as those produced when a continent overrides a divergent plate boundary. An example of this is in the eastern part of Africa along the East African Rift Zone (**Figure 4.2**).

Convergent boundaries are regions where plates are colliding as a result of compressional forces. These conditions produce the largest earthquakes. Three types of

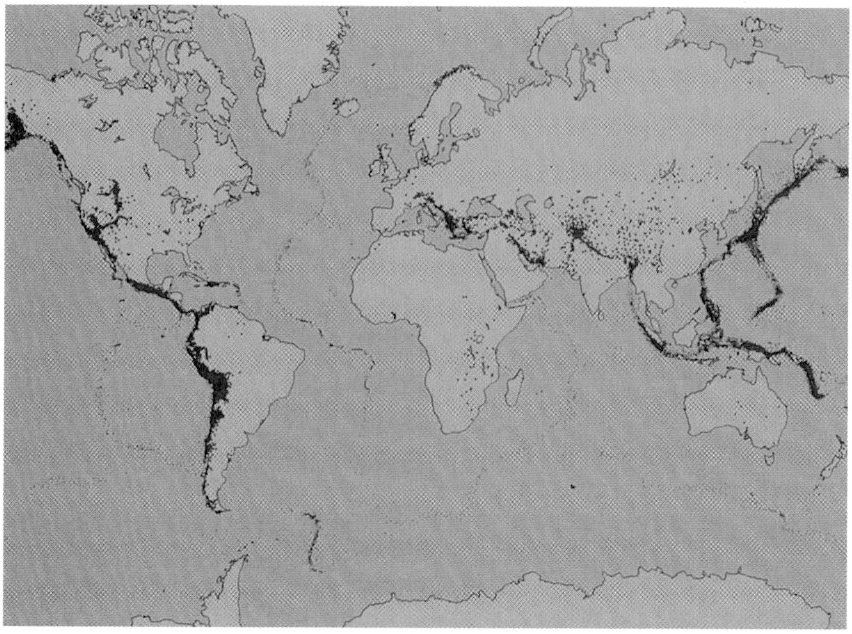

divergent boundary
An area where two or more lithospheric plates move apart from each other, such as along a mid-oceanic ridge.

convergent boundary
An area where two or more lithospheric plates are coming together, such as along the west coast of South America.

Figure 4.1 Global map showing earthquake epicenters for events between 1961 and 1967.

From *Physical Geology* by Dallmeyer. Copyright © 2000. Reprinted by permission of Kendall Hunt Publishing Co.

convergent configurations exist (**Figure 4.3**). A collision that involves two oceanic plates is an *ocean-ocean plate convergence* (**Figure 4.3a**). In this instance, one of the two plates is pushed under the other one, resulting in subduction of the downgoing slab. Once the slab starts its downward motion, gravitational forces pull the slab to greater depths. An example of this type of boundary is where the Caribbean Plate collides with the westward moving North American plate. The result is the string of volcanic islands that form the Lesser Antilles.

A convergence setting in which an oceanic plate collides with a continental plate is an *ocean-continent plate boundary* (**Figure 4.3b**). The oceanic plate, consisting of predominantly basalt, is denser than the continental granite so the oceanic plate is subducted. This subducted material can maintain its integrity as a solid slab to depths of several hundred kilometers, at which point it melts and is assimilated into the mantle.

Figure 4.2 Continental rifting in eastern Africa is caused by extensional forces shown by the arrows. The continent is moving past a set of divergent boundaries that exert pulling forces on the overlying material.

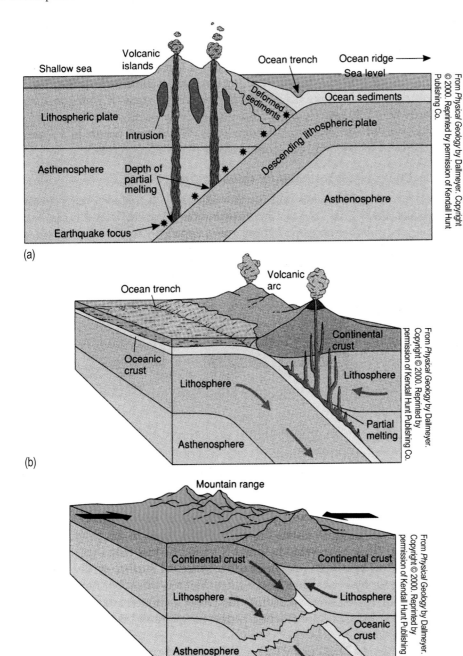

Figure 4.3 Different types of plate collisions. (a) ocean-ocean plates, (b) ocean-continent plates, (c) continent-continent plates.

The third example of plate collisions involves the *convergence of two continental plates* (**Figure 4.3c**). Each plate consists of the same general rock type—predominantly granite or granodiorite. As we have seen, continents rest atop the asthenosphere and denser mantle, so this type of plate convergence will not result in any significant subduction. An example is the collision of the Indian subcontinent with the lower edge of the Eurasian plate (**Figure 4.4**). Because no subduction takes place, the two colliding land masses basically rumple themselves up, producing the Himalayas, Earth's highest mountain range, which is located in South Asia. The regions surrounding this mountain range have experienced some of the most destructive earthquakes recorded over the past century; these have been especially catastrophic because of the poor construction standards and high population densities in that region of the world. (See **Box 4.1**, which describes the 2005 Pakistan earthquake.)

 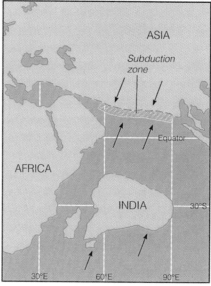 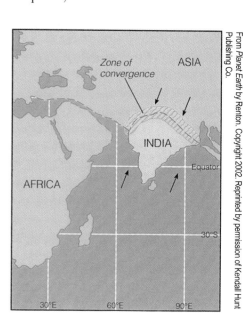

From *Planet Earth* by Renton. Copyright 2002. Reprinted by permission of Kendall Hunt Publishing Co.

Figure 4.4 Indian subcontinent moving north and colliding with the Eurasian continental plate. The lack of subduction of the continental masses produced the Himalayas. This collision began approximately 50 million years ago and continues today.

stress
The force being applied to a surface; forces can be compressional, extensional, or shearing.

strike-slip fault
A fault in which the motion of the two adjacent blocks is horizontal, with little if any vertical movement.

normal fault
A plane along which movement has occurred such that the upper block overlying the fault has moved down relative to the lower block.

| BOX 4.1 | The Pakistan Earthquake of October 2005 |

The Mediterranean-Trans-Asiatic belt runs in an east-west direction above the Mediterranean Sea stretching eastward through the upper Middle East into the lower portion of Asia. Stress in this region results from the African plate moving northward and colliding with the western portion of the Eurasian plate. There are also major strike-slip faults that run through Turkey into Iran and Iraq. The entire region experiences major earthquakes on a regular basis and the results are often devastating. Buildings and infrastructure here are old and poorly constructed, so a moderate amount of ground motion destroys villages and kills many people. Major earthquakes in this region of the world kill tens of thousands of people every decade or two.

The October 8, 2005, a magnitude 7.6 earthquake in Pakistan was related to major fault systems that are part of the Mediterranean-Trans Asiatic belt. The U.S. Geological Survey reported 80,361 people killed, more than 69,000 injured, and extensive damage throughout northern Pakistan. In some areas entire villages were destroyed and more than 32,000 buildings collapsed. The maximum intensity was VIII. An estimated 4 million people in the area were left homeless. Landslides and rock falls damaged or destroyed several mountain villages, along with roads and highways, thereby cutting off access to the region for several weeks. In the western part of the country there were reports of liquefaction and sand blows. Seiches, which are waves generated in closed bodies of water that produce a sloshing motion, were observed in West Bengal, India, and many places in Bangladesh.

Mechanics and Types of Earthquakes

When a force is applied to a surface, the result is termed **stress** (measured as force per unit area). Movement within the Earth occurs when the exerted stress exceeds the frictional force preventing movement from taking place. When blocks of rock move, slippage occurs along a surface or fault plane (**Figure 4.5**).

Faults

Orientation of the fault plane can range from nearly horizontal to vertical. The fault plane separates two adjacent blocks called the hanging wall and the footwall. It is easy to assign names to each of these blocks. Picture yourself standing on the fault plane. The block underneath your feet is the footwall; the block over your head is the hanging wall. Notice that the footwall always lies under the fault plane, while the hanging wall is above it.

A **strike-slip fault** is formed when two blocks slide one past the other with no vertical displacement (**Figure 4.6a**). In this instance a pair of parallel forces is acting one against the another, producing a shear couple. An example of a strike-slip fault is the San Andreas fault in southern California (**Figure 4.7**).

During the movement along a fault, if the hanging wall moves down relative to the footwall, the fault is classified as a **normal fault**. This type of fault is sometimes called a gravity fault, because the downward force of gravity causes the hanging wall to move down. Normal faults form when extensional or tensional forces are pulling on the two blocks (**Figure 4.6b**).

When the hanging wall moves up relative to the footwall, a **reverse fault** is produced (**Figure 4.6c**). Compressional forces squeezing two blocks together will generate a reverse fault, which has a relatively high angle. Reverse faults are commonly found at convergent plate margins.

When rocks do fracture, the initial point at which breakage occurs in the subsurface is called the **focus** (or hypocenter) of the earthquake. Energy radiates outward

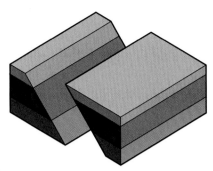

Figure 4.5 Block diagram showing the motion of a normal fault. The hanging wall moves down with respect to the footwall.

reverse fault
A plane along which movement has occurred such that the upper block overlying the fault has moved up relative to the lower block.

focus
The point in the subsurface where an earthquake first originates due to breaking and movement along a fault plane; sometimes referred to as a hypocenter.

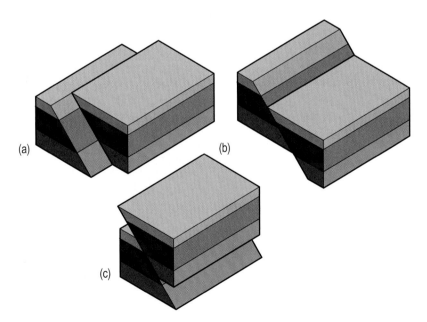

(a)

(b)

(c)

Figure 4.6 Block diagrams showing different types of fault motion: (a) a strike-slip fault, where motion is along the strike of the fault; (b) a normal or dip slip fault, where motion occurs down the dip of the fault; and (c) a reverse fault, where motion is in a reverse direction.

United States Geological Survey.

Figure 4.7 San Andreas fault is a strike-slip fault passing through central and southern California.

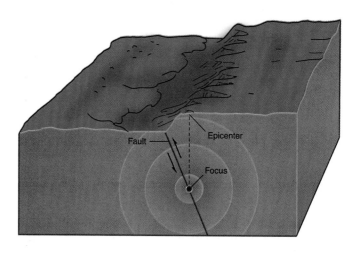

Figure 4.8 Location of the focus, epicenter, and fault plane. Notice how seismic energy radiates out from the focus in spherical paths.

epicenter
The point on Earth's surface directly above the focus. This is the position that is reported for the occurrence of an earthquake as a latitude and longitude value can be assigned to the point.

shallow-focus earthquake
An earthquake that has its focus located between the surface and a depth of 70 kilometers.

intermediate-focus earthquake
An earthquake that has its focus located between a depth of 70 kilometers and 300 kilometers.

deep-focus earthquake
An earthquake that has its focus located between a depth of 300 kilometers and roughly 700 kilometers.

in a spherical pattern from this point and travels in all directions (**Figure 4.8**). The point on the surface directly above the focus is termed the **epicenter**, a location that can be assigned latitude and longitude coordinates indicating where the earthquake occurred.

Earthquake foci have a wide range of depths. If an earthquake occurs at a depth between the surface and 70 kilometers, it is termed a **shallow-focus earthquake** (**Figure 4.9**). Shallow earthquakes are associated with the upper portion of Earth's crust, an area where rocks are brittle.

The subducted plate, which dips at an angle between approximately 30 and 45 degrees, is termed the Benioff zone. Focal depths lying between 70 and 300 kilometers in this zone are termed **intermediate-focus earthquakes**. Because these depths lie in the upper portions of the mantle, rocks here are becoming hotter and begin to bend. Intermediate-focus events are occurring in the cooler, subducted slab (Figure 4.9).

Deep-focus earthquakes have focal depths ranging from 300 kilometers to about 700 kilometers, which is the lower limit at which any brittle material exists. Material that is subducted below a depth of 700 km begins to undergo melting and is assimilated into the mantle.

Figure 4.9 Diagram showing subducted oceanic slab going underneath continental slab with shallow, intermediate, and deep earthquakes. Notice that the foci lie within the brittle, subducted slab.

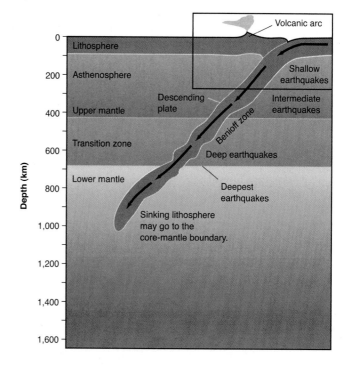

Seismic Waves

Energy radiates out from the focus of an earthquake (Figure 4.8) and is transmitted as seismic waves (sound and light waves are other examples of energy being moved through space). This energy produces two types of seismic waves: body waves and surface waves.

Body Waves

Body waves are transmitted through the body or interior of the Earth whereas surface waves move along boundaries, either on the Earth's surface or along boundaries within the Earth. Body waves consist of two types. **P-waves** are the primary waves that result from a push-pull action. They are also referred to as compressional or longitudinal waves, because the back and forth motion moves in the direction of wave propagation (**Figure 4.10a**). A compressional wave moves along by squeezing the material; the compressed particles then rebound or expand, which compresses the next particles adjacent to the ones that had been compressed. P-waves are the fastest moving of all the seismic waves, and they can travel through any type of material (solid, liquid, or gas). Velocities in rocks range between 5 and 8 km/sec, depending on the type of material the wave is moving through.

S-waves, also known as secondary waves, are so named because they are the second ones to arrive at a detection point. They travel at about 60 percent the velocity of P-waves and have a lower frequency and higher amplitude. S-waves are also referred to as shear or transverse waves because their actual motion causes particles to move at right angles to the direction of wave propagation (**Figure 4.10b**). Notice that this right-angle directed motion can be in both a vertical (SV) and a horizontal (SH) plane (producing both SV and SH waves, respectively). This combination of S-wave motions, particularly the horizontal component, is the reason that these waves impart considerable damage to buildings and other structures on Earth's surface. An important characteristic of S-waves is that they do *not* pass through liquids. This fact has allowed seismologists to determine that certain parts of Earth's interior are liquid because S-waves are not detected at all recording stations. Seismic recording stations that are located between distances of 103° and 143° from an earthquake focus will not record S-waves because the waves hit the liquid outer core and do not get transmitted any farther.

body wave
A seismic wave that is transmitted through the Earth. P-waves and S-waves are body waves.

surface wave
A seismic wave that moves along a surface or boundary.

P-wave
The primary wave, which is a compressional wave; this is the fastest moving of all the seismic waves, and arrives first at a recording station. P-waves are a type of body wave.

S-wave
The secondary wave, which is a shear wave that moves material back and forth in a plane perpendicular to the direction the wave is traveling. S waves do not travel through liquids. S-waves are a type of body wave.

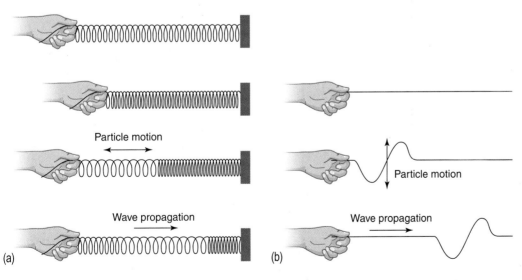

Figure 4.10 Seismic motion for P-waves is depicted in the upper set of drawings. Energy propagates in a push-pull fashion along the direction of movement (left to right). S-wave motion is in a direction perpendicular to the sense of movement (left to right). Vertical S-waves are shown.

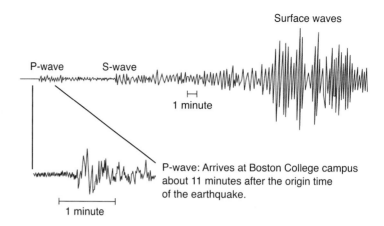

P-wave: Arrives at Boston College campus about 11 minutes after the origin time of the earthquake.

1 minute

Figure 4.11 Seismogram showing the arrivals of the P-, S-, and surface waves. Notice the changes in amplitudes of the various waves as they arrive at the detection station.

Surface Waves

The second main group of waves consists of surface waves, which include Rayleigh (R) and Love (L) waves, two types that are named after mathematical physicists who each did research into their respective characteristics. For shallow and some intermediate-focus earthquakes, surface waves have a high amplitude (**Figure 4.11**). Vertically directed R-waves make the ground ripple up and down, similar to the wave motion created when you jump into a swimming pool. The lateral motion caused by faster moving L-waves is similar to that seen when a sidewinder snake moves across the desert floor. As the snake moves forward, its body moves side to side (**Figure 4.12**). The key point about each of these surface waves is that they involve a rolling motion of particles and they move much slower than either type of body wave.

When an earthquake occurs, all these types of motion (P, S, R, and L) are generated so the energy is passed through an area at different velocities, different times, and varying amplitudes. The velocity relationship is $V_P > V_S > V_L > V_R$, but the amplitudes of these waves are essentially in reverse order. The rocks and buildings that are being affected undergo a great deal of mixed stresses that produce the damage we observed after the shock waves pass. Engineers can design buildings to withstand most of these forces, but the underlying geology (bedrock, loose soil, or sediment, for example) must also be considered in the design process.

Figure 4.12 A sidewinder rattlesnake displays a horizontal ripple motion along with forward progress similar to that of an L-wave.

seismometer
An instrument used to detect very small movements in the ground. This movement is amplified electronically to generate seismograms.

seismograph
A device that records the ground motion of an earthquake.

seismogram
The written or electronic record of ground motion detected by a seismometer.

Detecting, Measuring, and Locating Earthquakes

As seismic energy moves through rocks, it abruptly displaces them, sometimes on a large scale. As the ground moves, the motion is detected by a **seismometer**, a very delicate instrument that responds to small changes in the vertical and horizontal positions at a point. This movement is amplified before being sent to a recording device called a **seismograph**, which produces a record of the ground motion as a **seismogram** (**Figure 4.13**). Careful analysis of a seismogram can provide information about the strength or size of the earthquake and its distance from the recording station. An important parameter is to know the precise time related to the recorded signal so the source region of an event can be located.

There are numerous recording stations located throughout the world, each of which is capable of recording events of magnitude 6 or greater that occur anywhere in the world. Within the United States there are hundreds of stations, with a large number in earthquake-prone regions such as California and Alaska (**Figure 4.14**). These stations continually record all ground motion and report any unusual activity to the National Earthquake Information Center (NEIC) in Golden, Colorado, which issues reports and warnings of earthquake activity. Rapid computer analysis of earthquake data enables the NEIC to send information and warnings to all parts of the globe.

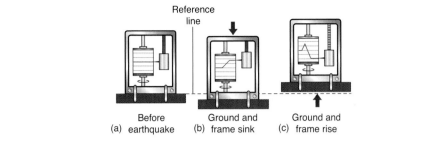

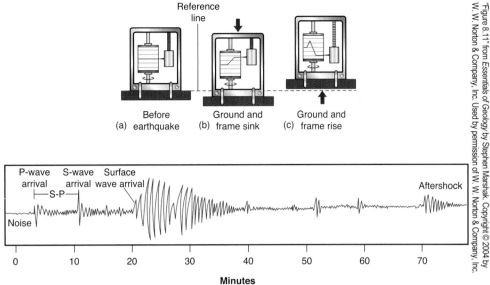

Figure 4.13 Seismometer is attached to the frame and moves upon detecting ground motion. Pen records motion as the frame and seismometer move vertically. Seismogram shows the arrivals of the P– and S– waves, along with the later surface wave. Approximately 66 minutes after the P-wave is detected, aftershocks are recorded.

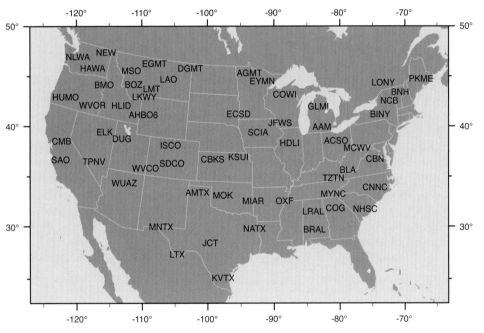

Figure 4.14 Selected seismic stations located throughout the continental United States. The three- and four-letter symbols are the unique code names for each reporting station.

Measuring the Size of an Earthquake

Several factors control the recorded size of an earthquake. The amplitude of the seismogram is related to the recording station's distance from the epicenter. The greater the distance, the less pronounced is the record. Also the magnitudes of the recorded motion depend on where the seismometers are placed. Loose soil moves more rapidly than solid rock and can give a false record of the true scale of the event. Seismometers are usually located on a solid base to minimize extraneous ground movement.

Earthquakes located near a recording station will generate a sudden jolt that will be detected by the seismometer as a spiked signal. Earthquakes occurring at a greater distance tend to produce longer period signals which have a lower amplitude (**Figure 4.15**). The effects of the Northridge, California, earthquake shown in Figure 4.15 are discussed in **Box 4.2**.

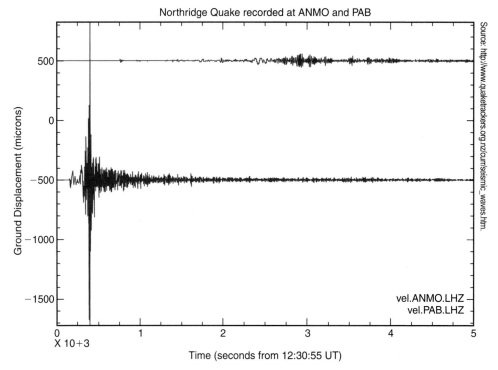

Figure 4.15 Recordings of the Northridge, California, earthquake of January 17, 1994. The upper record is from a recording station in Pablo, Spain; the lower record is from Albuquerque, New Mexico. Notice the greater amplitude for the signal recorded at the station that was much closer to the event.

Source: http://www.quaketrackers.org.nz/curr/seismic_waves.htm.

BOX 4.2   Northridge Earthquake of January 1994

Many of the residents of the Los Angeles, California, area had planned to take the Martin Luther King Day holiday on Monday, January 17, 1994. Unfortunately, Mother Nature had decided it was a day of work as the city of Northridge, a suburb on the northern edge of the city of Los Angeles, was rocked by a M = 6.7 earthquake at 4:31 A.M. The United States Geological Survey reported that 60 people were killed and more than 7,000 injured. More than 40,000 buildings were destroyed; numerous gas lines ruptured and several key freeway overpasses collapsed, blocking major traffic arteries in the Los Angeles area. The Northridge earthquake still ranks among the most costly natural disasters in U.S. history with more than $20 billion in damage. Major retrofitting of freeways in the Los Angeles area took place to minimize similar failures when the next large earthquake strikes the area.

Additional source for this earthquake: http://pubs.usgs.gov/of/1996/ofr-96-0263/

United States Geological Survey.

Box Figure 4.2.1 Parking deck at California State University Northridge.

Figure 4.16 is a seismic record of a minor earthquake in central California. The horizontal lines are recorded by a pen that marks paper on a drum that turns once every 15 minutes, hence four lines per hour. Notice the sudden onset of this signal, which tells us this event was close to the recording seismometer. The recorded ground motion for this event lasted about two and one-half minutes.

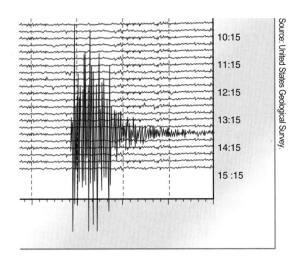

Source: United States Geological Survey.

10:15

11:15

12:15

13:15

14:15

15 :15

Figure 4.16 Magnitude 4.0 earthquake near Cloverdale, California, on January 10, 2000.

Magnitudes

Several different magnitude scales are used to describe the "size" of an earthquake. Various texts use several different styles of symbols to designate magnitudes. We will use the following symbols to designate magnitudes:

- local magnitude (ML)
- body-wave magnitude (Mb)
- surface-wave magnitude (Ms)
- moment magnitude (Mw)

The **local magnitude**, ML, is used for moderate-sized earthquakes measured close to the epicenter. This method uses the **Richter scale**, which was first developed in 1935 by Charles F. Richter of the California Institute of Technology for local earthquakes in California. Richter used the maximum amplitude of the S-wave and took into account the distance from the recording station to the epicenter to establish the magnitude of an event. It is an open-ended scale, so it can include the very smallest events or the largest possible ones that occur. The U.S. Geological Survey (USGS) reports that each year more than 10,000 earthquakes occur in the southern California area. Most of them are very low magnitude with only a few hundred measuring greater than magnitude 3.0, and only about 15 to 20 being greater than magnitude 4.0. If a large earthquake does occur, it will trigger smaller aftershocks (**Figure 4.13**) that can persist for several months.

Earthquake magnitude is a logarithmic scale signifying the size of the event. If we were the same distance from the epicenters of two different events, ground motion would be 10 times greater during an ML = 6 earthquake than that of an ML = 5 earthquake. This 10-fold increase in ground motion translates to approximately a 32-fold increase in the amount of energy produced by the larger event. This larger amount of energy explains why higher magnitude earthquakes are generally more destructive. A magnitude 6 earthquake would have more than 1,000 times the energy of an ML = 4 event (**Figure 4.17**).

Since the ML scale is technically applied only to local earthquakes in California, Richer along with a colleague, Beno Gutenberg, devised other scales to handle other types of events. They developed the body-wave scale (Mb) for deep-focus earthquakes and the surface-wave magnitude scale (Ms) for larger earthquakes that were over several hundred kilometers from the recording station.

Surface-wave magnitudes are not used for very large earthquakes (those with Ms ≥ 7.3). Brumbaugh (2010) explains the problems as they relate to surface waves with periods near 20 seconds. The volume of rock needed to generate such a wave is often unrealistic. A better measure was developed for very large earthquakes. The **moment magnitude** (Mw) is dependent on the shear strength of the rock, the area af-

local magnitude
A term that describes the size of an earthquake in an area near the epicenter; see Richter scale.

Richter scale
A measurement scale developed by Charles Richter to determine the size of earthquakes in California. This is commonly used to describe many types of earthquakes but it is more correctly used for small, localized events.

moment magnitude
A measure of an earthquake that is based on the area affected, the strength of the rocks involved, and the amount of movement along the primary fault.

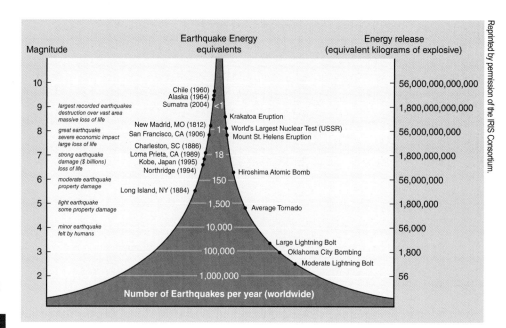

Figure 4.17 Earthquake energy as it relates to known events. Notice the small number of larger events compared with the overall number of earthquakes. Data reflect the annual occurrence of earthquakes.

fected by the fault rupture, and the average displacement along the fault. This scheme is used for large earthquakes having magnitudes greater than an Mw value of 8.

Table 4.1 shows a comparison of S-wave magnitude (Ms) and moment magnitude (Mw) values. We see that moment magnitudes are larger than S-wave magnitudes for each earthquake. Moment magnitude values are more realistic for great earthquakes because these quakes generate enormous amounts of energy and affect larger volumes of rock within the Earth. Variations in magnitudes exist because of the different types of seismic waves used to determine the magnitude (**Figure 4.18**).

A review of the 10 largest earthquakes in recorded history (Table 4.2) shows that all of them are related to active convergent plate boundaries, with nine of them occurring in the circum-Pacific Ring of Fire. The Assam, Tibet, event of August 1950 (along the India-China border) was caused by the collision of India with the Eurasian plate (Figure 4.4).

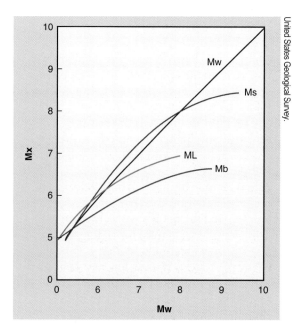

Figure 4.18 Different magnitude scales have a range of values when compared to each other. Mb = body-wave magnitude; ML = local or Richter magnitude; Ms = surface-wave magnitude; Mw = moment magnitude. Mx = Mw, ML, Ms, or Mb. Note the effective, upper limit for each magnitude.

TABLE 4.1	Comparison of Surface Magnitude Values with Corresponding Moment Magnitude Values		
Location	Date	S-Wave Magnitude	Moment Magnitude
Chile	May 1960	8.5	9.5
Alaska	March 1964	8.4	9.2

TABLE 4.2	Ten Largest Recorded Earthquakes Since 1900	
Date	Location	Magnitude (Mw Value)
May 1960	Chile	9.5
March 1964	Prince William Sound, Alaska	9.2
December 2004	Off west coast of northern Sumatra, Indonesia	9.1
November 1952	Kamchatka, east Russia	9.0
February 2010	Off shore Maule, Chile	8.8
January 1906	Off coast of Ecuador	8.8
February 1965	Rat Islands, Aleutian Islands	8.7
March 2005	Off west coast of northern Sumatra, Indonesia	8.6
March 1957	Andreanof Islands, Aleutian Islands, Alaska	8.6
August 1950	Assam, Tibet	8.6

Source: United States Geological Survey.

Measuring the Intensity of an Earthquake

The intensity of an earthquake is a measure of the damage caused by ground shaking and how buildings and other structures respond to the seismic energy. The observed damage is a subjective assessment because different observers might assign a variety of levels of severity to a given area depending on their interpretation and evaluation of the damage.

In 1902 Italian scientist Giuseppe Mercalli established a scale that assigned values to damaged areas based on how different materials responded to an earthquake. The scale was modified in 1931 by American seismologists Frank Neumann and Harry Wood, who refined the scale to better address building standards in the United States. The scale is based on 12 different levels of damage (values are assigned Roman numerals), with I being the lowest level of damage and XII being the highest (Table 4.3). The Modified Mercalli scale differs from magnitude scales because there

TABLE 4.3	Abbreviated Descriptions for Levels of the Modified Mercalli Intensity Scale

I.   Not felt except by a very few under especially favorable conditions.

II. Felt only by a few persons at rest, especially on upper floors of buildings.

III. Felt quite noticeably by persons indoors, especially on upper floors of buildings. Many people do not recognize it as an earthquake. Standing motor cars may rock slightly. Vibrations similar to the passing of a truck. Duration estimated.

IV. Felt indoors by many, outdoors by few during the day. At night, some awakened. Dishes, windows, doors disturbed; walls make cracking sound. Sensation like heavy truck striking building. Standing motor cars rocked noticeably.

V.   Felt by nearly everyone; many awakened. Some dishes, windows broken. Unstable objects overturned. Pendulum clocks may stop.

VI. Felt by all, many frightened. Some heavy furniture moved; a few instances of fallen plaster. Damage slight.

VII. Damage negligible in buildings of good design and construction; slight to moderate in well-built ordinary structures; considerable damage in poorly built or badly designed structures; some chimneys broken.

VIII. Damage slight in specially designed structures; considerable damage in ordinary substantial buildings with partial collapse. Damage great in poorly built structures. Fall of chimneys, factory stacks, columns, monuments, walls. Heavy furniture overturned.

IX. Damage considerable in specially designed structures; well-designed frame structures thrown out of plumb. Damage great in substantial buildings, with partial collapse. Buildings shifted off foundations.

X.   Some well-built wooden structures destroyed; most masonry and frame structures destroyed with foundations. Rails bent.

XI. Few, if any (masonry) structures remain standing. Bridges destroyed. Rails bent greatly.

XII. Damage total. Lines of sight and level are distorted. Objects thrown into the air.

Source: United States Geological Survey.

is no mathematical basis for the values given to the damage observed in an area. The Modified Mercalli Intensity Scale is more relevant for engineers and homeowners as they can relate to the impact of the earthquake.

The recorded intensity of a given earthquake can vary based on the location of buildings and other features. However, the recorded magnitude for the same event is constant and does not change with distance or location of the recording instruments. The U.S. Geological Survey has found that local responses to seismic energy are a function of the thickness of sediments overlying solid bedrock and the softness of the soil or rock near the surface. As seismic waves move from hard rock to softer material, they slow down and increase in amplitude. These larger waves produce more shaking and hence create more damage. In the eastern United States, earthquakes are felt over much greater areas than in the west (**Figure 4.19**).

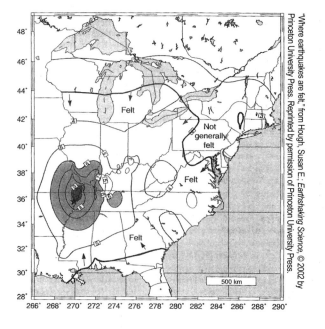

Figure 4.19 Isoseismal map of the December 16, 1811, earthquake in New Madrid, Missouri. Estimated magnitude was at least Ms $\geq$ 8.0, based on analyses of historical records and descriptions of the damage, which also were used to assign MM values as shown. The heavy line delineates regions where the quake was felt.

"Where earthquakes are felt," from Hough, Susan E.; *Earthshaking Science.* © 2002 by Princeton University Press. Reprinted by permission of Princeton University Press.

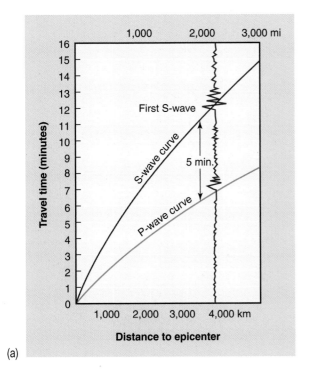

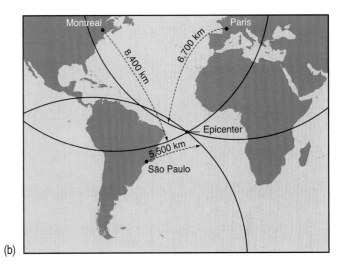

Figure 4.20 (a) Travel-time graph showing the P-wave and S-wave travel times as measured against the distance they have moved from the earthquake focus; (b) Using arrival data from three different stations, it is possible to determine the location of the earthquake epicenter.

Locating an Earthquake

Whenever an earthquake occurs, often several seismic stations record the event. In the case of a major earthquake, hundreds of stations produce seismic records. Today many stations are set up to digitally record earthquakes and then transmit the information to some central office where a determination is made regarding the location and magnitude of the earthquake. In the United States the primary reporting station is the National Earthquake Information Center in Golden, Colorado. If sufficient data are available at a single recording station, it is possible for scientists to make a quick determination of the distance and magnitude of a single earthquake from that station.

To get a quick fix on the location of an earthquake, arrival data are needed from a minimum of three recording stations. Each station will have a seismogram that shows the P- and S-wave arrivals of the event. The S-P time is noted and then converted to a distance using a travel-time curve (**Figure 4.20a**). When this distance is determined for a single station, the locus of all points located the same distance from the station is a circle. To be more precise in assigning a location for the epicenter, a similar procedure must be completed for two more stations. When this is done, the three circles should intersect at a single point, but because of minor differences in estimated rock velocities used to construct the travel-time graph, the lines usually establish a region of intersection where the earthquake occurred. As more data are analyzed from other stations, this region is eventually refined to more precisely define the point of the earthquake (**Figure 4.20b**).

Lessons from the Geologic Record

Earthquakes have occurred on Earth ever since crustal material was cold and brittle enough to break. Numerous examples exist of faults being preserved in rocks that are millions of years old. If a rock was faulted and later covered by sedimentary material,

Courtesy of David M. Best.

Figure 4.21 A high-angle normal fault shows a small amount of offset. The footwall is to the left.

the fault is preserved and we can reason that movement occurred prior to the later deposition of the overlying material (**Figure 4.21**). Undoubtedly, major earthquakes occurred throughout the world in the past, breaking up rock in many locations. This fracturing process allowed once solid material to become more easily eroded and removed, thereby erasing the record of earthquake activity. Whenever the broken material was rapidly covered by sediment and other debris, evidence of the fault was preserved.

Along the San Andreas Fault in California, geologists have dug a series of trenches across a section of the fault to detect areas where soil and sediment layers have been offset. By age-dating the layers using radiocarbon techniques, it is possible to determine when earthquakes occurred and thus generate a recurrence interval for a given area. With sufficient data they can extrapolate back in time to determine what happened and to make some educated predictions for future activity. However, because data are often not very continuous in terms of time, the recurrence intervals carry a significant amount of error.

Lessons from the Historic and Recent Record and the Human Toll

Earthquakes have occurred on Earth for hundreds of millions of years. Early reference to them was made by Chinese historians and by early Roman and Greek philosophers and naturalists. One of the best documented historic earthquakes was the Great Lisbon Earthquake of 1755 that devastated the capital city of Portugal. On November 1, All Saints' Day, when many people came to the city to attend church services, there was a great earthquake centered about 200 miles (325 km) off the coast of Portugal. Estimated magnitude values fall in the range $8.3 < Mw < 8.5$, making it one of the largest earthquakes ever to hit Europe. The damage suffered by the city was immense—more than 90,000 died from the earthquake, tsunamis, and a major fire that raged throughout the city. Nearby coastal regions, including Lisbon and its harbor, experienced several tsunamis that ravaged seaside towns and killed thousands more. Cities in Morocco suffered also, where more than 10,000 people were killed.

Densely populated areas suffer large losses of life in natural disasters. Table 4.4 shows that some of the most deadly earthquakes have occurred in China, the world's most populous country. Although China is not directly affected by subduction processes related to plate tectonics, the country is laced with major fault systems that obviously experience a great deal of stress. The resulting earthquakes have killed large numbers of people over the years. However, until recently, with an opening up of news agencies to the world, we often did not know the amount of devastation and the number of lives lost in these events.

On May 12, 2008, a major earthquake of magnitude 7.9, having a shallow focal depth of 19 km, occurred in the Sichuan province of southern China. Located about 1,600 km southwest of Beijing, the event occurred along a major fault system associated with movement of the Tibetan plateau against crustal material of southeastern China. Two weeks following the main event, an aftershock of magnitude 6 destroyed more than 70,000 buildings that had been weakened by the main event. Approximately 65,000 people were killed and an additional 23,000 people were reported missing. Major landslides dammed rivers, producing the potential for major flooding.

Earthquakes occur every day on Earth. They certainly disrupt the daily routine of those in the immediate area of the earthquake. Occasionally a local event can change the plans of an entire nation, such as what took place when the Loma Prieta earthquake of October 17, 1989 shook the Bay area around San Francisco, California (see **Box 4.3**). Millions of baseball fans were prepared to watch the third game of the World Series between the San Francisco Giants and the Oakland Athletics, only to have the game postponed and the outcome of the series delayed.

TABLE 4.4	Earthquakes That Have Caused More Than 100,000 Deaths Since A.D. 1100			
Year	Location	Magnitude	Deaths	Comments
1138	Syria		230,000	
1290	China		100,000	
1556	China	~8	830,000	Largest loss of life in one event
1908	Italy	7.2	70–100,000	Earthquake and tsunami
1920	China	7.8	200,000	Major fractures, landslides
1923	Japan	7.9	143,000	Tokyo destroyed by fire
1927	China	7.9	200,000	
1948	USSR	7.3	110,000	
1976	China	7.5	240,000	Some estimates are as high as 655,000
2004	Sumatra	9.1	230,000	Earthquake and tsunami

Source: United States Geological Survey.

BOX 4.3	The Loma Prieta Earthquake of October 1989

On October 17, 1989, at 5:04 P.M. (PDT), a magnitude 6.9 earthquake severely shook the San Francisco and Monterey Bay regions. The U.S. Geological Survey reported that the epicenter was located near Loma Prieta Peak in the Santa Cruz Mountains, approximately 14 km (9 mi) northeast of Santa Cruz and 96 km (60 mi) south-southeast of San Francisco.

As is true in many disasters, timing means a lot. The Great San Francisco Earthquake of 1906, the San Fernando quake of 1971, and the Northridge earthquake of January 1994, all struck in the early morning hours, sparing numerous people from imminent disasters. The Loma Prieta earthquake was no different in terms of timing. Baseball fans from the Bay Area were gathered in Candlestick Park in South San Francisco to watch Game 3 of the World Series between the San Francisco Giants and the Oakland Athletics.

More than 60,000 fans were in the relative safety of the stadium and off the highways. Many more were at home or in buildings watching the game so they were spared the disaster that struck the area. A portion of I-880, the Nimitz Freeway, collapsed, crushing cars and killing 47 people (this figure would have been much higher, given the rush hour timing of the earthquake, had not so many people been watching the World Series game). One section of the freeway was constructed on loose soil and the poor design of the vertical columns holding up the highest set of traffic lanes caused them to experience increased vertical resonance in the support columns. They all disintegrated under the downward force. Refer to Levy and Salvadori (1995) for details on the engineering issues.

The earthquake obviously caused the baseball game to be cancelled and, although the stadium was only slightly damaged by the seismic waves passing underneath it, the games were moved to Oakland after a 10-day postponement. The A's swept the Giants in four games.

United States Geological Survey. Photo by C. E. Meyer.

United States Geological Survey. Photo by H. G. Wilshire.

Box Figure 4.3.2 A section of I-880, the Cypress viaduct and Nimitz Freeway that disintegrated during the Loma Prieta earthquake. This road is usually heavily traveled at rush hour, and thousands were spared from death because of the scheduled World Series game that kept drivers off the freeways.

Box Figure 4.3.1 A section of the San Francisco-Oakland Bay Bridge collapsed during the Loma Prieta Earthquake of October 1989.

Additional Web sites for this event:
http://earthquake.usgs.gov/regional/nca/1989/index.php
http://pubs.usgs.gov/fs/1999/fs151-99/
http://pubs.usgs.gov/dds/dds-29/

Predicting and Forecasting Earthquakes

Regions that are likely to experience earthquakes on a regular basis are continually monitored for seismic activity. Scientists are aware of movement in the Earth and analyze the data for temporal and spatial patterns that might be useful in predicting future events. We do not currently have the ability to predict an earthquake with any degree of certainty. Far too many variables are involved to allow this to happen, so we view the occurrences of earthquakes as essentially random events in nature.

Short-Term Predictions

Several indicators can suggest that an earthquake could occur in a wide area. Among these are small movements in the ground or increases in recorded seismic activity in an area. Numerous sites are located along the San Andreas fault, where lasers are aimed across the fault to reflectors on the opposite side. Any minute changes in movements of either block of the fault are detected and can warn of impending larger scale movements.

Monitoring seismic activity might show that a large number of small events in a short time period have been recorded, producing a swarm. Such an increase in activity could be the precursor to a larger event in the immediate future. However, this activity cannot be termed a series of foreshocks to a larger event until the main shock takes place. Thus, we are not always able to interpret an increase in activity as a signal of bigger things to come. If no main shock occurs, then the earlier swarm would be just that and would not provide an indication of larger events to follow.

A key problem with making definitive, short-term predictions is that it can send society into a panic if such an alarm were sounded. People would have to evacuate their communities and find new, temporary living conditions. All businesses and schools would close, banks would not be available, abandoned areas would be subject to looters and criminals. Should the predicted event not happen, who would be held liable for all the problems that would follow?

There are documented cases of ill-founded predictions of major earthquakes hitting populated areas. In the mid 1970s there was an erroneous prediction of a great earthquake that was "supposed" to happen along the southeast coast of North Carolina. A similar erroneous, baseless prediction was made for the New Madrid seismic region of Missouri in late 1990. In each instance the general populace became unnecessarily alarmed about the potential for a major disaster. Needless to say the events never occurred but the experiences had been unnerving for people in those regions of the country.

Long-Term Predictions

Long-term predictions are much less precise but they do have some scientific merit. Information such as how often a certain magnitude event occurs or the location of past epicenters can be used to define regions likely to experience an increase in seismic activity. By keeping track of when large magnitude earthquakes occur, scientists can develop a time line that shows the **recurrence interval** of earthquakes. Collecting data over several decades allows researchers to produce a crude model to predict earthquakes. For example, if a magnitude 6 event occurred every eight to ten years in a region, the recurrence interval would be defined as about every nine years. The probability is that every nine or so years an earthquake of magnitude 6 or higher could occur. The likelihood that such an event would occur in a given year would be one year in nine (1/9) or approximately 11 percent.

Analyses of earthquake epicenters show where events have occurred and where they have not. Plots that display gaps of epicenters indicate areas that could be likely

recurrence interval
A measure of the elapsed time between events of a similar size, such as earthquakes, floods, or large storms.

seismic gap
An area in a faulted region where no seismic activity has occurred in recent time. These areas are undergoing stress buildup and could be the site of future earthquakes.

to experience an event. A **seismic gap** is caused by a lack of crustal motion that is often explained by fault planes that are locked in place, although the region has experienced activity in the past. Once the stresses are large enough to overcome the frictional forces holding the blocks or rock in place, an earthquake or series of earthquakes will most likely happen. If a long period has passed since any significant activity, a very large event could be waiting to happen because of the tremendous amount of strain accumulating in the rocks.

High-Risk Areas in and near the United States

In the United States and adjoining regions, several areas are considered hazardous in terms of earthquake activity. In 2006 the USGS reported that more than 75 million Americans in 39 states (25 percent of the country's population) are at risk from earthquakes.

The plate tectonic settings for Alaska, Washington, Oregon, and California cause those states to have a high risk factor. Over the past 300 years great earthquakes (magnitude 8 or higher) have been centered in each of these areas. In several instances tsunamis have inundated the coastlines of the states, causing significant damage. Although Hawaii rests in the middle of the Pacific plate, it has experienced great earthquakes and numerous tsunamis in the past several centuries.

In the area adjoining the Caribbean Sea, including Puerto Rico and the U.S. Virgin Islands, great earthquakes and tsunamis have killed thousands of people over the past several centuries. In addition, the entire east coast of the United States is at risk. In 1886 a major earthquake (Mb = 6.7) occurred near Charleston, South Carolina, causing significant damage to the nearby communities. This event ranks among the most widely felt earthquakes in the history of the United States. The earthquake was felt in at least 23 states, ranging as far west as Arkansas and as far north as Wisconsin and New York state. Rocks in the eastern and middle portions of the country are very old and not very highly fractured, so they transmit seismic energy more efficiently than the fractured rocks of the western United States.

New England and eastern Canada have been subjected to significant earthquake activity over the past three centuries with more than 10 earthquakes exceeding M = 5 (**Figure 4.22**). These earthquakes are related to the St. Lawrence River and an ancient rift valley that underlies the river. A second trend is evident that connects epicenters of events through New Hampshire northwestward into Quebec. These earthquakes are thought to occur along a landward extension of the Kelvin Fracture Zone in the North Atlantic Ocean.

Figure 4.22 Selected earthquake activity in New England and eastern Canada, 1627–1998. Larger stars are M > 5.5 events; smaller stars are M 2 and M 3 events.

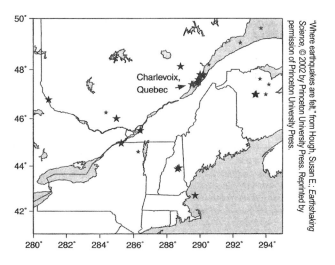

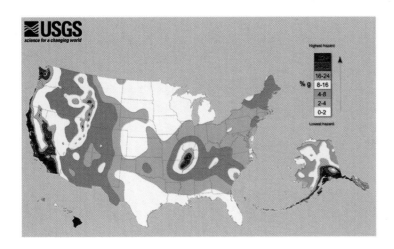

Figure 4.23 Relative shaking hazards in the continental United States. The highest amounts of shaking occur in the Aleutian Islands of Alaska and in California.

New Madrid, Missouri, and the surrounding area have an earthquake hazard risk as great as most of California (**Figure 4.23**). A series of three large earthquakes (Mw > 7.5) occurred between December 1811 and February 1812. The entire east coast was rocked by these events, with extensive damage being recorded over most of the eastern and areas of the central United States (refer to Figure 4.19). Fortunately, the region was relatively sparsely populated at the time, although some deaths were recorded.

Unexpected Events in the United States

In early 2008 several earthquakes occurred in Nevada, which is the third most active state in the United States, ranking behind California and Alaska. On February 21, 2008, a magnitude 6 event shook northeastern Nevada near the town of Wells, just south of the Utah border. Within two months the city of Reno experienced a magnitude 5.2 earthquake, which was followed by more than 200 aftershocks, including one of magnitude 4.7 soon after the main event.

Residents of southeastern Illinois and the surrounding region were awakened on April 18, 2008, by a magnitude 5.2 earthquake. This activity is associated with the Wabash Valley Seismic Zone, which is related to the New Madrid Seismic Zone to the southwest. Evidence exists that earthquakes having estimated magnitudes between 6.5 and 7.5 have occurred in this area in the past 20,000 years.

These two episodes demonstrate that seismic activity can recur in regions that are usually thought of as having little or no movement in the crust, although examination of a longer time frame shows that such events have indeed happened there.

Earthquake Hazard Mitigation

Many countries lie in regions that are affected by earthquakes, especially those on the edges of the circum-Pacific belt (Figure 4.1). Numerous densely populated communities are in danger of being affected by seismic activity that has historically devastated these same cities and towns, resulting in massive loss of property and lives. Communities tend to rebuild on the established sites. This is true following all types of natural disasters. Any attempt to lessen or mitigate future damage would be welcomed.

Of the world's 12 most populous countries in 2007 (**Table 4.5**), only Brazil, Bangladesh, and Nigeria are exempt from suffering significant seismic activity. The countries with the highest likelihood of major earthquakes have 4.05 billion people

TABLE 4.5	Twelve Most Populous Countries as of 2007. Only Brazil, Bangaladesh, and Nigeria Are Spared Major Earthquake Activity.
China	1.3 billion
India	1.1 billion
United States	301 million
Indonesia	235 million
Brazil	190 million
Pakistan	169 million
Bangladesh	150 million
Russia	141 million
Nigeria	135 million
Japan	127 million
Mexico	109 million
Philippines	91 million

Source: Information Please Almanac.

or roughly 60 percent of the world's population. The majority of these countries are developing regions that do not have rigid building standards in place to lessen the effects of earthquake damage. Some countries are almost totally within seismic zones, such as the Philippine Islands, Indonesia, and Japan, while others would be affected in only relatively isolated regions, such as the west coast of the United States or the eastern islands of Russia.

The loss of life and amount of damage inflicted on buildings are a function of several variables. The major factor controlling damage is the distance a populated area is from the epicenter of a quake. The focal depth of the source is important in determining how much motion is felt on the surface. Building construction style and engineering integrity respond directly to the energy that travels through an area. Buildings constructed on bedrock are more stable than those situated on loose or unconsolidated material.

However, problems can be reduced if buildings are designed to dampen seismic energy, much like shock absorbers are designed to lessen the bumps an automobile encounters on the road. This concept utilizes Newton's law of inertia, which states that a body (for example, a building) remains at rest unless a force (for example, seismic energy) is applied to it. To prevent a building from being destroyed, it should be "disconnected" or decoupled from the ground. Levy and Salvadori (1995) outline several good examples of ways to do this in earthquake-prone areas.

Mexico City experienced a major earthquake in 1985 that literally flattened scores of buildings. The city is constructed on ancient lake beds that provide a very weak geologic foundation. When the seismic energy passed through the area, it shook the ground as if the city rested on gelatin. Numerous buildings collapsed in a pancake fashion and many others crumbled (**Figure 4.24**). More than 9,500 people were killed and an estimated 30,000 were injured.

Regions that experience earthquakes must strengthen their building codes and attempt to retrofit existing structures that tend not to withstand seismic events well. Following the 1994 Northridge, California, earthquake, the state has spent billions of dollars to reinforce bridges and other structures to try to preserve the critical infrastructures that are often devastated by major earthquakes.

Earthquakes of equal or very similar magnitudes can produce extremely different effects, depending on where they occur. In December 2003 an earthquake of M = 6.6 killed 26,200 people in southeastern Iran, while a magnitude 6.7 event in Hawaii in October 2006 did not kill anyone. The primary cause of the large death toll in Iran was the poor construction standards used in many of the older buildings. The lack of reinforced concrete or other building material allows buildings to collapse very easily. Careful planning and the enforcement of strict building codes has reduced the effects of earthquakes in many developed countries.

Two major earthquakes in early 2010 produced major damage in the Western Hemisphere. The earthquakes in Haiti (**Box 4.4**) and Chile (**Box 4.5**) are among the most deadly and strongest in decades.

Figure 4.24 Damage was extensive as the General Hospital in Mexico City was destroyed following the September 19, 1985, earthquake that was centered off the west coast near the subduction zone, where the Cocos plate dives under the North American plate.

BOX 4.4   Haiti Earthquake of January 12, 2010

A major earthquake (M = 7.0) struck the country of Haiti on January 12, 2010, at 4:53 p.m. local time. The region lies south of the boundary of the westward moving North American plate and the eastward moving Caribbean plate. This left-lateral strike slip motion has produced numerous earthquakes to the east in the Dominican Republic, the country that lies on the eastern side of the island of Hispaniola.

Two major east-west trending, strike-slip fault systems occur in the area. The Septentrional fault system is found in northern Haiti and the Enriquillo-Plantain Garden fault system traverses the southern part of the country in a similar manner. It was the latter fault system that ruptured, producing the earthquake. The last major earthquake in the vicinity of Port-au-Prince occurred in 1770. Thus stresses in the rock have been accumulating for 240 years, even though the average movement along these faults is about 21 mm per year.

More than four dozen significant aftershocks rocked the area, creating very unsafe conditions for rescue workers struggling to free trapped people from the rubble. Eight days after the main shock an aftershock of M 5.9 occurred within ten km of the main shock.

The capital city of Port-au-Prince, with a population of more than two million people, suffered severe damage. Thousands of buildings were destroyed and at this time (late January 2010), an estimated 150,000 people died with projections reaching as high as 230,000 dead. More than 250,000 people were injured and more than 1.5 million people are homeless. The shallow depth of the event, about 13 km below the surface, added to the large amount of destruction. A worldwide effort has been mounted to raise relief money and assistance for an area that will take years to recover, as had been the case in South Asia following the December 2004 earthquake and tsunami.

Haiti, which is among the poorest on Earth, has experienced other major disasters in recent years including four hurricanes in 2008. The country lacks the resources and infrastructure to cope with such events and has been the recipient of worldwide aid to begin rebuilding.

The geological setting for this area is Haiti is very similar to that in southern California, where the right-lateral strike slip motion of the San Andreas fault system passes under heavily populated areas. In the next few decades possible movement along the southern San Andreas fault could possibly generate a similar catastrophe.

Box Figure 4.4.1 Newspaper headline the day of the earthquake. The death toll rose to more than 150,000 people within two weeks.

Box Figure 4.4.2 Epicenter of the Haiti earthquake of January 12, 2010 was located at a distance about 25 km west southwest of Port-au-Prince.

BOX 4.5 The Great Chilean Earthquake of February 27, 2010

On February 27, 2010 at 3:34 a.m. local time, the earth shook in the vicinity of Maule, Chile, located about 115 km north-northeast of the city of Conception. This event measured M 8.8, making it the fifth largest earthquake in recorded history. The region lies at the boundary between the eastward-moving Nazca plate and the westward-moving South American plate. Over the past forty years, the region has experience 30 earthquakes of magnitude 7 or greater. The epicenter for this event is located about 275 km north of the location of the May 1960 magnitude 9.5 earthquake, the largest earthquake in recorded history.

Given the proximity of the earthquake to the Pacific Ocean, the Pacific Tsunami Warning Center in Hawaii issued several tsunami warnings to alert more than fifty nations and territories bordering the Pacific Ocean. However, the warnings was canceled after it was determined that these areas would be spared from the destructive waves. Countries as far away as Japan and Russia were expected to receive tsunami about 21 hours following the main shock, but only a slight wave arrived in those areas. One contributing factor to the warning was the relatively shallow depth of the focus—only 35 km below the surface. More than 150 significant aftershocks were generated in the first two days following the main shock. These, of course, can be destructive from the standpoint of destroyed buildings that were weakened by the main shock. The aftershock region extended for more than 700 km along the Chilean coastline. By the end of the third day the number of fatalities reached more than 700 people. Undoubtedly the number will increase as reports come in from remote, outlying regions and more bodies are discovered in the rubble of thousands of collapsed buildings.

BOX TABLE 4.5		Two Largest Earthquakes in the History of Chile		
Date	Mw	Location and Depth	Death Toll	Notes
May 22, 1960	9.5	38.29S 73.05W Depth = 33 km (est.)	1,655	Largest earthquake in recorded history; epicenter located in the southern portion of the Nazca-South American plate boundary
February 27, 2010	8.8	35.846S 72.719W Depth = 35 km Conception 115 km Santiago 325 km	More than 730 (preliminary count as of March 3, 2010)	Fifth largest earthquake in recorded history; epicenter located in the south-central portion of the collision of the Nazca and South American plates

Mitigation involves reducing the risk that earthquakes create. Revising building codes to strengthen existing buildings, freeway bridges, and other structures lessens the damage they incur when they undergo shaking by seismic energy. Buildings experience damage because of the direct waves that pass through an area; if these do not destroy buildings, they can weaken them to the point that energy from aftershocks can finish the job of leveling the edifice.

Earthquakes will always occur because of the dynamics associated with plate tectonics, whether the events are the result of direct collision of plates or the result of volcanic activity that is also attributable to plate motion. Being aware of regions on Earth that are most likely to generate earthquakes will make people understand why they occur, although the actual timing of the events Is not predictable.

Summary

Earthquakes are a daily occurrence, happening somewhere almost at any given moment. Fortunately most of them are rather small in magnitude and do not affect people's lives. Seismically active regions are spread around the world and are associated with plate boundaries. Major earthquake areas include the circum-Pacific belt and the Mediterranean-Trans Asiatic belt. In the United States, the Pacific coast along with Alaska are most likely to be affected by large magnitude events.

Historically, millions of people have been killed by earthquakes, but in the United States, the numbers are quite small because of better building codes and a lower population density. Within the past 20 years California has been hit with several large earthquakes but fortunately the loss of life has been small. During that same time span, several hundred thousand people have died in other parts of the world, these areas have a higher population density and do not have the engineering standards used in the United States.

The ability to predict earthquakes does not exist. Earthquakes are random events that have numerous variables at work that cause them to occur where they do, when they do. The best way to lessen the impact and damage in any area is to plan for what in many areas is the inevitable and employ good construction practices, especially in areas where the population density is high.

References and Suggested Readings

Bolt, Bruce A. 2006. *Earthquakes.* 5th ed. New York: W. H. Freeman.

Brumbaugh, David S. 2010. *Earthquakes—Science and Society.* 2nd ed. Upper Saddle River, NJ: Prentice-Hall.

Doyle, Hugh. 1995. *Seismology.* Chichester, England: John Wiley.

Fradkin, Philip L. 1998. *Magnitude 8.* New York: Henry Holt.

Hough, Susan Elizabeth. 2002. *Earthshaking Science—What We Know (and Don't Know) About Earthquakes).* Princeton, NJ: Princeton University Press.

Levy, Matthys and Mario Salvadori. 1995. *Why the Earth Quakes.* New York: W. W. Norton.

Lutgens, F. and E. Tarbuck. *Essentials of Geology.* 9th ed. 2006, Upper Saddle River, NJ: Pearson Prentice-Hall.

Marshak, S. 2004. *Essentials of Geology.* New York: Norton.

Mogi, Kiyoo. 1985. *Earthquake Prediction.* Orlando, FL: Academic Press.

Prager, Ellen J. 2000. *Furious Earth—The Science and Nature of Earthquakes, Volcanoes, and Tsunamis.* New York: McGraw-Hill.

Ritchie, D. and A. E. Gates. 2001. *Encyclopedia of Earthquakes and Volcanoes.* New York: Facts on File.

Yeats, Robert S., Kerry Sieh, and Clarence R. Allen. 1997. *The Geology of Earthquakes.* New York: Oxford University Press.

Web Sites for Further Reference

http://earthquake.usgs.gov/
http://earthquake.usgs.gov/regional/states.php
http://quake.wr.usgs.gov/recent/index.html
http://vulcan.wr.usgs.gov/Glossary/Seismicity/description_earthquake.html
http://water.usgs.gov/wid/index-hazards.html
http://www.cumbavac.org/Earthquakes_Volcanoes.htm
http://www.geophys.washington.edu/seismosurfing.html
http://www.isc.ac.uk
http://www.iris.edu
http://www.ngdc.noaa.gov/seg/hazard/hazards.shtml
http://www.scec.org/

Questions for Thought

1. What are the three types of plate boundaries?

2. Which of the plate boundaries has the largest number of earthquakes and why?

3. Explain why the San Andreas fault system in California has earthquakes but not volcanic activity.

4. What are the differences between P-waves and S-waves?

5. How does a seismometer detect seismic activity?

6. Compare the Richter scale and the Mercalli scale. What does each measure?

7. How are earthquake epicenters located?

8. What reasons can you give for larger losses of life from earthquakes in China than in the United States?

9. How successful are scientists in predicting earthquakes?

10. How likely is an earthquake in your hometown?

Unstable Ground: Mass Movements and Subsidence

5

United States Geological Survey. Photo by R. L. Schuster.

La Conchita, California, has experienced two massive landslides in recent years. This one occurred in 1995. In 2005 another one affected the same slope after a series of rain storms saturated the slopes Ten people died in the latter event.

Key Terms

angle of repose
cause
collapse
dissolution
driving force
Factor of Safety (FoS)
landslide
liquefaction
mass movement (mass wasting)
normal force
regolith
resisting force
shear force
sinkholes
slide
slope failures
slump
subsidence
talus
trigger
undercutting

mass movement (mass wasting)
A term used more by geologists to describe the down-slope movement of material.

regolith
The layer of varied material that overlies unaltered bedrock and includes unconsolidated and fragmental particles.

landslide
A general term used to describe the down-slope movement of material under the force of gravity.

subsidence
A relatively slow drop in the ground surface caused by the removal of rock or water located under the surface.

sinkhole
A circular depression on the surface that forms from the collapse of material into an underlying void.

Although the ground we live on may appear to be stable and unchanging, it is in fact a dynamic place where the force of gravity is constantly acting on Earth's materials, over time causing the downward movement of rock and loose surface material. **Mass movement** (also called *mass wasting*) is the downward movement of Earth's surficial material from one place to another under the direct influence of gravity. Earth's surface consists mainly of slopes ranging from gentle to steep and often covered with several components of the surface. **Regolith** consists of all material lying above unaltered bedrock, including unconsolidated and fragmental material which breaks down to form soil. Although gravity acts constantly on all slopes, the strength and resistance of materials usually hold the slope in place. However, natural processes or human activity may destabilize a slope, causing failure and mass movement of material. This movement may be so slow that it is almost undetectable (known as *creep*), or be sudden and swift, as in devastating **landslides**. Mass movements do not always have to involve failure on a slope, however. **Subsidence** is the vertical (downward) motion of earth materials, and includes vertical movements such as the gradual sinking of the land surface when fluids are withdrawn, or the sudden collapse into subterranean voids creating **sinkholes**.

Mass movement is an important natural erosional process but can be a serious hazard where expanding populations are building homes on or near steep hillsides or over abandoned mine areas. Rapid downslope movements mostly involve landslides, which every year causes loss of life, property damage, or economic hardship to people around the world.

Landslides have occurred in all 50 States (**Figure 5.1**), and damage from these landslides exceeds $3.5 billion every year, while claiming 25 to 50 lives. In 1985, a massive landslide in Puerto Rico killed 129 people, making it the greatest loss of life

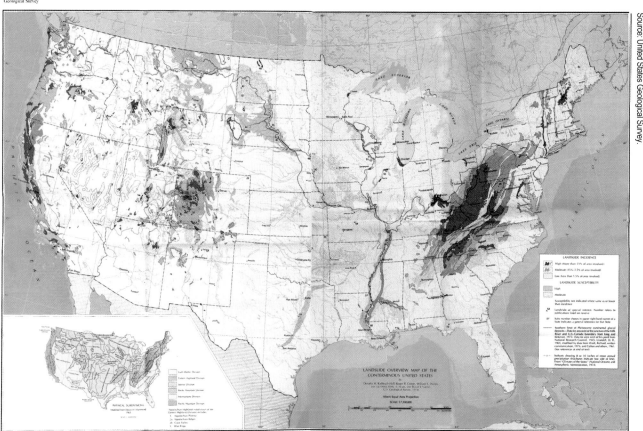

United States   Department of the Interior
Geological Survey

Figure 5.1 Landslides occur in all 50 states (Alaska and Hawaii not shown). Moist, unstable terrain in the eastern United States is especially affected by landslides.

from a single landslide in U.S. history. The Thistle, Utah landslide of 1983 caused an estimated $400 million in losses, making it the most expensive single landslide in U.S. history, while the 1997-98 El Niño rainstorms in the San Francisco Bay area produced thousands of landslides, causing over $150 million in losses. Landslides can also be accompanied by other hazards, such as earthquakes, floods, storm surges, severe storms, and volcanic activity. **Slope failures** are a secondary effect of wildfires that remove the vegetation holding soil and surface materials in place. Often they are more damaging and deadlier than the associated hazard event.

slope failure
The down-slope movement of an unstable area.

What makes many mass movements, such as landslides and collapse sinkholes, so hazardous is that they can occur at anytime and almost any place with little or no warning, and the event is over very quickly. Knowledge about the relationships between the local geology and mass movement processes can aid in better development planning and a reduction in vulnerability to such hazards. Therefore, we will look at the various types of mass movements, their underlying causes, factors that affect ground and slope stability, and what we can do to reduce the vulnerability and risk of living on unstable ground.

Landslides: Mass Movements on Slopes

Earth's surface materials that make up slopes comprise a complex physical, chemical, and biological system. These materials are a mixture of rock, regolith, and soil, with a variable amount of water and organic material. The amount of water and organic material (mostly vegetation) may vary on a slope from season to season and year to year. This variability depends upon such factors as the nature and extent of precipitation, rates of materials added or removed to the slope by deposition or erosion, and activities that affect the vegetation, such as human impact or fire. These complex interactions involving the strength and composition of earth materials are responsible for the different shapes of slopes, and how slopes may fail under the influence of gravity. The resulting mass movement plays an important part in the erosional process, moving material downslope from higher to lower elevations.

The term landslide is used in the general sense to describe a wide variety of mass movement landforms and processes involving the downslope transport of earth material under gravitational influence. Landslides have a great variety of shapes (morphologies), rates of motion, and types of movement, and can range in size from a small area to a region of many square kilometers. Landslides play a significant role in the evolution of the landscape, and are among the most widespread natural hazards on Earth.

Slope Processes and Stability: The Balance Between Driving Forces and Resisting Forces

Two types of forces are involved in mass movements on slopes: **driving forces** and **resisting forces**. Driving forces are those that promote movement, while resisting forces are those that tend to prevent movement. The main *driving force* responsible for mass movement is *gravity*. Gravity acts everywhere on the Earth's surface, pulling everything in a vertical direction toward the center of the Earth. On a flat surface the force of gravity acts directly downward, so, as long as the material remains on the flat surface, it will not move. On a slope, the force of gravity can be divided into two components: one acting perpendicular to the slope (the **normal force**) and one acting parallel to the slope (the **shear force**) (**Figure 5.2**). The normal force holds the material in place on the slope, whereas the shear force moves the material down the slope.

driving force
A force that produces down-slope movement caused by gravity.

resisting force
A force that tends to prevent downhill movement.

normal force
A force that is acting perpendicular to a surface.

shear force
A force that acts parallel to a surface.

On and below the ground surface, resisting forces, such as friction and particle cohesion, act in opposition to the driving forces. When the driving forces acting downhill exceed the resisting forces, mass movement results.

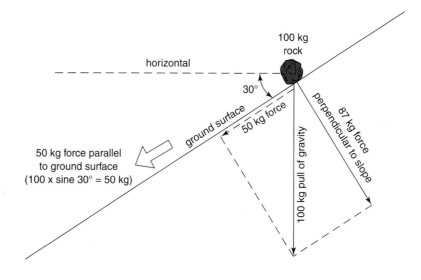

Figure 5.2 Forces acting on a 100 kilogram boulder resting on a 30° hillslope.

Factor of Safety (FoS)
The ratio of resisting force to driving force.

The **Factor of Safety (FoS)** can be defined for a slope as:

$$\text{FoS} = \text{resisting forces/driving forces.}$$

If the resisting forces are greater than the driving forces, the FoS is greater than 1 and the slope is stable. If the driving forces are greater than the resisting forces, the FoS is less than 1 and slope failure occurs. The magnitudes of the resisting and driving forces can change as a result of weather conditions or slope configuration.

Slopes are in a continuous state of dynamic equilibrium by constantly adjusting to new conditions caused by changes in the driving and resisting forces. Although we view mass wasting events as disruptive and often destructive processes, it is simply a way for a slope to adjust to new and changing conditions. For example, when a building or a road is constructed on a hillslope, the equilibrium of the slope is upset and the slope may become unstable. This could lead to slope failure as the slope adjusts to the new set of conditions (additional mass) placed upon it. The role of gravity is to level out all slopes and create a horizontal landscape.

Causes and Triggers That Influence Slope Stability

Rock, regolith, and soil remain on a slope only when the driving forces are unable to overcome the resisting forces keeping the material in place. Long before a landslide occurs, slopes undergo various processes that weaken rock material and gradually make the surface more susceptible to the pull of gravity. Eventually, the strength of the slope is weakened to the point that some event (known as a *trigger*) allows it to suddenly cross from a state of stability to one of instability.

cause
A controlling factor that makes a slope vulnerable to failure.

trigger
An event that disturbs the equilibrium of a slope, causing movement to occur.

Following a landslide, geologists and engineers often look for clues (the contributing factors) to explain why the slope failed. The reasons for slope failure can be broken down into two categories: *causes* and *triggers* (**Table 5.1A-B**). The difference between causes and triggers of landslides is subtle, but important to understand. **Causes** are considered to be the controlling factors involved in making the slope vulnerable to failure in the first place; in other words, they contribute to the slope becoming weak and unstable. A **trigger** is the single event that temporarily disturbs the equilibrium of the slope and initiates mass movement, resulting in slope failure. Triggers that initiate movement include earthquakes, heavy rainfall, and volcanic activity. The causes are the explanation for why a landslide occurred in a certain location, and include geological factors, morphological factors, and factors associated with human activity. Landslides most often have multiple causes that combine to make a slope vulnerable to failure, and then a trigger finally initiates the movement.

TABLE 5.1A	Common Landslide Causes

1. Natural Geological Causes

   a. Weak or sensitive materials
   b. Weathered materials
   c. Sheared, jointed, or fissured materials
   d. Adversely oriented discontinuity (bedding, schistosity, fault, unconformity, contact, and so forth)
   e. Contrast in permeability and/or stiffness of materials

2. Natural Morphological Causes

   a. Tectonic or volcanic uplift
   b. Glacial rebound
   c. Fluvial, wave, or glacial erosion of slope toe or lateral margins
   d. Subterranean erosion (solution, piping)
   e. Deposition loading slope or its crest
   f. Vegetation removal (by fire, drought)
   g. Thawing
   h. Freeze-and-thaw weathering
   i. Shrink-and-swell weathering

3. Human Causes

   a. Excavation of slope or its toe
   b. Loading of slope or its crest
   c. Drawdown (of reservoirs)
   d. Deforestation
   e. Irrigation
   f. Mining
   g. Artificial vibration
   h. Water leakage from utilities

TABLE 5.1B	Common Landslide Triggers

1. Physical Triggers

   a. Intense rainfall
   b. Rapid snowmelt
   c. Prolonged intense precipitation
   d. Rapid drawdown (of floods and tides) or filling
   e. Earthquake
   f. Volcanic eruption
   g. Thawing
   h. Freeze-and-thaw weathering
   i. Shrink-and-swell weathering
   g. Flooding

Source: USGS Fact Sheet 2004-3072 and Circular 1325-508.

In most cases it is relatively easy to determine the trigger after the landslide has occurred, but the causes leading up to slope failure can be much more difficult to identify, because once a landslide occurs, any pre-existing evidence is usually destroyed.

Common Controlling Factors (Causes) in Slope Stability

The following is a discussion of the more common controlling factors that influence slope stabilities, and the role they play in making them vulnerable to failure. Although they are discussed separately, most of them are interrelated and can collectively affect a slope's stability.

The Role of Slope Materials

Some rocks are inherently more stable than others. Massive, uniformly-textured rocks such as granite, basalt, and quartzite, have interlocking crystals and grains that give them nearly equal strength in all directions. Such rocks will hold their position against gravity, and form steep slopes only when they become fractured and jointed. Over time, chemical and physical weathering processes produce unconsolidated regolith, and change its resistance to gravity. Freezing and thawing of water in cracks, or shrinking and swelling of clay minerals decreases friction and cohesiveness and promotes slope failure.

Unconsolidated earth materials such as soil, clay, sand, and gravel are more susceptible to movement. Clay becomes plastic and weak when it absorbs water, causing it to act as a lubricant for mass movements. For sediment coarser than clay, stability on slopes is influenced by grain shape (whether angular or round) and grain roughness. The presence of plant root networks and mineral are very effective in holding material in place, thus promoting stability.

The Role of Water

The water content in slope material strongly influences its stability in several ways, and is a critical factor in many causes of mass movements, as well as being a trigger. As a driving force, slope material can become saturated with water (after heavy rains, melting snow, or rising groundwater) by filling pores and fractures. This increases the weight (load) of material on the slope and increases the stress acting parallel to the surface (water weighs 1 kg per liter or 8.35 lbs per gallon). However, this extra weight of water is probably less important than the reduction of the resisting forces of the material. Water percolating through slope material helps decrease friction between grains and contributes to a loss of cohesion among particles.

Another aspect of water that affects slope stability is fluid pressure, which in some cases can build in such a way that water in the pores can actually support the weight of the overlying material. When this occurs, friction is greatly reduced, and thus the resisting forces holding the material on the slope is also reduced, causing slope failure. Water can also reduce rock strength by circulating through the pores of some rocks and dissolving soluble cementing materials, such as calcium carbonate. This process reduces cohesion as well. Another effect of water is that it can soften layers of shale, and even cause some types of clay minerals to expand, reducing friction between rock layers. Clay consists of platy particles that easily slide over each other when wet. This lubricating effect of water with clay is the reason why clay beds are frequently the slippery layer along which overlying material slide down-slope.

angle of repose
The natural angle of a slope that forms in a pile of unconsolidated material.

undercutting
A process whereby a slope or hillside has supporting material removed by erosion.

The Role of Slope Angle

Slope angle is a major factor that influences mass movement. Commonly, the steeper the slope, the less stable it is, and therefore, steep slopes are more likely to experience mass movement than gentle slopes. The steepest angle that a slope can maintain without collapsing is called its **angle of repose** (**Figure 5.3**). For dry, unconsolidated materials, the angle of repose increases with increasing grain size, but usually lies between about 33° and 37° on naturally formed slopes. At this angle, the resisting forces of the slopes material counterbalance the force of gravity. Stronger material such as massive bedrock can maintain a steeper slope. If the slope angle is increased, the rock and debris will adjust by moving downslope.

A number of processes can over steepen a slope and make it unstable. **Undercutting** by stream and wave action is a common process that removes the base of a slope, increasing the slope angle, and increasing the driving force acting parallel to the slope. Waves pounding against the base of a cliff, especially during storms, often result in mass movements along the shores of oceans and lakes. Human activities also often create over-steepened and unstable slopes that become prime sites for mass movement. Mass movement occurs when grading and cutting into the slope too steeply increases the downhill forces and the slope is then no longer strong enough to maintain the steep angle. Such actions by humans is analogous to undercutting by streams and has the same result, explaining why so many mountain roads and building sites are plagued by frequent mass movements.

The Role of Vegetation

Plants protect regolith and soil against erosion, and contribute to the stability of slopes (**Figure 5.4**). Vegetation helps absorb rainfall and leads to a decrease in water saturation of the slope material that would otherwise lead to a loss of shear strength. The root systems of plants also help stabilize the slope by binding soil particles together and holding the soil to bedrock. Where vegetation is lacking, mass movement is enhanced. The removal of vegetation by natural events (fires, drought) or human processes (timber, farming, or development) frequently results in unconsolidated materials moving down-slope.

The Role of Overloading

Overloading occurs by the additional weight of material being added to a slope. As mentioned earlier, water can become a driving force by increasing the weight of slope material after prolonged periods of rain. However, overloading is most often the result of human activity, and typically results from the dumping, filling, or piling up of material on a slope (**Figure 5.5**). The additional weight created by the overloading increases the water pressure within the material, which in turn decreases the stability of the slope. If enough material is added to the slope, it will eventually fail, sometimes with tragic consequences. Water can become a driving force by increasing the weight of slope material after prolonged periods of rain by completely filling pore spaces. The water increases the weight of the sediment and thus increases the driving force and causes many mass movements to occur during or shortly after prolonged rainfall.

The Role of Geologic Structures

Plate tectonic forces can reorient rocks after their formation, causing them to be rotated and tilted. If the rocks underlying a slope dip in the same direction as the slope, mass movement is more likely to occur than if the rocks are horizontal or dip in the opposite direction of the slope. When rocks dip in the same direction as the slope, water can

Figure 5.3 The angle of repose is the natural angle of a slope formed by a pile of material.

Figure 5.4 Vegetation on slopes is effective in holding the soil and regolith in place as root systems help stabilize the material. Whenever grasses and trees are removed, soil and regolith can readily move downslope.

Figure 5.5 Unstable slopes are created when tailings and other material from mining operations are piled up. These slopes are unable to develop soil layers that can support vegetation, so in addition to being unsightly they are likely to move if a triggering mechanism occurs nearby.

Figure 5.6 Bedding planes dipping in slope direction lead to slope failure.

Figure 5.7 Terraces are cut in an open-pit mine near Bisbee, Arizona, in order to help stabilize the steep walls of the pit.

percolate along the bedding planes and decrease the cohesiveness and friction between adjacent rock layers (**Figure 5.6**). This is particularly important when clay layers are present, because they become slippery when wet.

Even if the rocks are horizontal or dip into the opposite direction of the slope, other structural weaknesses (such as joints, foliations, and faults) may dip in the same direction as the slope. The water percolating through the rock structures weathers and expands the openings further until the weight of the overlying rock causes it to fail.

The Role of Human Activity

There are many ways in which human activities can affect slope stability and promote landslides. One is by the clearing of stabilizing vegetation during logging operations, exposing sloping soil to rain. Many types of construction projects can lead to over-steepening of slopes. Highway roadcuts, quarrying, or open-pit mining, and construction of homes on benches cut into hillsides are among the most common activities that can cause problems (**Figure 5.7**). Where dipping layers of rock are present, removal of material at the lower end of the layers may leave large masses of rock unsupported.

In the 1870s throughout Europe there was a large demand for slate, which was used to make blackboards. To meet this demand, slate quarry miners near Elm, Switzerland dug slate quarry at the base of a steep cliff containing excellent planar foliation that dipped toward the quarry. By September 1881, the quarry was 180 m long and 60 m into the hill below the cliff, and a fissure above the cliff had opened to 30 m wide. Falling rocks were frequent in the quarry and almost continuous noises were heard coming from the overhang above the quarry. On September 11, 1881, a 10 million m³ mass of rock above the quarry suddenly fell to the quarry floor, and produced an avalanche moving at 180 km/hr that traveled over 2 km, burying the village of Elm and killing 115 people.

Slopes cut into unconsolidated material at angles higher than the angle of repose without planting stabilizing vegetation becomes extremely unstable. The very act of building a house above a naturally unstable or artificially steepened slope adds weight to the slope, thereby increasing the stress acting parallel to the slope. Other activities connected with the presence of housing developments on hillsides include watering the lawn, use of septic tanks, or an in-ground swimming pool from which water can seep slowly. These are all activities that increase the water content of the soil and can render the slope more susceptible to movement. Even homes built on apparently solid rock could be prone to destruction (**Figure 5.8**).

Common Triggers of Slope Failure

The factors discussed thus far all contribute to slope instability and are considered causes that make the slope vulnerable to failure. A sudden triggering event may then happen that initiates movement on the unstable hillside. Movement could eventually occur without a trigger if

Figure 5.8 This home in Valparaiso, Chile, is situated on solid rock. However, earthquakes are a common occurrence in Chile, and this home could easily collapse down the hill.

slope conditions became more unstable over time. However, sudden, rapidly moving landslides usually involve a triggering mechanism. Water, volcanic activity, and seismic activity are three common landslide trigger mechanisms that can occur either singly or in combination (refer to Table 5.1a-b).

Water as a Trigger

Rainfall. When a slope becomes saturated with water, the increased mass of the water, and its lubricating effects, create instability. Intense or prolonged rainstorms act as a trigger of the mass movement. Porous soil and layer of underlying rock allow the pore pressure to increase, thereby decreasing the adhesion of the grains. Failure results. Heavy rainfall occurs during tropical storms and hurricanes, short-lived, severe thunderstorms, and prolonged rainfall related to slow-moving weather systems. Flooding and landslides are commonly the result (**Figure 5.9**).

Snowmelt. In regions that experience prolonged cold weather that allows snow packs to increase in size, seasonal melting will produce large amounts of water in relatively short time periods. Melting can be accelerated by short periods of unseasonably warm weather. In addition to direct runoff of melt waters, water can percolate into the subsurface and saturate the soil above the zone of permanently frozen ground (permafrost). If there is any slope to the saturated layer, the fluids and solid material will rapidly move downhill as increased pore pressure reduces the shear strength of the soil. Depending on the amount of water present and the angle of the slope, rapid or slow mass movement can occur.

River and Wave Undercutting. Erosional processes caused by moving water in streams, or wave action along a shoreline, can remove material at the base of a slope. If sufficient removal occurs, the slope is no longer supported and slope failure occurs. A large-scale example of this is seen at Niagara Falls along the border of the United States and Canada (**Figure 5.10**). Water cascading over the falls erodes the relatively soft underlying shale layers. The upper rocks are no longer supported and collapse. Erosion there is occurring at the rate of about one meter per year.

Volcanic Activity as a Trigger

Some of the largest and most destructive landslides known have been associated with large volcanic eruptions. These occur either in association with a violent eruption of the volcano itself, or as a result of mobilization of its weak deposits that are formed as a consequence of volcanic activity. Volcanic eruptions produce *seismic shaking* similar to high-energy explosions and earthquakes, and can trigger flank collapses that create fast-moving debris avalanches. Hot volcanic lava or ash eruptions can melt snow and ice on a volcano summit at a rapid rate, causing a flow of rock, soil, ash, and water that accelerates rapidly on the steep slopes of volcanoes. These volcanic debris flows (known as *lahars*) can reach great distances, once they leave the flanks of the volcano, destroying anything in its path (see volcanic hazards in Chapter 4). The 1980 eruption of Mount St. Helens triggered a massive debris avalanche on the north flank of the volcano, which is the largest landslide in recorded history (**Figure 5.11**). The 1985 Nevado del Ruiz eruption in Colombia created pyroclastic flows that melted

Figure 5.9 These houses were destroyed by landslides associated with Typhoon Morakat, which hit Taiwan in August 2009.

Figure 5.10 Undercutting occurs as water flows over Niagara Falls. The softer, underlying rock (shale) is eroded, thereby removing support for the more resistant overlying dolomite, which subsequently falls into the area below.

Figure 5.11 A massive debris avalanche in the Coldwater Lake area near Mount St. Helens, Washington.

ice and snow at the summit, forming lahars up to 50 meters thick. These massive flows traveled more than 100 kilometers down several river valleys, destroying many houses and towns. The town of Armero was completely covered by the lahar, killing approximately 23,000 people, making it the second-deadliest volcanic disaster in the 20th century (after the 1902 eruption of Mount Pelée, which killed 29,000 people).

Seismic Activity as a Trigger

United States Geological Survey.

Figure 5.12 The ground in Anchorage, Alaska, was affected greatly by the M 9.2 earthquake of March 1964.

Earthquake activity can cause potentially unstable slopes to fail as seismic energy passing through rock shakes the material. Consolidated, solid rock, such as granite is not affected by the energy passing through it. However, loose, unconsolidated material or mud can be turned into a liquid if sufficient water is present. A combination of vertical and horizontal stresses creates failure in slopes when the earthquake trigger occurs. In regions where water is common, pore pressure is increased by the seismic waves, and liquefaction occurs.

Most of the monetary losses due to the 1964 Great Alaska Earthquake were caused by widespread slope failures and other ground movement (**Figure 5.12**). In other areas of the United States, such as California, Oregon, and the Puget Sound region of Washington, slides, lateral spreading, and other types of ground failure due to moderate to large earthquakes have been experienced in recent years. Worldwide, landslides caused by earthquakes kill more people and damage more structures than in the United States.

Classification and Types of Mass Movements

Often, the term landslide is used in the general sense to describe many types of slow to rapidly descending masses of material down a slope. However, the term landslide does not tell us anything about the processes involved in the movement. Most geologists prefer the term *mass movement*, because slope movements can occur on the ocean floor and not just on land. Also, many slope movements occur by mechanisms such as falling, flowing, and creeping rather than by sliding and more descriptive terms such as *rockfall*, *debris flow*, and *rockslide* are used. Geologists and civil engineers distinguish different types of mass movement based on three characteristics:

- **Type of Material.** The type of material involved in a mass movement depends upon whether the descending mass began as unconsolidated material or as bedrock. If regolith dominates, terms such as *debris*, *mud*, or *earth* are used in the description. The term *earth* refers to material that is composed mainly of sand-sized or finer particles, and debris is composed of coarser fragments. If the massive bedrock breaks loose and moves downslope, the term *rock* is used as part of the description.
- **Type of Motion.** The type of motion describes the way material moves down the slope. The most common motions described are *fall*, *slide*, or *flow*. Falls move unimpeded through the air and land at the base of a slope. Slides move in contact with the underlying surface. Flows are plastic or liquid movements in which the mass breaks up during movement. Other movements include topples and lateral spreads.
- **Rate of Movement.** The movement of slope material occurs at a wide range of speeds. Some mass movement occurs so slowly that it may take years before the movement is noticeable from the downslope displacement of trees, fences, and walls. Other mass movements occur very rapidly and reach speeds in excess of 200 kilometers per hour. Water also affects the rate of movement. Commonly, the higher the water content, the faster the rate of movement.

BOX 5.1	The Nile Landslide of Southern Washington October 11, 2009

At approximately 6 am on the morning of October 11, 2009, a massive landslide moved down a hill in a lightly populated area in southwest Washington state about 35 km northwest of the city of Yakima. The slide was a rotational slide in an area thought to lie on the side of a large anticlinal fold. The slide covered about 80 acres and earliest estimates indicated more than 1 million cu meters of material came down the slide. The Naches River, lying at the base of the slide, was partially dammed and more than 800 m of State Highway 410 was destroyed. More than 12 m of soil and debris covered the highway. Traffic was diverted to a small county highway located on the facing side of the river (to the right of the image) while a temporary bypass was constructed. Flooding and the landslide damaged 30 homes in the immediate area and more than 600 people were inconvenienced by the landslide.

Courtesy of Washington State Department of Natural Resources.

Box Figure 5.1.1 More than 1 million cu m of material slid down this hillside, destroying about 800 m of highway and temporarily damming the Naches River.

Courtesy of Washington State Department of Natural Resources.

Box Figure 5.1.2 Landslide debris covers State Highway 410, creating a major detour for travelers going between Mount Rainier National Park and the city of Yakima.

Landslide Hazards

Since the hazards associated with landslides are highly variable and depend on the material involved and mechanism of movement, the common types of landslides are discussed below, along with their associated hazards. These definitions are based mainly on the work of Varnes (1978) and USGS Fact Sheet 2004-3072, which discuss landslide types and processes. The type of movement describes the actual internal mechanics of how the landslide mass is displaced: fall, topple, slide, spread, or flow. Thus, landslides are described using two terms that refer respectively to material and movement (that is, rockfall, debris flow, and so forth).

Slides

Slides refer to mass movements where there is a distinct zone or surface of weakness that separates the overlying slide material from more stable underlying material. The two major types of slides are rotational slides and translational slides, which are characterized by the surface of rupture.

slide
The movement of material along a curved or flat plane.

Rotational Slide. This is a slide in which the surface of rupture is curved concavely upward and movement is roughly rotational about an axis that is parallel to the ground surface and perpendicular across the slide (**Figure 5.13**). Rotational slides occur most frequently in homogeneous materials and are the most common landslides occurring in "fill" materials. They are usually associated with slopes ranging from 20 to 40 degrees, and travel extremely slowly (less than 0.3 meter or 1 foot every 5 years) to moderately fast (1.5 meters or 5 feet per month) to rapid. Most are triggered by intense and/or sustained rainfall or rapid snowmelt that can lead to the saturation of slopes and increased groundwater levels within the mass.

slump
A mass movement in which generally unconsolidated material moves down a hillside along a curve, rotational subsurface plane.

A **slump** is an example of a small rotational landslide and commonly occurs in unconsolidated sediments and in some weaker rocks. Slumping is commonly caused by erosion at the base of a slope, which removes support for the slope material. The erosion may be natural, such as cutting away of the base of a coastal cliff by waves, undermining of a river bank by stream flow, or the result of human activity, such as road construction. When slope failure occurs, the slump block rotates downward, a *scarp* (cliff) is formed at the head of the slope and the toe moves outward over the slope below. Notice that the toe provides the primary resisting force, so when it is upset by erosion or removal, the material generally moves downhill.

Slumping is an especially serious hazard where structures are built on cliffs above a shoreline when wave energy erodes the base of the cliffs, causing slope failure (especially during storms). In many cases, homes built along the cliffs are destroyed and the scarp progresses inland, posing a hazard for the next line of houses on the cliff.

The Great Alaska Earthquake of March 27, 1964 (M 9.2, see Chapter 4) caused serious land displacement in the city of Anchorage, which is mostly underlain by the Bootlegger Cove Clay. This former marine clay layer originally held saltwater in its pores that was later replaced by freshwater, making this area an unstable *quickclay*. The ground shaking that accompanied the earthquake caused the clay to liquefy and the ground fail, producing a series of slump scars. Some of the worst damage was in the subdivision of Turnagain Heights built on a flat-topped bluff some 30 m above the level of Cook Inlet (**Figure 5.14**). Slumping began in the overlying glacial deposits as soon as the underlying clay lost its cohesion. Within minutes, the flat bluff area was changed into a mass of rotated slump blocks, covered with twisted trees and destroyed homes.

Translational Slide. In this type of slide, the movement occurs along a roughly planar surface with little rotation or backward tilting (**Figure 5.15**). A *block slide* is a

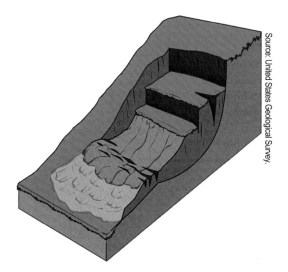

Source: United States Geological Survey.

Figure 5.13 A rotational slide has a curved plane along its base, giving the sense of rotation of the material.

United States Geological Survey.

Figure 5.14 Severe slumping occurred in the Turnagain Heights region of Anchorage due to the March 27, 1964, earthquake. A combination of clay and water created unstable slopes that collapsed.

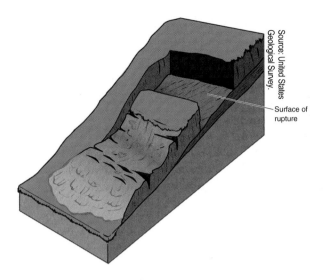

Source: United States Geological Survey

Surface of rupture

Figure 5.15 A translational slide often has a solid block of rock sliding down a surface, which has been translated to a new, lower position on the hillside.

Courtesy of David M. Best.

Figure 5.16 Material at the base of the cliff forms a talus slope. Weathered rock falls downhill and builds up a slope.

translational slide in which the moving mass consists of a single unit or a few closely-related units that move downslope as a relatively coherent mass. Often called *rock-slides*, the sliding surface is commonly a bedding plane, but rockslides can develop on other surfaces such as fractures that cut across layered rocks. One of the most common types of landslides, translational slides are found throughout the world. They generally occur close to the surface, and can range from small (residential-lot size) failures to very large, regional landslides that are kilometers wide. Movement may initially be slow (1.5 meters per month) but many are moderate in velocity (1.5 meters per day) and some extremely rapid. With increased velocity, the landslide mass of translational failures may disintegrate and develop into a debris flow. They are triggered primarily by intense rainfall, rise in ground water within the slide due to rainfall, snowmelt, flooding, or other inundation of water resulting from irrigation, or leakage from pipes or human-related disturbances, such as undercutting. These types of landslides can be earthquake-induced. Occasionally the translating block will ride on a cushion of air, which greatly reduces the frictional forces. These situations are preceded downslope by an air blast that often flattens trees.

Falls

Falls are abrupt movements of masses of geologic materials, such as rocks and boulders, that become detached from steep slopes or cliffs. Rocks can separate along potential planes of weakness such as joints, fractures, or bedding planes. When material falls under the force of gravity, it comes to rest and accumulates at the base of the hillside or slope. The material at the base is termed **talus** (**Figure 5.16**). If the buildup is significant, the result is a talus slope.

Rockfalls. Rockfalls are abrupt, downward movements of rock, earth, or both, that detach from steep slopes or cliffs (**Figure 5.17**). The falling material usually strikes the lower slope at angles less than the angle of fall, causing bouncing. The falling mass may break on impact, or may begin rolling on steeper slopes, and may continue until the terrain flattens. They are common worldwide on steep or vertical slopes, in coastal areas, and along rocky banks of rivers and streams. The volume of material in a fall can vary substantially, from individual rocks or clumps of soil to massive blocks thousands of cubic meters in size. They are very rapid to extremely rapid,

talus

The natural pile of material that builds up at the base of a cliff or hillside by material falling from above.

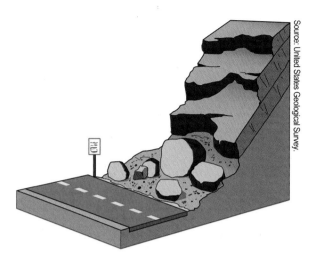

Figure 5.17 A rockfall is often a free-fall of rock to the surface, where it can break into smaller pieces or roll along the surface.

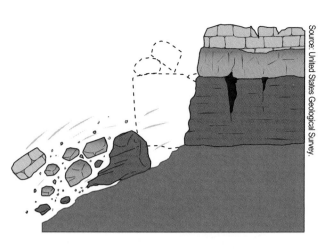

Figure 5.18 Topples result from rocks rotating about a pivot point at the base of a cliff and the rocks basically tilt over. The dashed blocks designate the position of the rocks prior to collapsing.

free-fall, bouncing and rolling of detached soil, rock, and boulders. The rolling velocity depends on slope steepness. Rockfalls are commonly triggered by undercutting of slope by natural processes such as streams and rivers or differential weathering (such as the freeze/thaw cycle), human activities such as excavation during road building and (or) maintenance, and earthquake shaking or other intense vibration.

Topples. Toppling failures are distinguished by the forward rotation of a unit or units about some pivotal point below or low in the unit (**Figure 5.18**). Toppling is sometimes driven by gravity exerted by the weight of material upslope from the displaced mass. Toppling can also be caused by water or ice expanding in cracks or voids in the mass. Topples can consist of rock, debris (coarse material), or earth materials (fine-grained material). They are known to occur globally, often prevalent in columnar-jointed volcanic terrain, as well as along stream and river courses where the banks are steep. Velocities can range from extremely slow to extremely rapid, sometimes accelerating throughout the movement, depending on distance of travel.

Flows

Flows are mass movements in which the material behaves like a viscous fluid. In many cases, mass movements that start as falls, slides, or slumps are transformed into flows further downslope. Flows have broad characteristics and encompass wettest to driest, and slowest to fastest types of mass movement.

Debris Flow. A debris flow is a form of rapid mass movement in which a combination of loose soil, rock, organic matter, air, and water mobilize as a slurry that flows downslope (**Figure 5.19**). Debris flows typically consist of less than 50 percent fine grained material that has been carried along by extreme fluid flow generated by heavy rainfall or snowmelt. Debris flows are common in areas that have experienced wildfires, and are also associated with volcanic deposits.

These types of flows can exhibit a range of viscosities, being thin and watery or thick with sediment and debris, and are usually confined to the dimensions of the steep gullies that facilitate their downward movement. Movement across the surface is relatively shallow, and the runout is both long and narrow, sometimes extending for kilometers in steep terrain. The debris and mud usually terminate at the base of the slopes and create fanlike, triangular deposits called debris fans, which may also be unstable. They can be rapid to extremely rapid, depending on consistency and slope angle. Debris flows are often triggered by intense surface-water flow, due to heavy

Source: United States Geological Survey.

Figure 5.19 A debris flow is characterized by a low viscosity combination of soil, rock, and water that easily spreads out across the surface.

precipitation or rapid snowmelt, that erodes and mobilizes loose soil or rock on steep slopes. Debris flows also commonly mobilize from other types of landslides occurring on steep slopes, are highly saturated, and consist of a large proportion of silt- and sand-sized material.

Debris and Rock Avalanche. Debris avalanches are large, extremely rapid flows formed when an unstable slope collapses and the resulting fragmented debris is rapidly transported away from the slope (**Figure 5.20**). In some cases, snow and ice will contribute to the movement if sufficient water is present, and the flow may become a debris flow and/or a lahar. They occur worldwide in steep terrain environments and on very steep volcanoes, where they may follow drainage courses. Rock avalanches can occur whenever shear rock walls collapse. Some large avalanches have been known to transport material blocks as large as 3 kilometers in size, several kilometers from their source. They are rapid to extremely rapid and can travel close to 100 meters per sec. Numerous debris and rock avalanches occurred in Alaska following the Great Alaskan earthquake (**Figure 5.21**).

Source: United States Geological Survey.

Figure 5.20 A debris avalanche consists of a viscous mixture of rock, soil debris, and a small amount of fluid that travel a short distance before becoming its own impedance.

United States Geological Survey. Photo by Austin Post.

Figure 5.21 Rockslide avalanche on Sherman Glacier following the Great Alaska earthquake. The source was from the area marked by the fresh scar on Shattered Peak (middle distance). The debris displays flowlines and terminal digitate lobes.

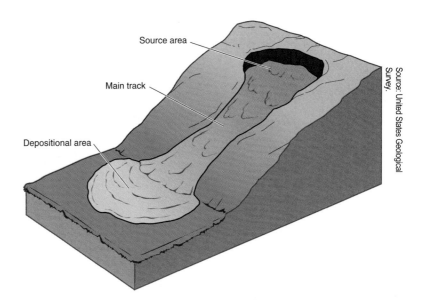

Source: United States Geological Survey.

Figure 5.22 Earthflows are relatively fluid masses that include a mixture of soils and weathered bedrock.

Earthflow. Spring and early summer are prime times for earthflows, when the upper layers of the regolith trap water. This reduces friction and grain cohesion, which allows the mass to make a slow downhill movement. The Earthflows can occur on gentle to moderate slopes, generally in fine-grained soil, commonly clay or silt, but also in very weathered, clay-bearing bedrock (**Figure 5.22**). The mass in an earthflow moves as a plastic or viscous flow with strong internal deformation. Slides or lateral spreads may also evolve downslope into earthflows. Earthflows range from very slow (creep) to rapid and catastrophic. A *mudflow* is an earthflow consisting of material that is wet enough to flow rapidly and that contains at least 50 percent sand, silt, and clay-sized particles. Newspaper reports often refer to mudflows and debris flows as "mudslides."

Creep. Creep is the imperceptibly slow, steady, downward movement of slope-forming soil or rock. Evidence of creep is seen in curved tree trunks, tilted walls or fence posts, or small ridges or ripples on the surface of the soil. The driving forces acting on the slope are sufficient enough to permanently deform the material but not large enough to cause shear failure. There are three types of creep:

1. Seasonal, in which changes in soil moisture and temperature allow movement by expansion of the slope surface (**Figure 5.23**),
2. Continuous, where the driving force always exceeds the strength of the material (resisting force), and
3. Progressive, where slopes are reaching the point of failure as other types of mass movements affect the slope.

Figure 5.23 Creep occurs when a surface is expanded by the addition of moisture or frost action. Dessication or thawing causes the expanded surface to fall back under the force of gravity, with the result that a particle beginning at point 1 is repositioned at point 2 and so forth.

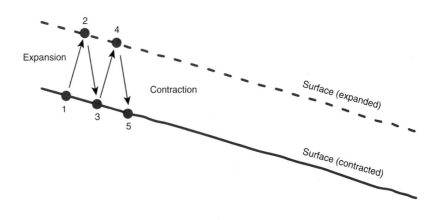

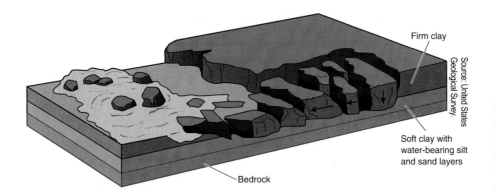

Source: United States Geological Survey.

Figure 5.24 Lateral spreads result from the horizontal movement of a relatively firm layer of material moving over a deeper layer of water-rich silt and sand. Very little slope is required for this to occur.

Creep is widespread around the world and is probably the most common type of mass movement, often preceding more rapid and damaging types of landslides. Creep can occur over large regions (tens of square kilometers) or simply be confined to small areas. It is difficult to define the boundaries of creep since the event itself is so slow, and surface features representing perceptible deformation may be lacking. The rates of movement range from very slowly to extremely slowly, usually less than 1 meter per decade. Wherever seasonal creep is occurring, rainfall and snowmelt are the typical triggers, whereas other types of creep could have numerous causes, including chemical or physical weathering, leaking pipes, poor drainage, or destabilizing types of construction.

Lateral Spreads

Lateral spreads are easily identified because they usually occur on very gentle slopes or flat terrain (**Figure 5.24**). Movement is caused by horizontal tension that pulls the surface material and produces shear or tensile fractures. **Liquefaction** is the primary cause of lateral spreads, developing when loose sediments such as sand and silt become saturated. Water holds the particles in suspension so any movement will cause the liquid and suspended solids to move rapidly across the surface. Usually two layers are present; the overlying one is a more cohesive layer lying above material that can undergo liquefaction. If liquefaction occurs, the upper layer then responds by subsiding, rotating, or disintegrating, and begins to either flow or liquefy. The initial failure occurs in a small area that spreads rapidly as liquefaction increases.

Lateral spreads are known to occur where there are liquefiable soils. The area affected may start small in size and have a few cracks that spread quickly, affecting areas hundreds of meters in width. Rate of movement may be slow to moderate; however, if the event is triggered by an earthquake, movement will be very rapid. The ground may then slowly spread from a few millimeters per day to tens of meters per day. Triggers that destabilize the weak layer include liquefaction of a lower weak layer by earthquake shaking. Other causes could be:

- Natural or anthropogenic overloading of the ground above an unstable slope;
- Saturation of underlying weaker layer due to precipitation;
- Snowmelt, and (or) ground-water changes;
- Liquefaction of underlying sensitive marine clay following an erosional;
- Disturbance at base of a riverbank/slope; and
- Plastic deformation of unstable material at depth such as thick deposits of buried salt.

liquefaction
A condition that exists when an overabundance of liquid, usually water, is present.

Submarine and Subaqueous Landslides

Submarine and subaqueous landslides include rotational and translational landslides, debris flows and mudflows, and sand and silt liquefaction flows that occur primarily underwater in coastal and offshore marine areas or in lakes and reservoirs. The failure

Source: United States Geological Survey.

Figure 5.25 The harbor at Kodiak, Alaska, was totally overrun by a tsunami that was generated by the Great Alaskan Earthquake of March 27, 1964. The city is located near the major fault that moved during the event.

of underwater slopes can result from overloading by rapid sedimentation, release of methane gas in sediments, storm waves, current scour, or earthquake seismic shaking. Subaqueous landslides pose problems for offshore and river engineering, jetties, piers, levees, offshore platforms and facilities, and pipelines and telecommunications cables, as well as producing tsunami hazards.

On November 18, 1929 a magnitude 7.2 earthquake occurred about 250 km south of Newfoundland near the Great Bank. The entire region felt the event, with reports of ground motion as far away as New York and Montreal. The earthquake triggered a large submarine slump estimated at 200 cubic kilometers of material. At the time, major undersea transatlantic cables stretched from eastern Canada to Great Britain. An analysis of numerous cable breakages was used to determine that the slump moved across the ocean floor at 95 kilometers per hour as it rolled down the continental slope. The event also produced a tsunami that killed 28 people along the Canadian coast.

The much larger 1964 Alaska earthquake (M 9.2) resulted in almost instantaneous catastrophic failure of the steep submerged shore in the harbor of Kodiak, Alaska (**Figure 5.25**) The submarine slide retrogressed beyond the shoreline, submerging areas of coastal land and harbor facilities, and almost 75 million m³ of shoreline of Valdez harbor disappeared into the sea.

Lessons from the Geologic Record

Slopes have always existed on Earth, and are constantly changing as a result of plate tectonic and erosional processes. Since mass movement is an erosional process that removes surface material, ancient landslide deposits are not as frequently found in the ancient rock record. However, the world's largest landslides discovered are prehistoric and are important to geologists in recognizing danger zones for future development. The following are examples of some of the largest mass movements recorded in the geologic record. Landslides of this magnitude have never been witnessed in historic times, and if they where to occur in today's populated areas, the destruction would be devastating.

Heart Mountain Detachment

Heart Mountain is located in northwestern Wyoming and represents the largest known rockslide on Earth. It covers about 1300 sq km of surface area and has a thickness of several kilometers. This volume of rock rests on a gentle slope of 1° to 2°. The leading edge of the land mass moved a minimum of 40 km while the entire mass covered more than 3400 sq km. Erosion has removed most of the oldest part of the slide sheet, which completed its move about 48 million years ago. How could such a massive amount of material move? Although many hypothetical models have been put forward, geologists who have studied the mountain and surrounding area in detail concluded that volcanism held the key to the source of motion. Frictional heating of underlying limestone generated free CO_2 along with water vapor that served to reduce friction along the major fault plane.

Mount Shasta

One of the largest landslides triggered by volcanic activity is a debris avalanche formed about 300,000 years ago in northern California. The most likely cause of the hummocky terrane was a major eruption of 4317-m-high Mount Shasta in the Cascade

Range of northern California (**Figure 5.26**). This debris avalanche today extends 43 km westward from the base of the volcano and has an estimated volume of 26 km³. This huge deposit and its moon-like surface was not recognized as a landslide until after the 1980 eruption of Mount St. Helens (described below), which resulted in a similar, but smaller, disruption of the Earth's surface.

Figure 5.26 Hummocky hills in the foreground of Mount Shasta are thought to have formed from a massive landslide that occurred about 300,000 years ago.

Lessons from the Historical Record and the Human Toll

Landslides can cause flooding by forming landslide dams that block valleys and stream channels, allowing large amounts of water to back up. This causes backwater flooding and, if the dam fails, subsequent downstream flooding. Also, solid landslide debris can upset any stream equilibrium by adding volume and density to otherwise normal stream flow. New debris in a channel can produce blockages and diversions, creating flood conditions or localized erosion. Landslides can also cause overtopping of reservoirs and/or reduced capacity of reservoirs to store water.

Usoi Landslide Dam and Lake Sarez, Pamir Mountains, Tajikistan

The world's largest and highest historic landslide dam was formed by the earthquake-triggered Usoi rock slide-rock avalanche, which dammed the Murgab River in the Pamir Mountains of southeastern Tajikistan in 1911. The country lies to the north of Afghanistan and Pakistan, sharing its southern border with those countries. Today the resulting 600-m-high dam impounds 53-km-long, 550-m deep Lake Sarez, a natural dam that is twice as high as the world's highest man-made dam (Nurek Dam, also in Tajikistan). The dam has not been overtopped and river inflow from the Murgab River and seepage through the dam, in the form of several large outlet springs, appear to be in equilibrium now. If the dam were to fail, hundreds of thousands of people downstream would be adversely affected.

Gros Ventre, Wyoming

One of the largest landslides to occur in recent times, the Gros Ventre (pronounced GROW vaunt) slide, took place on June 23, 1925 in the area east of Jackson Hole valley in Wyoming. The landslide was caused by a geologic setting in which sedimentary sandstone layers rested on top of clay layers. Following several weeks of heavy rainfall, the friction holding up the limestone layers was reduced by the water and caused almost 40 million cubic meters of rock to slide down the north face of Sheep Mountain (**Figure 5.27**). The slide crossed the Gros Ventre River and moved up the opposing side of the valley. The landslide created a dam over 60 m high and 400 m wide, creating Lower Slide Lake. This remained in place for only about two years, when on May 18, 1927, part of the dam failed following heavy rains that melted snowpack in the region. Flood waters reached more than 5 m in depth and downstream destroyed the town of Kelly, located 10 km downstream, killing several people. Water continued to travel for almost 30 more kilometers.

Figure 5.27 The Gros Ventre slide in June 1925 moved 40 million m³ of material that dammed the Gros Ventre River until it was breached two years later by heavy rainfall and snow melt.

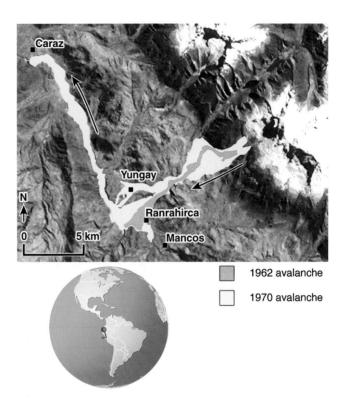

Figure 5.28 The 1970 avalanche was much farther reaching than that of 1962, although the two had a parallel course in the upper portions of the valley. Source: Data from the World Glacier Monitoring Service, Zurich, Switzerland and based on figure by UNEP's DEWA/GRID-Europe, Geneva, Switzerland and *Hugo Ahlenius, UNEP/GRID-Arendal.*

United States Geological Survey. Photo by George Plafker.

Figure 5.29 Peru Earthquake May 31, 1970. Lower part of the Huascaran debris avalanche of May 31,1970. The combined Yungay and Ranrahirea debris lobes cover an area of about 8 kilometers and probably contain close to 50 million cubic meters of material.

Nevados de Huascarán, Peru

Nevados de Huascarán (elevation 6786 m) is a peak that is situated in the central Andes Mountains of Peru. The summit is covered with glacial ice throughout the year. Between 1962 and 1970 two catastrophic rock avalanches occurred, killing more than 22,000 people (**Figure 5.28**). In January 1962, a massive slab of glacial ice and rock fell without warning and without any apparent triggering mechanism. The resulting avalanche, consisting of 10 million m³ of material, rapidly covered the valley at the base of the mountain, killing 4,000 people in a town at the base. The larger city of Yungay was spared any disaster, as the flow was stopped by a hill on the valley floor.

This section of South America experiences many high-magnitude earthquakes. On May 31, 1970, a M 7.7 earthquake occurred on the subduction zone about 135 km away from the Nevados de Huascarán (**Figure 5.29**). The duration of ground motion was 45 seconds, long enough to shake loose a large block of material from the Nevados de Huascarán between 5,400 and 6,600 m elevation. The ensuing rock and ice avalanche (estimated at a minimum of 50 cubic meters) slid across the glacier at the mountain base, reaching velocities exceeding 300 km per hour. The avalanche gained speed as it dropped 4 km in elevation over a distance of 16 km downslope.

Almost 18,000 residents of the city of Yungay died as the material covered the valley floor and moved up the opposite side of the valley, where an additional 600 people died. The city has been reoccupied but, given its tenuous location, it will undoubtedly be in the path of another catastrophic mass movement in the future.

Thistle, Utah

On April 14, 1983, a massive landslide in Utah County, Utah moved part of the mountain and blocked two creeks, forming an earthen dam (**Figure 5.30**). The Thistle landslide began moving in the spring in response to ground-water buildup from heavy rains the previous September and melting snowpack from the winter of 1983. Abnormally high precipitation in 1982-84, related to a strong El Niño Southern Oscillation, caused thousands of landslides in mountain areas of the western United States. Within a few weeks, the landslide dammed the Spanish Fork River, consequently obliterating U.S. Highway 6 and the main line of the Denver and Rio Grande Western Railroad.

The town of Thistle was inundated by the floodwaters rising behind the landslide dam. Eventually a drain system was engineered to drain the lake and avert a potential disaster. The landslide reached a state of equilibrium across the valley, but fears of reactivation caused the railway to construct a tunnel through bedrock around the slide zone, at a cost of millions of dollars. The highway likewise was realigned around the landslide. When the lake was drained, residual muck partially buried the town, and virtually no one returned to Thistle. It is now a ghost town. Total costs (direct and indirect) incurred by this landslide exceeded $400 million, making this the most costly single landslide event in American history.

Utah Geological Survey.

Figure 5.30 Heavy rains and a melting snow pack caused the most costly landslide in the history of the United States in spring 1983. The town of Thistle, Utah, was flooded by a lake that formed when the Spanish Fork River was dammed by the landslide.

Madison County, Virginia

The Appalachian Mountains of eastern North America have experienced major rainfall events over the past several decades. Some of the storms have been related to tropical cyclones, such as Tropical Storm Agnes in 1972, which affected West Virginia and Pennsylvania (see Chapter 9). Other storms such as a very intense rain storm in June 1995 are isolated events in time. However, both types of storms generate large amounts of rain that produce landslides.

Madison County, Virginia, located in the north central part of the state, received 30 inches of rain in a 16-hour period in late June 1995. Although local hillsides are covered in heavy vegetation, the amount of rain was sufficient to saturate the ground and produce hundreds of debris flows that devastated the region. This event was only the most recent of several that have hit the area. In 1969 the remnants of Hurricane Camille (a tropical storm when it struck Virginia) saturated an area about 150 km south of Madison County with the result being 150 people killed and more than $100 million (in 1969 dollars) damage. In November 1985 three days of storms struck Virginia and West Virginia, producing flooding and debris flows that killed 70 people and generated $1.3 billion in property damages.

Every year, numerous landslides occur worldwide and cause thousands of casualties and billions of dollars in monetary losses, making landslides a serious natural hazard. These annual costs are approximately equal to the damage caused by earthquakes over a 20-year interval. Yet losses are increasing in the United States and worldwide as development expands under pressures of increasing populations. Improper construction techniques lead to many of the problems. In most instances, losses are due to the public's lack of recognition or concern of the dangers. The resulting encroachment of developments into hazardous areas, expansion of transportation infrastructure, deforestation of landslide-prone areas, and changing climate patterns may lead to continually increasing losses.

Landslide Prediction and Mitigation

The potential for landslides to occur in an area can be assessed by examining the geology and geologic setting to determine if conditions could result in some type of mass movement. Several factors need to be considered including the actual location of an area and what types of human influence might add to the instability of the landscape. Past landslide history must be determined to develop a possible set of recurrence frequency curves.

Mass movements are inevitable where the terrain has a slope and where weather and other conditions can contribute to downhill movement. The effects of landslides can be reduced by preventing construction on or near hazardous zones. Governments can reduce the effects of landslides by establishing regulations and policies that minimize or reduce commercial and residential development. Governmental agencies that have control over the development of areas that could experience landslides should carefully monitor requests to use those areas. Prior to any construction, the services and knowledge of engineering geologists, civil engineers, and geotechnical professionals should be sought in order to properly assess the potential for future land failures.

The potential hazards of landslides can be mitigated if construction does not occur on steep slopes or on areas that have experienced landslides. Stabilization of slopes can be increased by constructing walls to reinforce the hillside (**Figure 5.31**). The removal of surface water from the top of a slide area increases stability, as the water cannot percolate into the subsurface and contribute to slope failure. Channels at the tops of slopes and drainage pipes leading from the lower portions of retaining walls direct water out of the soil and away from unstable areas.

Courtesy of David M. Best.

Figure 5.31 The retaining wall was built following the 1995 slope failure at La Conchita, California.

collapse
The sudden sinking of the surface into a subsurface void.

Subsidence and Collapse: Vertical Mass Movement

Under the steady pull of gravity, unsupported material at the Earth's surface can subside or collapse owing to subsurface vertical movement. Land subsidence is a gradual settling of the Earth's surface, while **collapse** is the sudden sinking into voids. Subsidence and collapse can be natural processes, but mainly they are due to human actions. For hundreds of years, humans have removed large amounts of Earth's underground resources: groundwater, oil, coal, metal ores, and rock. Removal of these materials creates many potential problems underground. When fluids are removed (example: water, oil, natural gas), the pores in which they are found are emptied. When solids are removed (example: coal, salt, ores), large void spaces are created. In both of these cases, support is decreased for the surface materials above.

We continue to withdraw increasing volumes of these fluids and solids from the subsurface to meet the demands of our growing population and industrial economy. Subsidence and collapse are a serious global hazards and growing more common due to land development expanding into areas that are more vulnerable. In the United States, an area roughly the size of New Hampshire and Vermont combined, has been directly affected. More than 80 percent of the identified subsidence and collapse in the United States is a consequence of our removal of groundwater resources. The growing demand for groundwater in many communities lowers the water table and reduces pore pressure, thus causing more subsidence. In some areas, old underground mines or caves have collapsed, causing damage to new developments on the surface. The increasing exploitation of land and water resources threatens to accelerate existing subsidence and collapse problems and initiate new ones.

Common causes of land subsidence from human activity are pumping water, oil, and gas from underground reservoirs; **dissolution** of limestone aquifers (sinkholes); collapse of underground mines; drainage of organic soils; and initial wetting of dry soils. Land subsidence occurs in every region of the United States.

dissolution
The natural dissolving of a substance by chemical reactions.

The Effect of Removal of Fluids from under the Surface

In the subsurface, pore space exists between grains of soil, rock, and other buried materials. Void space also exists in fractures within the rock. These voids can collect fluids such as water, oil, or gas. Whenever the fluids are lost, either through natural flow processes or by human extraction, the ground can respond by collapsing. The fluids exert a force (pore pressure) that helps keep the grains apart. Removal of the fluid removes this pore pressure, and compaction results.

The single greatest cause of ground collapse, termed subsidence, is the overpumping of ground water. If the level of the water table in the subsurface is near the ground surface, collapse is relatively rapid and nonreversible. The largest user of ground water is agriculture, especially in the arid regions of the western and southwestern United States.

When the ground collapses, the pore space is compacted due to a collapse of the sand, silt, and clay. The compaction reduces or eliminates the interconnected pores that once allowed water to flow through the material.

Compaction also occurs in municipal regions as cities use ground water as a primary water source for their inhabitants. Areas surrounding Las Vegas, Nevada, a large city located in a desert, have experienced subsidence as more people occupy a region that has little surface. Water is drawn from Lake Mead and from ground water. Similar conditions have developed in southern Arizona as expanding agriculture has used increasing amounts of ground water. Extensive fissuring and subsidence has lowered the landscape between Phoenix and Tucson and has caused large dessication cracks in the ground.

Two major agricultural valleys in California, the San Joaquin and the Santa Clara Valleys, have subsided more than 15 m since widespread pumping of ground water began in the early 1900s. The conditions only worsen as more people migrate to these areas.

Removal of oil and natural gas began in a field between Galveston and Houston, Texas, in 1917. For several decades these fluids along with water were removed from the subsurface, causing the ground to subside as much as 2.5 m. The sinking of the land increased as several million people in the area began using ground water as their water source. Readjustment of the land has reactivated old faults and created new ones in the region. Coastal sea water has moved landward into low-lying areas that dropped during the subsidence process.

The Natural Removal of "Solid" Rock

Rock is generally considered to be solid—"hard as rock" is the saying. However, certain types of rock, especially sedimentary rock consisting of soft mineral grains or grains that can be chemically eroded, are prone to removal by natural means. Water can erode soft sediments, causing collapse of any overlying material. The chemical solutioning of limestone produces a landscape that forms by the collapse of surface material into large subsurface voids.

Limestone is a common sedimentary rock whose primary mineral is calcite ($CaCO_3$). The atmosphere naturally contains CO_2 which combines with rain water or surface water and forms carbonic acid (H_2CO_3). The naturally occurring reaction between these two compounds, calcite and carbonic acid, is

$$CaCO_3 \quad + \quad H_2CO_3 \quad \rightarrow \quad Ca^{++} + 2HCO_3^-$$

Calcite   carbonic acid   soluble compound

Measuring the pH of a liquid determines its acidity. Normal rainwater has a pH of 5.6 (neutral pure water has a pH of 7). Therefore, falling rain has sufficient acidity to dissolve limestone. In addition, surface water from lakes and streams can flow through cracks in limestone and react with the rock below the surface.

The soluble compound in the equation above is now moved around by any excess water present and taken away from the source area. When the water leaves an area, such as by a period of drought or perhaps a drop in the water table level, the dissolved rock produces a void that we term a cave or cavern (**Figure 5.32**). This cave can expand in volume through time as more dissolving of the rock takes place. Eventually the roof of the cave will not have sufficient support to hold it in place and it will collapse. The ground surface now lies within a depression on the surface. The collapse of subsurface voids can have a negative effect on the landscape, particularly if it occurs in developed areas.

Subsurface mining of minerals such as halite (common salt, NaCl), and the excavation of coal, have produced numerous examples of ground failure through subsidence. Once the surface has collapsed the area is more susceptible to erosion.

About 25 percent of the conterminous United States is underlain by limestone and other dissolvable rocks, including salt. The landscape that forms from the processes described earlier, or by the dissolving of salt in water, are termed *karst topography*. The name karst comes from the Karst region along the coast of the Adriatic Sea along Croatia and Slovenia in southern Europe.

Figure 5.32 The solutioning of limestone can result in intricate cave formations. Natural chemical reactions have dissolved limestone and produced these subsurface voids.

Courtesy of David M. Best.

Predicting and Mitigating Subsidence Hazards

In addition to removal of dissolvable rock, excessive excavation of salts or extraction of ground water obviously increase the likelihood of collapse. Such excavations must be done thoughtfully in order to minimize the possible failure of the surface.

In such areas it is important to map the rock units along with the flow of water, and identify fracture patterns in the rock. Once the potential hazard zones are recognized, these areas should be placed off limits for future development. Continuous monitoring is necessary to forecast any sudden changes that might occur in the near-surface geologic conditions. The withdrawal of ground water and other fluids should be closely controlled in order to minimize upsetting the balance between the forces of the fluids and those of the rocks.

Summary

Mass movements result when driving forces exceed resisting forces. The major driving force is gravity, assisted primarily by water. Resisting forces include friction and the shear strength of the material. Rock structures, such as bedding and joints, also play an important role in facilitating mass movement, especially when these features are inclined toward a valley or excavation. The removal of material on slopes, either by human activity or natural events such as wildfires, increases the likelihood of slope failure. The causes of mass movement can be natural (geological or morphological) or human-caused. Triggering events upset the equilibrium and led to slope failure. Often the causes include several conditions that result in mass movement.

Mass movements are mainly characterized as slides, falls, or flows, depending on the nature of the movement. Translational slides occur when solid rock slides along a flat plane. Rotational slides (or slumps) occur when sediment or poorly consolidated rock moves down along a curved surface. In falls, material moves through the air to land as a pile on the ground. Slides are movements along a well defined surface. Flows can be mudflows, debris flows, or avalanches, depending on the amount of water versus solid material being moved.

Subsidence is the vertical lowering of the land surface, while collapse is the relatively fast or sudden opening of the land surface and movement of surface material into cavities below. Removal of rock in the subsurface through chemical action or anthropogenic mining creates unstable situations in which the overlying rock collapses, either slowly in large-scale subsidence or by rapid collapse.

References and Suggested Readings

Browning, J. M., 1973, Catastrophic rock slides. Mount Huascarán, north-central Peru, May 31, 1970. *American Association of Petroleum Geologists Bulletin. v.* 57, p. 1335–1341.

Coch, N. K., 1995, *Geohazards, Natural and Human.* Upper Saddle River, NJ: Prentice Hall, 481 pp.

De Graff, J. V. and S. G. Evans. 2002. Catastrophic Landslides: Effects, Occurrences, and Mechanisms, *Reviews in Engineering Geology.* Boulder, CO: Geological Society of America

Highland, L. M., and Bobrowsky, Peter, 2008, The landslide handbook—A guide to understanding landslides: Reston, Virginia, U.S. Geological Survey Circular 1325, 129 p.

Jibson, R. W. 2005. Landslide Hazards at La Conchita, California. Reston, Virginia: *U.S. Geological Survey* Open-File Report 2005-1067.

Risley, J.,Walder, J. and Denlinger, R. 2006, Usoi Dam Wave Overtopping and Flood Routing in the Bartang and Panj Rivers, Tajikistan. *U.S. Geological Survey* Water-Resources Investigations Report 03-4004

U.S. Geological Survey, 1997. *Debris-Flow Hazards in the United States.* USGS Fact Sheet 176-97.

U.S. Geological Survey, 1999. *Real-Time Monitoring of Active Landslides—Reducing Landslide Hazards in the United States.* USGS Fact Sheet 091-99.

U.S. Geological Survey, 2004. *Landslide types and processes.* USGS Fact Sheet 2004-3072.

U.S. Geological Survey, 2005. *Landslide hazards—a national threat.* USGS Fact Sheet 2005-3156.

Varnes, D. J., 1978. Slope movement types and processes, *in* Schuster, R. L., and Krizek, R. J., eds., Landslides-analysis and control: National Research Council, Washington, D.C., Transportation Research Board, Special Report 176, p. 11–33.

Web Sites for Further Reference

http://geochange.er.usgs.gov/sw/changes/anthropogenic/subside/
http://landslides.usgs.gov/
http://landslides.usgs.gov/learning/
http://ndrd.gsfc.nasa.gov/
http://pubs.usgs.gov/of/2001/ofr-01-0276/
http://www.usgs.gov/hazards/landslides/
http://3dparks.wr.usgs.gov/landslide/index.html

Questions for Thought

1. How does the orientation of rock features such as bedding and joints affect the development of mass movement?

2. Distinguish among the types of mass movement called falls, slides, and flows.

3. Give several examples of how human activities can cause mass movement.

4. Many landslides occur after rainfalls. Why?

5. Describe the conditions that result in a debris avalanche.

6. How do rockslides and slumps differ?

7. Describe five different problems resulting from surface subsidence.

8. Explain why subsidence is an especially serious problem in coastal areas.

9. Describe ways in which human activities have resulted in sinkhole development, subsidence, and collapse in limestone areas such as Florida.

10. Explain the basic chemistry involved with the dissolving of limestone by natural processes.

Tsunami: The Utmost in Destruction from the Sea

6

Key Terms

advisory
drawdown
information bulletin
runup
seismic sea wave
submarine landslide
tsunami
warning
watch
wavelength

The December 26, 2004 tsunami was extremely destructive in South Asia and adjoining regions, as more than 280,000 people were killed and several million injured, as the wave trains traveled across the Indian Ocean and nearby seas.

Tsunami have existed on Earth for millions of years, but it took the events of December 26, 2004, to remind the world of their existence and destructive power. Off the west coast of the island of Sumatra in Indonesia, two oceanic plates converged. The resulting subduction generated a M 9.1 earthquake, the third largest in recorded history. The vertical movement of the ocean floor displaced thousands of cubic kilometers of water, generating a tsunami that raced across the adjoining oceans, killing an estimated 280,000 people. Those people who lived near the source had very little time to react to the event. Once officials learned what had happened, countries farther away were notified and thus were able to warn residents of the impending disaster. Coastal regions in East Africa did not experience the large loss of life that occurred in Sumatra, Sri Lanka, India, and Thailand because of the warnings. Obviously if a warning network had been in place throughout the region around the Indian Ocean, more inhabitants would have been informed in time to save many lives.

What Is a Tsunami?

Prior to December 2004, few people were familiar with the word *tsunami* (soo-NAH-mee). The word is derived from two Japanese characters: *tsu* meaning harbor, and *nami* which means wave (the word is both singular and plural in Japanese, as is the word *deer* in English). A tsunami (also termed **seismic sea wave**) is a series of waves that are generated by a large displacement of sea water. The energy moves out from the source, producing a series of waves that have long wavelengths. The time between the wave pulses can be as much as an hour or more.

Throughout the long documented history of Japan, these waves have occurred rather frequently. Almost 200 tsunami have been recorded hitting the islands. The location of Japan on the northwest edge of the Pacific Ocean is a focal point for these destructive waves as they travel across the Pacific. The shoreline of the Japanese islands consists of many harbors and inlets. When a seismic sea wave hits these indentations, the energy is rapidly focused into a small area. Water can move inland more than one kilometer. Other areas bordering the Pacific Ocean have also been struck by these waves, including Alaska, Hawaii, Guam, and California.

Tsunami commonly occur once or twice a year throughout the world as active tectonic plates produce large magnitude earthquakes that displace massive amounts of water. Although the majority of tsunami result from the vertical displacement of the sea floor or adjoining areas, strike-slip motion is thought to produce about 15 percent of all tsunami. Energy from strike-slip mechanisms tends to remain close to the source, thus not producing large-scale disasters. As populations have increased in the areas surrounding the Pacific, more people have settled in low lying, coastal areas that are in harm's way when these waves strike the shoreline. Coastal inhabitants are aware of the potential disasters and have developed escape routes to allow them to evacuate low areas (**Figure 6.1**).

Characteristics of Tsunami

Tsunami are most often generated in deep water when vertical movement occurs along a fault or water that is displaced by a landslide. Many of the plate boundaries in the Pacific Ocean are elongate compressional regions that produce vertical offset when they collide. The initial vertical motion projects water upward, which then collapses, creating a splash that generates waves in all directions. This upward motion is not very large, perhaps on the order of several meters. A massive volume of water is moved. Successive waves move out from this point, separated by a distance measured in tens or hundreds of kilometers. A similar condition is produced when you drop a stone into a pond. The waves radiate from the point where the water was displaced. If you were on a ship in the open ocean when a tsunami passed underneath, it would be difficult to detect it. In addition to the long **wavelengths** (the distances between

tsunami
A series of giant, long wavelength waves produced by the displacement of large amounts of ocean water by an earthquake, volcanic eruption, landslide, or meteorite impact.

seismic sea wave
A large ocean wave that is produced by a major disturbance in the ocean, such as an earthquake or a landslide.

wavelength
The distance separating two adjacent crests (or troughs) on a wave form.

Figure 6.1 Evacuation routes are set up in areas that can experience tsunami. These routes direct people to higher ground and a safer environment.

the crests of adjacent waves), the amplitude of the wave is very small, on the order of a meter or two. However, ocean buoys can be set in deep water to detect these waves and transmit the information to warning centers.

Tsunami typically have velocities in the range of several hundred kilometers per hour. Water depth is the controlling factor of velocity. In very deep water tsunami can travel more than 800 kilometers per hour, the same as a jet airliner. As the energy approaches more shallow water, its velocity drops drastically, but the height of the water increases rapidly because the same amount of energy is present.

Tsunami differ from wind-generated waves in that tsunami have a much longer period, the time separating adjacent crests of the waves, and they also have much larger wavelengths. The periods of tsunami range from about 5 minutes to as many as 60 or more minutes. Normal, wind-generated waves typically have periods of between 5 and 20 seconds. Wavelengths for tsunami range between tens and hundreds of kilometers, while wind-generated waves average around 100 to 200 meters (**Figure 6.2**).

Tsunami have been referred to in the past as tidal waves, although their genesis is not related to tides at all. Perhaps the misnomer came about because, as tsunami approach the shore, they appear similar to breaking waves. Tsunami come ashore and continue moving inland; they do not recede in a normal ebb-and-flow sense that is typical of normal tides.

Figure 6.2 Wind-generated waves have a much shorter period and a shorter wavelength than tsunami.

Life Cycle of a Tsunami

Tsunami are relatively short-lived events. The forces that produce them are rather quick—an earthquake in the ocean, a submarine volcanic explosion, a landslide that displace material into a body of water, or a meteorite impacting the ocean. The United States Geological Survey (USGS) recognizes four stages of a tsunami: initiation or creation, splitting, amplification, and runup.

The first stage, initiation, is when some energy source displaces water. Vertical movement on the sea floor caused by an earthquake pushes water upward, and an accompanying down drop can pull water down (**Figure 6.3a**). A volcanic explosion could displace water both vertically and horizontally, depending on how the volcano erupted. Vertical displacement in deep water results on a relatively small uplift of water. This can occur over several tens or hundreds of kilometers of distance, displacing enormous volumes of water.

The uplift and subsequent collapse of water at the surface then splits the uplifted column, at which point water is traveling radially outward (**Figure 6.3b**). If the point of origin for the source of the tsunami is near shore, two distinct tsunami are created.

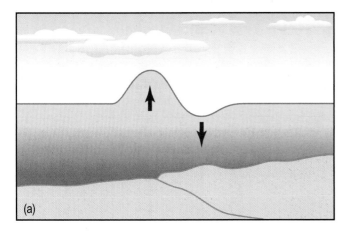
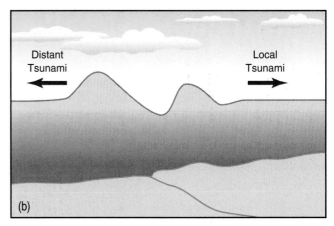

Figure 6.3 Distant and local tsunami. (a) Movement along a submarine fault causes water to rise in one area and drop in an adjacent area (wave heights are exaggerated in terms of water depth). Actually surface uplift is perhaps one meter. (b) During the splitting stage water collapses, producing a local tsunami that affect the nearby coast and a distant tsunami that traverses the open ocean.

drawdown
A lowering of water level at the beach when the trough of a tsunami comes ashore.

runup
The height to which water comes up onto the shoreline, either from natural wave action or the occurrence of a tsunami.

Figure 6.4 Runup has an appearance similar to this wave, but it continues inland for a kilometer or more, covering the flat coastal areas.

The one that travels out into deep water is termed the distant tsunami, while the one closer to the coast is the local tsunami. The local tsunami reaches land faster because of the shorter travel distance and does the initial damage. The velocity with which the tsunami travels is related to the square root of the water depth. For this reason, the distant tsunami travel much faster than the local tsunami because it passes through shallow water near the coast.

As the tsunami approaches more shallow water near the continent, the energy is now focused in a much smaller volume. The amplitude of the waves increases dramatically as the velocity and wavelength decrease. As the wave approaches the shore, if the trough portion arrives first, this gives an appearance of the tide and water going out to sea. This is termed **drawdown**. Shortly after the water recedes, the next crest in the wave train arrives, producing **runup**.

Water arrives as a surge as the sea level constantly rises (**Figure 6.4**). The amount of runup can be variable, because it is controlled by offshore seafloor topography and the outline of the coast. Areas with a shallow offshore topography were overrun by water, while those beaches with a steep offshore seafloor were not inundated as severely. Strong currents develop as massive volumes of water move in and out of confined areas. Tsunami generally have several major runup events separated by periods of drawdown. In the December 2004 tsunami, beaches that are close to each other experienced extremely different effects.

Propagation of Tsunami Through the Oceans

The origin of most tsunami is in deep water, although some can occur along the continental shelf. Because the wavelength of tsunami is much greater than the depth of the water through which they travel, the velocity of the wave is only dependent on the water depth (D). So the velocity V is determined by

$$V = \sqrt{gD}$$

where g is the gravitational constant (9.8 meters per second) and D is the water depth, measured in meters.

The average depth of the world's oceans is 4,000 meters. Therefore the velocity of a tsunami traveling in water that deep is

$$V = \sqrt{9.8 \times 4000} = 198 \text{ meters per second or } 713 \text{ kilometers per hour}$$

This is about the average speed of most commercial airplane flights. If you were on the ocean in a small boat, the wave would move so fast you could not detect it. In addition in deep water the vertical movement of the water might be one or two meters, no different that most waves in the open ocean. Obviously in deeper water velocities can approach 850 kilometers per hour or higher.

At sea tsunami are basically harmless waves. Very little energy is lost as they travel through the open ocean. When the energy reaches shallow water near the shoreline, the waves rise up and become very destructive. They may have an appearance of normal wind-generated waves (**Figure 6.5**), but when tsunami hit the shoreline, they continue to move inland in a relentless flow of water.

Frequency of Tsunami

The National Geophysical Data Center (NGDC), a division of the National Oceanic and Atmospheric Administration (NOAA), maintains a database of tsunami activity since 2000 B.C.E. to the present. A key part of the destruction of tsunami is the runup

TABLE 6.1	Comparative Amounts of Tsunami and Runup Events, by Region. More than 13,000 Runup Events Have Been Recorded.	
Region	Percentage of Tsunami	Percentage of Runup Events
Pacific Ocean	63	82
Mediterranean Sea	21	4
Indian Ocean	6	9
Caribbean Sea	4	4
Atlantic Ocean	5	2
Black Sea	1	4
Red Sea	< 1	4

Source: National Geophysical Data Center, NOAA.

of water on land adjacent to the ocean. Numerous sources of information have been examined to develop a better understanding of where tsunami and runup events occur (Table 6.1). Eighty-four percent of all tsunami have occurred in the Pacific Ocean and the Mediterranean Sea. Major tsunami that affect a large area occur about once per decade.

Extensive documentation exists, especially in Japan and the Mediterranean areas, where inhabitants have recorded the occurrence of natural disasters. The first recorded tsunami occurred off the coast of Syria in 2000 B.C.E. Since 1900 (the beginning of instrumentally located earthquakes), most tsunami have been generated in Japan, Peru, Chile, New Guinea, and the Solomon Islands. However, the only regions that have generated remote-source tsunami affecting the entire Pacific Basin are the Kamchatka Peninsula, the Aleutian Islands, the Gulf of Alaska, and the coast of South America. All of these are very active areas of tectonic activity. Hawaii, because of its location in the center of the Pacific Basin, has experienced tsunami generated in all parts of the Pacific.

Regions around the Mediterranean and the Caribbean have experienced numerous, locally destructive tsunami. A small subduction zone exists where the African Plate is colliding with the Eurasian Plate. The Caribbean Plate is surrounded by the North American and South American Plates on its eastern edge and the North American and Cocos Plates on its western edge.

Only a few tsunami have been generated in the Atlantic and Indian Oceans. In the Atlantic Ocean, there are no subduction zones at the edges of plate boundaries to generate such waves, except small subduction zones under the Caribbean and Scotia Arc, which is located at the southern end of South America. Subduction is active along the eastern border of the Indian Ocean, where the Indo-Australian plate is being driven beneath the Eurasian plate at its eastern margin. Thus, most tsunami generated in this region are propagated toward the southwest shores of Java and Sumatra, rather than into the Indian Ocean. The December 2004 Indonesian tsunami was anomalous in that it affected all the countries in the immediate and distant areas because of the size of the earthquake that generated the seismic sea wave (Figure 6.6).

© JupiterImages Corporation.

Figure 6.5 Wind-generated waves tend to break near the shore, whereas tsunami produce a rapid rise in water level.

© capturefoto, 2010. Under license from Shutterstock, Inc.

Figure 6.6 Damage caused by the December 2004 tsunami was far-reaching. Areas such as this one near Banda Aceh, Indonesia, were overrun by water that stripped the landscape clean.

Historical Occurrences of Tsunami

Throughout recorded history tsunami have affected millions of people, especially in Asia and in areas surrounding the Mediterranean Sea. Table 6.2 lists selected tsunami that have been generated by several different sources. The major events are all associated with sites that border the oceans. Also significant tsunami were generated by the three largest earthquakes in recorded history—Chile 1960 M 9.5, Alaska 1964 M 9.2, and Indonesia 2004 M 9.1. These events were each caused by active subduction of oceanic plates under adjoining continental or oceanic plates.

TABLE 6.2	Significant Tsunami That Have Occurred Since 1755			
Date	Location	Source Mechanism	Death Toll	Interesting Facts
November 1, 1755	Lisbon, Portugal	Earthquake estimate M 9	20,000 to 30,000	Portugal, Spain, and Morocco hit by tsunami; tsunami crossed the Atlantic Ocean and struck West Indies.
August 27, 1883	Krakatoa, Indonesia	Volcanic eruption	36,000	Collapse of top of volcano generated large tsunami that reached the English Channel.
June 15, 1896	Japan	Earthquake	26,000	Largest wave 38 meters (70 ft. reported)
November 18, 1929	Grand Banks, Newfoundland	Submarine landslide	29	Rate of motion calculated by determining breaks in transatlantic transmission cables.
April 1, 1946	Hawaii	Earthquake M 8.1	164	Pacific Tsunami Warning System was set up after this; 159 lives lost in Hawaii; 5 in Alaska.
July 9, 1958	Lituya Bay, Alaska	Landslide	Less than 10	Waves reached more than 500 meters up the slope opposite the slide, the largest recorded runup in history.
May 22, 1960	Chile	Earthquake M 9.5	1000 killed in Chile	Largest earthquake in recorded history generated a tsunami that traveled across the Pacific Ocean; 61 died in Hawaii.
March 27, 1964	Alaska	Earthquake M 9.2	132	Second largest earthquake in recorded history generated a major tsunami that killed 132, including 12 in Crescent City, CA who went to watch the incoming wave.
August 16, 1976	Philippine Islands	Earthquake M 7.9	7,000	Moro Gulf region hit by tsunami; 700 km of coastline hit.
September 1, 1992	Nicaragua	Earthquake M 7.6	170	Slow moving fault generated weak seismic waves that were not felt, so people were unprepared for the tsunami.
July 12, 1993	Hokkaido, Japan	Earthquake M 7.8	202	Hundreds missing; repeated warning went out but some ignored; can been seen on YouTube.
July 17, 1998	Papua New Guinea	Earthquake M7.1 triggered submarine landslide	3,000	Epicenter of M 7.1 event just off northwest coast. No time to generate a warning. In less than 10 minutes the tsunami hit the shoreline.
December 26, 2004	Off west coast of Sumatra, Indonesia	Earthquake M 9.1	280,000	Third largest recorded earthquake in history generated a tsunami that traveled around the globe.

Figure 6.7 (a) Scotch Gap lighthouse was a reinforced concrete structure (b) that was demolished by a large tsunami produced by a magnitude 8.1 earthquake in Alaska in 1946.

Unimak, Alaska, April 1, 1946

The source earthquake was located near Unimak Island. The resulting energy produced a tsunami that destroyed the Scotch Gap lighthouse, which was built in 1940 and situated about 15 meters above the mean tides in the area (**Figure 6.7**). Five people who were on duty in the station at the time of the tsunami died as a result of the waves hitting the structure.

Chile, May 22, 1960

The largest recorded earthquake in history (M 9.5) produced a massive tsunami that actually traveled around the globe. The majority of its destruction was in the Pacific Ocean and surrounding coastal regions. More than 1,000 died in coastal regions of Chile, including 200 in the community of Isla Chiloe, Chile (**Figure 6.8**).

Alaska, March 27, 1964

On Good Friday, the second largest recorded earthquake occurred near Prince William Sound, Alaska. This M 9.2 event generated a major tsunami train that traveled around the Pacific Ocean and affected numerous seaside localities. Coastal cities such as Valdez and Seward, Alaska, were overrun by the water. The port area of Seward was destroyed, with ships, vehicles, and general material being tossed about (**Figure 6.9**).

Figure 6.8 The coastal community of Isla Chiloe, Chile, was overrun by water from the series of tsunami produced by the May 22, 1960, Chilean Earthquake.

Figure 6.9 An overturned ship and a crumpled chemical truck, along with other debris, show the force exerted by the tsunami associated with the March 27, 1964, earthquake in Alaska.

BOX 6.1	An Experience of Living Through the Good Friday Great Alaskan Earthquake of March 27, 1964

The week was coming to a close for Darrell Boomgaarden and everyone in the area was preparing for the upcoming Easter weekend. As he was making his way across the U.S. Navy base in Kodiak, delivering newspapers along his route, he remembers the events that would be history making. "My strongest memory was the sound, which I heard about three to five minutes before the ground motion hit. I could hear it bouncing off the nearby hills as I said to myself, 'What is that?'"

- And then the motion hit. As he looked down the street, all the lights were moving and cars were bouncing up and down. "I couldn't believe concrete could move like it did, and it seemed to keep going on forever. There were several minutes of motion."
- To Darrell the houses looked fine and there were no poles that appeared to have fallen. The residential area didn't seem too different. After completing his paper route, he headed home for dinner. About one hour later the siren went off. The base was evacuated, which was challenging as it dark by this time. Everyone went up the mountain to safer ground, to a location about 20 feet above the water. They all spent the night there.
- Following the M 9.2 earthquake that occurred at 5:36 pm local time (the epicenter was about 455 km to the northeast in Prince William Sound), several tsunami struck the town of Kodiak and did significant damage (Figure 1). The townspeople consisted mainly of fishermen and cannery workers. The oil-fired power plant on the base was destroyed, which meant there was no electric power. The naval base was inundated with oil-filled sea water. All food in the commissary was lost. Darrell's father was the officer-in-charge of ground control at the naval air station, so he had to ensure the runway and its radar were not going to be flooded. Thousands of sandbags were used to protect the runway from incoming water.

United States Geological Survey.

Box Figure 6.1.1 Damage along the shoreline in Kodiak following the earthquake of March 27, 1964.

- Slowly, basic services were restored. A seaplane tender ship was sent to the naval base to generate power. There was no television in the region, so news was slow to reach everyone.
- It was several days before Darrell made it into town. "I could not believe that boats could get moved that far inland—several hundred yards from their docks onto land." School was canceled for two weeks, and teenagers joined in the effort to clean up the area. Everything was covered in oil.
- Darrell later learned that several of his friends died when they were driving along the shoreline after dark. Their car was hit by a tsunami, which they were unable to see. Recovery in the region took many months and the events of Good Friday, 1964, still live in the minds of those who survived.

Hokkaido, Japan, July 12, 1993

On July 12, 1993, the island of Hokkaido and a small island offshore experienced the effects of a magnitude 7.8 earthquake. In a matter of minutes the region was hit by one of the largest tsunami in the history of Japan (**Figure 6.10**). The coastline of Okushiri Island was inundated, resulting in the loss of more than 120 lives, and property damage of $600 million.

Figure 6.10 Damaged caused by a tsunami and fires that followed destroyed a significant part of Okushiri Island following a nearby major earthquake in July 1993.

Figure 6.11 The north shore of Papua New Guinea is relatively flat with some tidal marsh areas and river deltas. It was flooded in the 1998 tsunami.

Figure 6.12 Banda Aceh, Indonesia, was one of the most devastated cities along the Sumatra coast, following the December 26, 2004, tsunami.

Papua, New Guinea, July 17, 1998

The southwestern Pacific Ocean is an area that has a great deal of tectonic activity, as colliding plates produce numerous large earthquakes. Such was the case in mid-July 1998 when a magnitude 7.1 earthquake occurred just north of the coast of Papua, New Guinea (**Figure 6.11**). Analysis after the event showed that a major **submarine landslide** was triggered by the earthquake. A tsunami formed very quickly following the main event, sending several waves onto the island. Wave heights ranged between 10 and 15 meters and pushed landward, destroying several villages. More than 2,000 people died.

submarine landslide
The collapse of land material either underwater or from the land that slides into water, producing a massive wave.

Sumatra, Indonesia, December 26, 2004

December 26, 2004, began very quietly for many vacationers who were visiting the beaches of Sumatra, Thailand, India, and nearby countries. Unknown to many people, one of the largest earthquakes in history occurred, displacing trillions of gallons of water in a tsunami wave that ended up going all the way around the globe.

At 7:58 local time, a tsunami generated by a huge (M 9.1) earthquake began its journey across the Indian Ocean. Within about 15 minutes, it struck Sumatra, killing over 130,000 people. Damage in Sumatra was complete—trees were ripped out of the ground by their roots, entire towns were leveled, and water flooded the land to a depth of up to 25 m (**Figure 6.12**). Within two hours, the tsunami had struck Sri Lanka and Thailand, claiming another 40,000 lives.

In all, at least 280,000 people died in one of the worst natural disasters in human history. The waves dramatically pointed out the vulnerability of coastlines to tsunamis, and the devastation caused when little warning is available.

Tsunami Prediction, Warning, and Possible Mitigation

Tsunami are often associated with large-magnitude earthquakes that occur in or near oceans. When a large earthquake occurs in a region near the ocean, instruments alert scientists about the potential for tsunami to be formed.

In the Pacific Ocean the Pacific Tsunami Warning Center located at Ewa Beach, near Honolulu, Hawaii , becomes the focal point of activity when an earthquake could potentially generate a tsunami. This facility is operated by the National Weather Service (NWS) and serves as the headquarters for the Operational Tsunami Warning System, a group that coordinates the monitoring and reporting of seismic activity around the Pacific Ocean, the Indian Ocean, and the Caribbean Sea. The West Coast/Alaska Tsunami Warning Center, located in Palmer, Alaska, monitors the northern Pacific Ocean, including Alaska and the Aleutian Islands. It also oversees tsunami activity along all coastal regions of the United States (except Hawaii), the coastal provinces of Canada, Puerto Rico, and the Virgin Islands.

Warnings are only as good as the people who heed them. If people decide to rush to the shore to watch waves come in, they will obviously be placing their lives in danger; such was the case in Crescent City, California. In March 1964 people went to the shore to watch tsunami generated by the great earthquake in Alaska, resulting in 12 deaths.

In order to mitigate property damage, fewer people must live near the shoreline, especially in areas where there is very little relief above sea level. Oncoming waves that are as much as 15 m or higher will move rapidly onshore and flood flat-lying areas that are only a few feet above sea level. All property will be inundated and mostly likely totally destroyed from the onslaught of water.

Tsunami Detection Devices

The National Oceanic and Atmospheric Administration (NOAA) operates a network of stationary tsunami detection buoys in several oceans throughout the world. The system that employs the recording technology is the Deep-ocean Assessment and Reporting of Tsunamis (DART). These buoys are primarily deployed in the Pacific Ocean, but there are several in the western Atlantic Ocean and in the Gulf of Mexico (**Figure 6.13**). Only six buoys were operational in the Pacific prior to the 2004 Indonesian event, and none were located in the Indian Ocean. Therefore there was no means to notify nations around the Indian Ocean of the impending disaster coming their way.

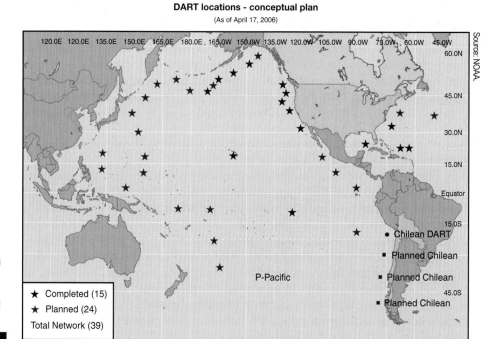

DART locations - conceptual plan
(As of April 17, 2006)

Figure 6.13 Tsunami can be detected in the Pacific and Atlantic Oceans as well as the Caribbean Sea and Gulf of Mexico by detection buoys maintained by the National Oceanic and Atmospheric Administration.

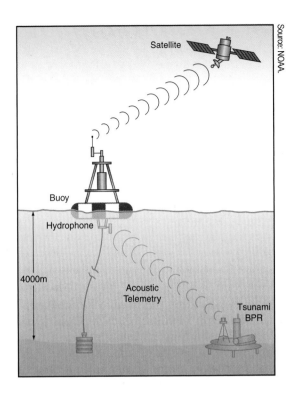

Figure 6.14 The real-time tsunami system operated by NOAA is capable of detecting a tsunami as it passes through the deep waters of the ocean. Information is transmitted via satellite to the warning centers.

The buoys float on the water surface and receive a signal from a source located on the seafloor (**Figure 6.14**). As a tsunami passes the sensor, there is a rapid change in pressure which is sent to the hydrophone underneath the buoy. The instruments are programmed to recognize this change and then transmit a signal to a warning center, which then sends the appropriate information to areas in the possible path of the tsunami.

If a tsunami is created, the first information available is related to the earthquake and its source area. This information gives scientists a general idea of the area that could be affected. As the tsunami travels across the ocean, its movement is tracked by data furnished from the recording buoys. As more data become available, the path of the tsunami can be projected more accurately, thus allowing steps to be taken to reduce the destruction ahead of the waves.

What Happens During a Tsunami Warning?

Within seconds of an earthquake's occurrence, seismographs located around the Pacific Basin send data on the earthquake to the Pacific Tsunami Warning Center in Hawaii. Computer programs determine a preliminary magnitude for the earthquake. If the magnitude is large enough, the system alerts the warning center staff that a large earthquake has occurred. Because of this, all warning center staff must live on the grounds of the warning center, with one staff member on the grounds at all times. Based on the magnitude of the earthquake, the staff may choose to issue an information bulletin, an advisory, a watch, or a warning.

- An **information bulletin**. These are informational statements to scientists and disaster specialists that an earthquake has occurred. They generally mean that a tsunami has not occurred, and are intended to prevent needless evacuation.
- An **advisory**. These are issued when it is believed that a tsunami may have been generated. They are issued to areas outside the current warning or watch area to inform disaster managers to remain alert and wait for further expansion of the warning or watch areas.

information bulletin
Notification to scientists and researchers that an earthquake has occurred, but not necessarily a tsunami was generated.

advisory
An announcement that is issued when a tsunami could result from a large earthquake that occurs near a coastal region.

watch
An alert that a tsunami could potentially occur in an area.

warning
The highest alert level that is given when a tsunami has been formed and its arrival is highly possible.

- A **watch**. Tsunami watches are issued based on seismic information, without confirmation that a tsunami has been generated. The size of the watch area depends on the magnitude of the earthquake. NOAA sets guidelines which allow tsunami warning centers to use the following rules of thumb for issuing a watch:

   For earthquakes over magnitude 7.0, the watch area is 1 hour tsunami travel time outside the warning zone.

   For all earthquakes over magnitude 7.5, the watch area is 3 hours tsunami travel time outside the warning zone.

   The watch will either be upgraded to a warning in subsequent bulletins or will be cancelled, depending on the severity of the tsunami. Watches are issued to provide disaster managers with the earliest possible indication that a tsunami may be on the way.

- A **warning**. This is the highest level of tsunami alert. Initially, a warning center may issue a tsunami warning for areas close to an earthquake before confirming that a tsunami has been generated, in order to give maximum time for disaster managers to evacuate the coastline. In successive statements, however, warning systems issue warnings only when tsunami generation has been confirmed (either through DART buoys or shoreline observation). Tsunami warning centers use these NOAA rules for issuing a warning:

   Earthquakes over magnitude 7.0 trigger a warning covering the coastal regions within 2 hours tsunami travel time from the epicenter.

   When the magnitude is over 7.5, the warned area is increased to 3 hours tsunami travel time.

   As water level data showing the tsunami are recorded, the warning will either be cancelled, restricted, expanded incrementally, or expanded in the event of a major tsunami.

   Initial warnings are issued based on the magnitude of the earthquake. Changes are made to previous warnings as size and location of the earthquake, along with tide gauge and DART information, are updated.

Tsunami Safety Rules

In case you are ever in an area where there is a threat of tsunami, heed the following tsunami safety rules from the West Coast/Alaska Tsunami Warning Center Home Page: http://wcatwc.arh.noaa.gov/safety.htm:

- A strong earthquake felt in a low-lying coastal area is a natural warning of possible, immediate danger. Keep calm and quickly move to higher ground away from the coast.
- All large earthquakes do not cause tsunami, but many do. If the quake is located near or directly under the ocean, the probability of a tsunami increases. When you hear that an earthquake has occurred in the ocean or coastline regions, prepare for a tsunami emergency.
- Tsunami can occur at any time, day or night. They can travel up rivers and streams that lead to the ocean.
- A tsunami is not a single wave, but a series of waves. Stay out of danger until an "all clear" is issued by a competent authority.
- Approaching tsunami are sometimes heralded by a noticeable rise or fall of coastal waters. This is nature's tsunami warning and should be heeded.
- A small tsunami at one beach can be a giant a few miles away. Do not let the modest size of one make you lose respect for all.
- Tsunami can strike every coastline in the Pacific. All tsunami are potentially dangerous even though they may not damage every coastline they strike.
- Never go down to the beach to watch for a tsunami! **If you can see the wave, you are too close to escape.** Tsunami can move faster than a person can run!

- During a tsunami emergency, your local emergency management office, police, fire and other emergency organizations will try to save your life. Give them your fullest cooperation.
- Homes and other buildings located in low-lying coastal areas are not safe. Do **not** stay in such buildings if there is a tsunami warning.
- The upper floors of high, multi-story, reinforced concrete hotels can provide refuge if there is no time to quickly move inland or to higher ground.
- If you are on a boat or ship and there is time, move your vessel to deeper water (at least 200 meters). If there is concurrent severe weather in the area, it may be safer to leave the boat at the pier and physically move to higher ground.
- Damaging wave activity and unpredictable currents can affect harbor conditions for a period of time after the tsunami's initial impact. Be sure conditions are safe before you return to your boat or ship or to the harbor.

Summary

Tsunami are perhaps the most destructive of all natural disasters. They are produced by large magnitude events that occur in or near oceans: earthquakes, volcanic eruptions, submarine landslides, or meteorite impacts. Tsunami (or seismic sea waves) travel very rapidly across open oceans because their velocities are related directly to the depth of the water through which the energy is traveling. Velocities of 800 kilometers per hour are commonly recorded in deep oceans. Over the past sixty years there have been several major tsunami that have killed hundreds of thousands of people, the largest having occurred in December 2004 off the coast of Indonesia.

Tsunami typically involve a series of waves that hit the coastline. The largest wave is generally not the first in the series. The devastation produced by these waves is far reaching when tsunami hit the flat shorelines. Countries around the Pacific Ocean are very likely to experience tsunami, because of the number of large earthquakes associated with active plate subduction around the Circum-Pacific region. Detection systems are in place in several oceans, but there is still a need for a system in the Indian Ocean, an area that is likely to experience tsunami because of the active tectonics in the regions surrounding it.

References and Suggested Readings

Bryant, Edward. 2008. *Tsunami: The underrated hazard.* Springer. Berlin and New York.

Duchamp, L. Timmel. 2007. *Tsunami.* Aqueduct Press. Seattle, WA

Dudley, Walter C., and Min Lee. 1998. *Tsunami!* Second Edition, University of Hawaii Press, Honolulu, Hawaii.

Kling, Andrew. A. 2003. *Tsunami.* San Diego: Lucent Books, 112 p.

National Science and Technology Council. Tsunami risk reduction for the United States: A framework for action. A joint report of the subcommittee on disaster reduction and the United States group on Earth Observations, December 2005.

Svarney, Thomas E. and Patricia Barnes-Svarney. 2000. *The Handy Ocean Answer Book.* Farmington Hills, MI. Visible Ink Press. 570 p.

U.S. Geological Survey. Tsunami hazards—a national threat. USGS Fact Sheet 2006-3023.

Web Sites for Further Reference

http://www.prh.noaa.gov/itic/
http://wcatwc.arh.noaa.gov/safety.htm
http://www.pmel.noaa.gov/tsunami/
http://www.usgs.gov/
http://woodshole.er.usgs.gov
http://www.tsunami.noaa.gov
http://walrus.wr.usgs.gov/tsunami/

Questions for Thought

1. What is the origin of the word *tsunami*?

2. Why does Japan experience so many tsunami?

3. How do tsunami and wind-generated waves differ?

4. Explain the difference between runup and drawdown.

5. How does vertical movement associated with earthquakes on the ocean floor generate tsunami?

6. How does a change in water depth affect the velocity of a tsunami?

7. Why is Hawaii prone to having a large number of tsunami hit it?

8. Explain how landslides on the continents or in the oceans create tsunami.

9. Explain how a tsunami detection buoy provides information on possible tsunami.

10. Give an example of how a tsunami warning system saves lives.

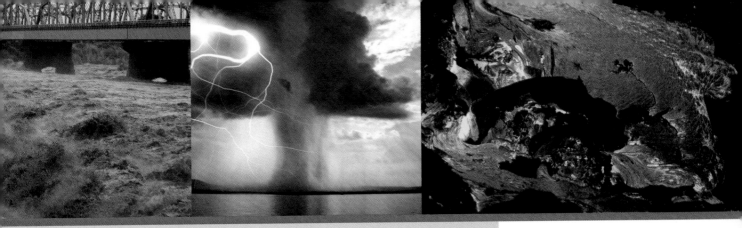

Target Earth

7

NASA.

The potential for asteroids to strike Earth is always present. This image was taken by spacecraft Galileo of Asteroid 951 Gaspra. This image is a combination of images taken with the violet, green and near-IR filters. The colors are highly exaggerated to bring out subtle differences in surface properties. Bluish regions represent fresher rock, while reddish regions are composed of regolith materials. Gaspra is an irregular shaped body about 19 × 12 × 11 km. The illuminated portion in this image extends about 18 km from lower left to upper right. The sun is shining from the right.

Key Terms

asteroid
asteroid belt
astronomical unit (AU)
Barringer Crater
Big Bang Theory
comet
Kuiper Belt
meteor
meteorite
meteoroid
Oort Cloud

Big Bang Theory
The idea that the Universe formed from an initial point mass that exploded about 13.7 billion years ago and moved outward in an ever-expanding fashion.

Astronomers estimate that the Universe began about 13.7 billion years ago. The **Big Bang Theory** postulates that the Universe first existed as a single, extremely dense point of matter. It exploded, spreading material throughout space, and has been continually expanding. Astronomical observations also show that other galaxies are moving away from this central point. Countless billions of objects are moving through space in hundreds of billions of galaxies that exist. Space research has discovered that other galaxies contain stars that have solar systems and planets. The existence of life in other parts of the Universe is an ongoing debate that will be part of the basis for further space exploration.

Earth as Part of Our Solar System

Calculations show that our Sun and the Solar System formed about 4.6 billion years ago (**Figure 7.1**). The Solar System is divided into two groups of planets: the inner or terrestrial planets that are Earth-like, and the giant or Jovian planets (Figure 7.1). The four inner planets, those closest to the Sun, are Mercury, Venus, Earth, and Mars. These four have several similar characteristics including their average densities and their general compositions. The giant planets—Jupiter, Saturn, Uranus, and Neptune—are much larger than the inner planets and have characteristics that are indicative of their frozen, gaseous compositions.

Earth is the third planet from the Sun and the only one in our Solar System that has an environment capable of sustaining life. Discussions later in the text (see Chapter 14) provide information about the biosphere, atmosphere, hydrosphere, and geosphere in which life exists. The composition of Earth is discussed in Chapter 2.

As Earth travels around the Sun and moves through space, the planet encounters extraterrestrial objects that are moving through the Solar System. Throughout its existence Earth has been bombarded by particles ranging in size from microscopic dust to large masses weighing billions of tons. A combination of metallic, stony, and icy material has contributed to the growth of the planet and the addition of water and chemical elements that have been incorporated into its composition.

Figure 7.1 A schematic view of the solar system from afar.

Classification of Space Objects

Extraterrestrial objects in space include asteroids (occasionally referred to as minor planets) and comets. These objects are produced from different sources, each of which is millions of kilometers from Earth.

Asteroids

An **asteroid** is a relatively small body that consists of rocky or metallic material. Several million asteroids lie in the **asteroid belt**, a region located between the orbits of Mars and Jupiter (**Figure 7.2**). They range in size from a few centimeters to more than 900 km in diameter. Of the estimated 750,000 to 1.7 million objects larger than 1 km, more than 200 of them have diameters exceeding 100 km, including the three largest (Ceres, diameter 933 km; Pallas, diameter 523 km; and Vesta, 501 km). Ceres was the first asteroid discovered, having been observed by the Italian astronomer Giuseppe Piazzi in 1801.

When astronomers determined the distances of the planets from the Sun, the position of the asteroid belt corresponded to the orbital distance that a possible planet could have had. The ring of material extends from about 300 million km to about 600 million km, or 2 AUs to 4 AUs from the Sun (1 AU equals one **astronomical unit**, equivalent to the distance separating the Sun and Earth, 150 million km or 93 million miles). About half of these asteroids lie near a distance of 2.8 AUs. Depending on the distance from the Sun, material takes between three and six years to complete a revolution around the Sun. The material is spread out in a manner that prevents it from coagulating and forming a planet because of the strong gravitational forces of Jupiter. If all the mass of the asteroids were brought together, it is estimated that the diameter of the planet would be approximately 1,500 km (about one-half that of our Moon).

As asteroids circle the Sun, sometimes they are moved out of their orbits, either by changes in the gravitational forces of other objects, or by collisions with each other. Asteroids can be grouped into three general types of bodies that revolve around the Sun in a generally elliptical orbit. The belt is divided into three parts: the inner belt, main belt, and the outer belt. Compositions of material within the entire belt vary,

asteroid
A rocky or metallic body ranging in size from a few centimeters to almost 1000 km in size.

asteroid belt
A region around the Sun lying between Mars and Jupiter that is the source of asteroids.

astronomical unit (AU)
The distance of the Earth from the Sun, averaging 150 million km or 93 million miles; used to describe large distances between bodies in space.

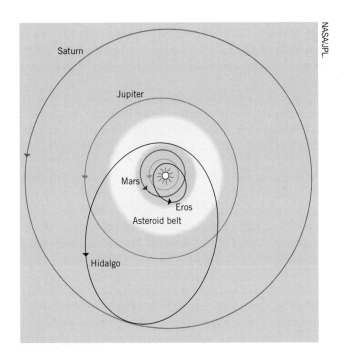

NASA/JPL

Figure 7.2 The asteroid belt lies between Mars and Jupiter and contains the majority of asteroids. Two notable exceptions are shown—Eros and Hidalgo, whose orbits lie outside the asteroid belt.

NASA/JPL

Figure 7.3 Asteroid Ida, measuring about 55 km in length, has its own moon. Dactyl is about 1.5 km in diameter and moves through space under the gravitational force of Ida.

TABLE 7.1	Types of Asteroids Include Those Rich in Carbon, Silicates, and Metals. The Predominant Chemical Composition Determines the Type of Asteroid.	
Type of Asteroid	**Approximate Percentage of Total**	**Composition and Location**
Carbonaceous C-type	75%	Gray to dark, composition similar to that of the Sun; low reflectivity; found in outer regions of main belt
Silicaceous S-type	17%	Appear bright with high reflectivity; metallic iron with Fe and Mg silicates; dominate inner belt
Metallic M-type	8%	Relatively bright; predominately Fe and Ni; common in middle belt region

Source: NASA.

but overall the less dense material lies the greatest distance from the Sun. These compositional differences are indicative of the variety of elements present in the Solar System (Table 7.1)

Numerous spacecraft have returned images of asteroids. In 1993 the spacecraft Galileo photographed the first asteroid that has its own moon. Asteroid Ida, a 55-km long mass, moves through space with its moon Dactyl, which is 1.5 km in diameter (**Figure 7.3**). Dactyl could have been a piece of Ida that was thrown out following a collision of Ida with another object or Dactyl could have been captured by the gravitational force of Ida.

Meteoroids, Meteors, and Meteorites

meteoroid
An relatively small object that moves through space; most range in size from dust to less than 10 m in size; composed of rock, metal, or ice.

meteor
A rapidly moving body that passes into the atmosphere and begins to burn up, producing what is sometimes referred to as a "shooting star".

meteorite
A meteor that hits Earth's surface

The term **meteoroid** is used to describe any small object that moves through space. Usually meteoroids, objects that are less than 10 meters, are composed of rock, metal, or ice. Upon entering Earth's atmosphere, the object is termed a **meteor** and becomes a **meteorite** when it strikes Earth's surface. The most common meteorites, known as ordinary chondrites, consist of rock, and appear to be relatively unchanged since the Solar System formed. About 86 percent of all recorded meteorites are classified as stony meteorites.

Meteoroids as small as a few microns (millionths of a meter) are much more common than larger ones. These very small ones continually strike the atmosphere and float to Earth after burning up. The resulting dust falls to Earth and adds an estimated 100,000 tons of mass to the planet each day. Stony-iron meteorites, on the other hand, appear to be remnants of larger bodies that contain heavier metals and lighter rocks separated into different layers. The objects remain more intact during their fall to Earth.

As meteoroids travel through space toward Earth, their flight patterns have changed in response to Earth's gravitational attraction. Racing along at velocities of up to 17 km per second, asteroids continue on their path and often do not head directly toward Earth or enter the atmosphere. Many glance off the upper atmosphere and continue their journey into outer space.

If an object does enter the atmosphere, it becomes a meteor. Ranging in size from dust particles to tens of meters in diameter, they begin to burn up in the upper portions of the atmosphere. The resulting fire produces what is termed a "shooting star," a streak of light visible at night as the meteor moves across the sky. The duration of the streak of light is very short due to the high velocity of entry.

Several times each year, Earth's orbit places it in the path of a comet that crossed the orbit years ago. The sky is filled with many streaks of light as clouds of material from the comet generate numerous showers of light. The Perseid meteor shower each August and the Leonid meteor shower each November are examples of such events. These are so named because the meteor showers appear to come from each of these respective constellations.

Formation of Impact Craters

When an asteroid strikes the surface of Earth, a crater is formed. Craters can be classified as either simple or complex. A simple crater has a generally circular shape and a raised rim produced by ejected material that is thrown out by the impact (**Figure 7.4a**). The depth and diameter of the crater are dependent on the velocity and mass of the asteroid.

A complex crater is formed from a very large asteroid impacting the surface. Because more material is thrown into the lower atmosphere, some of this falls back into the crater, producing a central peak (**Figure 7.4b**) These larger asteroids also generate extremely high temperatures and pressure conditions that form new minerals only found near such impacts.

Surface Impacts

Throughout its history Earth has been struck by extraterrestrial objects. Depending on the size of the object, the resulting impacts can produce small to large craters. More than 170 documented impacts have been recognized on Earth, with 57 located in North America (**Figure 7.5** and **Table 7.2**). Because of the dynamic weathering processes on Earth, many of the older impacts have been obliterated or highly altered. However, **Barringer Crater** (also known as Meteor Crater) in northern Arizona is extremely well preserved. The climate of the surrounding region is arid, which has minimized the physical and chemical alteration of the rock and the crater (**Figure 7.6**). The crater is more than 1 km in diameter and cuts more than 180 m into the sedimentary rock units on the surface. Material thrown out formed a rim ranging in height between 30 and 60 m above the flat-lying sediments.

Barringer Crater
The best preserved impact crater on Earth; located in northern Arizona, it has a diameter of almost 1 km and is about 180 m in depth.

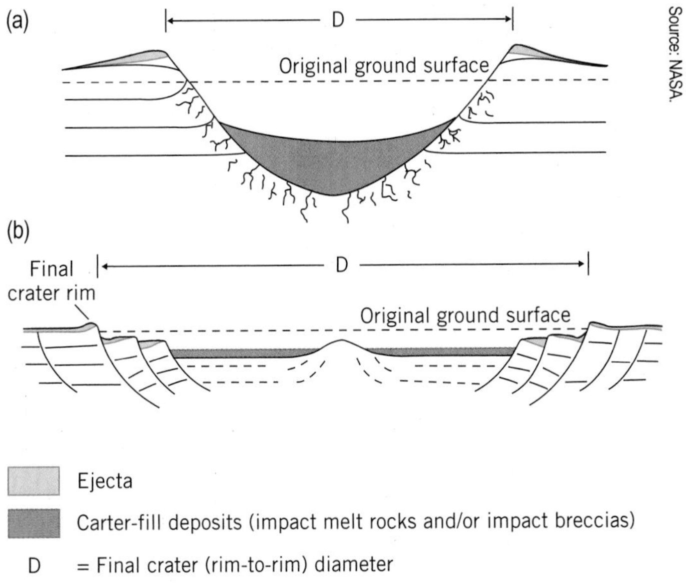
(a)
D
Original ground surface
(b)
Final crater rim
D
Original ground surface
Ejecta
Carter-fill deposits (impact melt rocks and/or impact breccias)
D = Final crater (rim-to-rim) diameter
Source: NASA.

Figure 7.4 (a) Simple impact crater is basically a lare depression that is infilled with some material produced by the impact. (b) A complex crater is larger and has a raised peak in the center that formed from material falling back into the center.

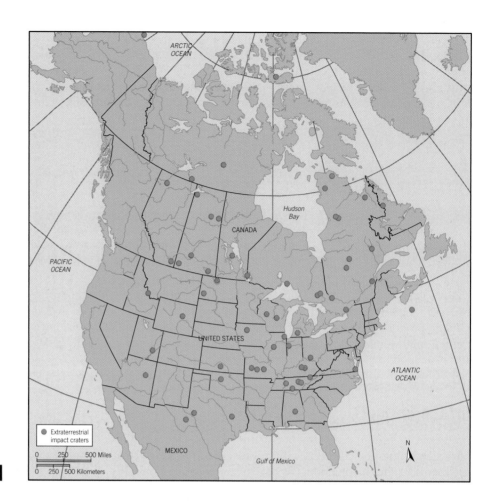

Figure 7.5 Locations of impact features in Canada and the United States (refer to Table 7.2). Surface and subsurface information has been used to provide locations.

TABLE 7.2	Impact Craters in North America						
Crater Name	Location	Latitude	Longitude	Diameter (km)	Age (Ma)*	Exposed	Drilled
Ames	Oklahoma, U.S.A.	N 36° 15′	W 98° 12′	16.00	470 ± 30	N	Y
Avak	Alaska, U.S.A.	N 71° 15′	W 156° 38′	12.00	3 − 95	N	Y
Barringer	Arizona, U.S.A.	N 35° 2′	W 111° 1′	1.18	0.049 ± 0.003	Y	Y
Beaverhead	Montana, U.S.A.	N 44° 36′	W 113° 0′	60.00	∼ 600	Y	N
Brent	Ontario, Canada	N 46° 5′	W 78° 29′	3.80	396 ± 20*	N	Y
Calvin	Michigan, U.S.A.	N 41° 50′	W 85° 57′	8.50	450 ± 10	N	Y
Carswell	Saskatchewan, Canada	N 58° 27′	W 109° 30′	39.00	115 ± 10	Y	Y
Charlevoix	Quebec, Canada	N 47° 32′	W 70° 18′	54.00	342 ± 15*	Y	Y
Chesapeake Bay	Virginia, U.S.A.	N 37° 17′	W 76° 1′	90.00	35.5 ± 0.3	N	Y
Chicxulub	Yucatan, Mexico	N 21° 20′	W 89° 30′	170.00	64.98 ± 0.05	N	Y

TABLE 7.2	Impact Craters in North America (Continued)						
Crater Name	Location	Latitude	Longitude	Diameter (km)	Age (Ma)*	Exposed	Drilled
Clearwater East	Quebec, Canada	N 56° 5′	W 74° 7′	26.00	290 ± 20	Y	Y
Clearwater West	Quebec, Canada	N 56° 13′	W 74° 30′	36.00	290 ± 20	Y	Y
Cloud Creek	Wyoming, U.S.A.	N 43° 7′	W 106° 45′	7.00	190 ± 30	N	Y
Couture	Quebec, Canada	N 60° 8′	W 75° 20′	8.00	430 ± 25	Y	N
Crooked Creek	Missouri, U.S.A.	N 37° 50′	W 91° 23′	7.00	320 ± 80	Y	N
Decaturville	Missouri, U.S.A.	N 37° 54′	W 92° 43′	6.00	< 300	Y	Y
Deep Bay	Saskatchewan, Canada	N 56° 24′	W 102° 59′	13.00	99 ± 4	N	Y
Des Plaines	Illinois, U.S.A.	N 42° 3′	W 87° 52′	8.00	< 280	N	Y
Eagle Butte	Alberta, Canada	N 49° 42′	W 110° 30′	10.00	< 65	N	Y
Elbow	Saskatchewan, Canada	N 50° 59′	W 106° 43′	8.00	395 ± 25	N	Y
Flynn Creek	Tennessee, U.S.A.	N 36° 17′	W 85° 40′	3.80	360 ± 20	Y	Y
Glasford	Illinois, U.S.A.	N 40° 36′	W 89° 47′	4.00	< 430	N	Y
Glover Bluff	Wisconsic, U.S.A.	N 43° 58′	W 89° 32′	8.00	< 500	Y	Y
Gow	Saskatchewan, Canada	N 56° 27′	W 104° 29′	4.00	< 250	Y	N
Haughton	Nunavut, Canada	N 75° 22′	W 89° 41′	23.00	39	Y	N
Haviland	Kansas, U.S.A.	N 37° 35′	W 99° 10′	0.01	< 0.001	Y	N
Holleford	Ontario, Canada	N 44° 28′	W 76° 38′	2.35	550 ± 100	N	Y
Ile Rouleau	Quebec, Canada	N 50° 41′	W 73° 53′	3.80	< 300	Y	N
Kentland	Indiana, U.S.A.	N 40° 45′	W 87° 24′	13.00	< 97	Y	Y
La Moinerie	Quebec, Canada	N 57° 26′	W 66° 37′	8.00	400 ± 50	Y	N
Manicouagan	Quebec, Canada	N 51° 23′	W 68° 42′	100.00	214 ± 1	Y	Y
Manson	Iowa, U.S.A.	N 42° 35′	W 94° 33′	35.00	73.8 ± 0.3	N	Y
Maple Creek	Saskatchewan, Canada	N 49° 48′	W 109° 6′	6.00	< 75	N	Y
Marquez	Texas, U.S.A.	N 31° 17′	W 96° 18′	12.70	58 ± 2	N	Y
Middlesboro	Kentucky, U.S.A.	N 36° 37′	W 83° 44′	6.00	< 300	Y	Y

(Continued)

TABLE 7.2	Impact Craters in North America (Continued)						
Crater Name	**Location**	**Latitude**	**Longitude**	**Diameter (km)**	**Age (Ma)***	**Exposed**	**Drilled**
Mistastin	Newfoundland/Labrador, Canada	N 55° 53′	W 63° 18′	28.00	36.4 ± 4	Y	N
Montagnais	Nova Scotia, Canada	N 42° 53′	W 64° 13′	45.00	50.50 ± 0.76	N	Y
New Quebec	Quebec, Canada	N 61° 17′	W 73° 40′	3.44	1.4 ± 0.1	Y	N
Newporte	North Dakota, U.S.A.	N 48° 58′	W 101° 58′	3.20	< 500	N	Y
Nicholson	Northwest Territories, Canada	N 62° 40′	W 102° 41′	12.50	< 400	N	N
Odessa	Texas, U.S.A.	N 31° 45′	W 102° 29′	0.16	< 0.05	Y	Y
Pilot	Northwest Territories, Canada	N 60° 17′	W 111° 1′	6.00	445 ± 2	Y	N
Presqu'ile	Quebec, Canada	N 49° 43′	W 74° 48′	24.00	< 500	Y	N
Red Wing	North Dakota, U.S.A.	N 47° 36′	W 103° 33′	9.00	200 ± 25	N	Y
Rock Elm	Wisconsin, U.S.A.	N 44° 43′	W 92° 14′	6.00	< 505		
Santa Fe	New Mexico, U.S.A.	N 35° 45′	W 105° 56′	6–13	< 1200	N	N
Saint Martin	Manitoba, Canada	N 51° 47′	W 98° 32′	40.00	220 ± 32	N	Y
Serpent Mound	Ohio, U.S.A.	N 39° 2′	W 83° 24′	8.00	< 320	Y	Y
Sierra Madera	Texas, U.S.A.	N 30° 36′	W 102° 55′	13.00	< 100	Y	Y
Slate Islands	Ontario, Canada	N 48° 40′	W 87° 0′	30.00	~ 450	Y	N
Steen River	Alberta, Canada	N 59° 30′	W 117° 38′	25.00	91 ± 7*	N	Y
Sudbury	Ontario, Canada	N 46° 36′	W 81° 11′	250.00	1850 ± 3	Y	Y
Upheaval Dome	Utah, U.S.A.	N 38° 26′	W 109° 54′	10.00	< 170	Y	Y
Viewfield	Saskatchewan, Canada	N 49° 35′	W 103° 4′	2.50	190 ± 20	N	Y
Wanapitei	Ontario, Canada	N 46° 45′	W 80° 45′	7.50	37.2 ± 1.2	N	N
Wells Creek	Tennessee, U.S.A.	N 36° 23′	W 87° 40′	12.00	200 ± 100	Y	Y
West Hawk	Manitoba, Canada	N 49° 46′	W 95° 11′	2.44	351 ± 20	N	Y
Wetumpka	Alabama, U.S.A.	N 32° 31′	W 86° 10′	6.50	81.0 ± 1.5	Y	Y
Whitecourt	Alberta, Canada	N 54° 00′	W 115° 36′	0.036	< 0.0011	Y	N

* pre-1977 K-Ar, Ar-Ar and Rb-Sr ages recalculated using the decay constants of Steiger and Jager (1977). Ages in millions of years (Ma) before present.

Source: Earth Impact Database, Planetary and Space Science Centre, University of New Brunswick, Canada (www.unb.ca/passc).

Comets

A **comet** contains a nucleus, or central core, which averages 5 to 10 km in diameter. The core consists mainly of frozen CO_2 and water ice, along with some solid materials. As they near the Sun, comets begin to be torn apart by solar energy that heats the frozen material and produces gases that stream away from the core and the Sun. Gases adjacent to the core produce a coma, a bright zone surrounding the nucleus (**Figure 7.7**). Two tails are produced, one composed of dust and the other from ions formed from the breakdown of gas particles torn apart by solar winds. The strong solar winds push the tail away from the Sun (**Figure 7.8**). The orbits of comets are often very ellipitical.

The source of 90 percent of comets is the **Oort Cloud**, named for the Dutch astronomer Jan Oort, who discovered it in 1950. It is a large, spherical region in space that stretches from 20,000 AU to 50,000 AU from the Sun. The Oort Cloud is considered to be part of our Solar System. Within the Oort Cloud are several trillion icy bodies. If the orbit of a body is disturbed by the passing of another object, the icy body can leave the Oort Cloud and come under the gravitational attraction of the Sun and the larger outer planets. This causes the icy body to begin a journey into our Solar System; the body is now a comet, speeding toward the Sun (**Figure 7.9**).

Michael Collier/Stock Boston.

Figure 7.6 Barringer Crater in northern Arizona is the best preserved impact feature on Earth. Its estimated age is 49,000 years.

comet
Body of icy and rocky material that moves around the Sun; most comets originate from the Oort Cloud.

Oort Cloud
A far-reaching, spherically shaped body of material located between 50,000 and 100,000 AU from the Sun; the major source of comets.

Figure 7.7 A comet as it approaches the Sun displays two tails pointing away from the Sun.

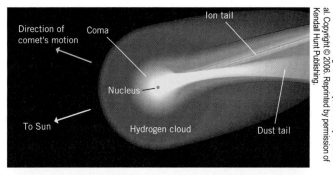

Direction of comet's motion

Coma

Ion tail

Nucleus

To Sun

Hydrogen cloud

Dust tail

From *Discovering Astronomy*, 5th Ed by Shawl et al. Copyright © 2006. Reprinted by permission of Kendall Hunt Publishing.

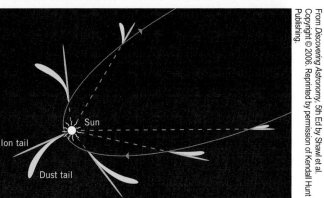

Sun

Ion tail

Dust tail

From *Discovering Astronomy*, 5th Ed by Shawl et al. Copyright © 2006. Reprinted by permission of Kendall Hunt Publishing.

Figure 7.8 The tails of a comet point away from the Sun as solar winds push material outward from the Sun.

Source: NASA.

Figure 7.9 Image of a comet. The tail points away from the Sun as solar winds drive material away from the core of the comet.

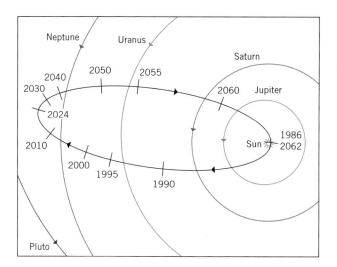

Figure 7.10 Halley's Comet returns to cross Earth's orbit about every 75 or 76 years. Its last close encounter was in 1986.

Kuiper Belt
A region beyond the orbits of Neptune and Pluto that is a source of short-period comets.

Both long- and short-term comets enter the Solar System. Long-term comets come from the Oort Cloud and have periodicities of more than 200 years. Short-term comets can have a return period of less than 200 years, causing them to return and be sighted again from Earth. Comets that come from the **Kuiper** (KY-per) **Belt**, a region lying just outside the orbit of Pluto, tend to be short-term comets. The gravitational effect of Neptune on objects in the Kuiper belt causes an occasional body to move out of its orbit and head toward the Sun.

Halley's Comet

The most famous and brightest of all comets is Halley's Comet. The earliest recorded observations of the comet were made more than 2,000 years ago in China. Subsequent observations and study showed the comet to have a very predictable return rate. The comet was studied by Edmond Halley, an English astronomer (later named Royal Astronomer) who predicted its return, based on an observed return cycle of 75 or 76 years. The comet was named in his honor following his death in 1742 (he predicted its return in 1758, which it did). Its most recent passage near Earth occurred in 1986 (**Figure 7.10**).

Other calculations revealed that Halley's Comet was the same one that appeared in 1066 before the Battle of Hastings. Its appearance then was interpreted by the Saxons as a bad sign prior to being defeated by the Normans.

Spacecraft observations of Halley's Comet in 1986 indicated that its core is about 15 km long and 8 km wide. This nucleus contains icy chunks that are thought to be covered with carbon compounds, causing the core to appear dark.

Comet Shoemaker-Levy 9

In March 1993 Gene and Carolyn Shoemaker along with David Levy discovered the comet bearing their name (this was the ninth comet the team discovered). It was soon determined that the comet was in orbit around Jupiter and had the potential to strike the planet. Jupiter's huge gravitational forces broke the comet into at least 21 separate pieces that collided with the planet over a six-day period beginning on July 16, 1994 (**Figure 7.11**). The series of collisions represented the first observed bombardment of two bodies in our solar system. The spectacular series of events was described by the Shoemakers and Levy to a worldwide television audience of millions of viewers throughout the world. It should be noted that following the untimely death of Gene Shoemaker in Australia in 1997, Carolyn Shoemaker has continued to discover asteroids, and is now credited with having found more than 800 asteroids and 32 comets.

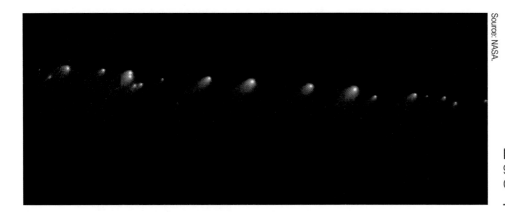

Figure 7.11 Comet Shoemaker-Levy 9 as taken with the Wide-Field/Planetary Camera on the Hubble Space Telescope.

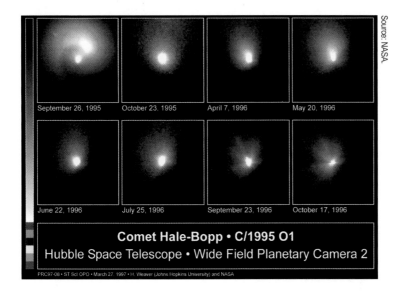

Figure 7.12 Images of Comet Hale-Bopp over a 13 month period as observed from the Hubble Space Telescope.

Comet Hale-Bopp

New comets are discovered almost every year by astronomers. On July 23, 1995 two amateur astronomers, Alan Hale from New Mexico and Thomas Bopp of Arizona, independently discovered the comet to receive their names. Comet Hale-Bopp was sighted outside the orbit of Jupiter at a distance of 7.15 AU (roughly 660 million miles from the Sun) (**Figure 7.12**).

A Recent Collision with Earth

Tunguska, Siberia, experienced an explosion of catastrophic proportions on June 30, 1908. Scientists have hypothesized that a fragment of a comet exploded about 8 km above the ground surface, generating a shockwave that flattened more than one thousand square kilometers of thick forest. Given the remoteness of the area, almost twenty years passed before researchers were able to visit the region and determine the source of the event. No evidence has ever been located of an impact crater.

Continued exploration of space has included our ability to intercept extraterrestrial bodies. One example is the crashing of a probe into Comet Tempel 1 in the summer of 2005 (**Figure 7.13**). Observations made of the actual impact aided scientists in describing the terrane and overall appearance of the comet.

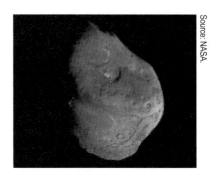

Figure 7.13 Tempel 1 is a comet that orbits the Sun about every 5.5 years. It was impacted on July 4, 2005 by a component of the NASA Deep IMPACT Probe. Its dimensions are estimated at 14 km by 8 km.

BOX 7.1	The story below, published by the Near-Earth Object Program Office of NASA and the Jet Propulsion Laboratory, describes a recent event that shows how vulnerable earth could be to a potential impact.

Small Asteroid 2009 VA Whizzes by the Earth

Don Yeomans, Paul Chodas, Steve Chesley NASA/JPL Near-Earth Object Program Office November 9, 2009

A newly discovered asteroid designated 2009 VA, which is only about 7 meters in size, passed about 2 Earth radii (14,000 km) from the Earth's surface Nov. 6 at around 16:30 EST. This is the third-closest known (non-impacting) Earth approach on record for a cataloged asteroid. The two closer approaches include the 1-meter sized asteroid 2008 TS26, which passed within 6,150 km of the Earth's surface on October 9, 2008, and the 7-meter sized asteroid 2004 FU162 that passed within 6,535 km on March 31, 2004. On average, objects the size of 2009 VA pass this close about twice per year and impact Earth about once every 5 years.

   Asteroid 2009 VA was discovered by the Catalina Sky Survey about 15 hours before the close approach, and was quickly identified by the Minor Planet Center

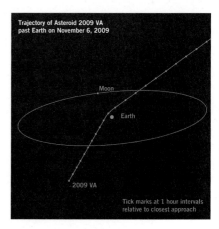

Trajectory of Asteroid 2009 VA Past Earth on November 6, 2009.

in Cambridge MA as an object that would soon pass very close to the Earth. JPL's Near-Earth Object Program Office also computed an orbit solution for this object, and determined that it was not headed for an impact. Only thirteen months ago, the somewhat smaller object 2008 TC3 was discovered under similar circumstances, but that one was found to be on a trajectory headed for the Earth, with impact only about 11 hours away.

Effects of Impacts
Mass Extinctions

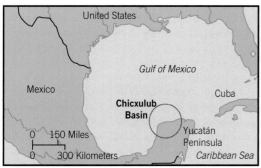

Figure 7.14 The Chicxulub Basin stretches for approximately 180 km and is the result of the imact of the Chicxulub asteroid 65 million years ago.

Major impacts of meteorites, in addition to producing scars on the landscape, have killed large amounts of flora and fauna. Megaimpacts, such as the one about 65 million year ago, caused global loss of life. The Chicxulub impact along the edge of the Yucatan peninsula is often related to the rapid dying of the dinosaurs that dominated the Earth at the time (**Figure 7.14**). The impact threw billions of tons of debris into the atmosphere, including a significant amount of iridium, a rare element on Earth. Studies of the boundary between the Cretaceous and Tertiary periods (refer to the geologic time scale in Chapter 1) align the deposition of iridium with the time boundary.

   In the distant geologic past undoubtedly impacts struck the continents (refer to Table 7.2). Because the majority of Earth's surface was covered with water, impacts struck the oceans, generating megatsunami. However, the geologic evidence of these events is more difficult to read in the geologic record.

Environmental Changes

Several stages of major impacts on the environment follow a large asteroid impact. The most immediate change is the resulting fireball that increase surface and atmospheric temperatures in the surrounding region. These temperatures declined over a period of several days.

Because the impact threw massive amounts of dust and ash into the atmosphere, incoming solar radiation is decreased, causing global temperatures to drop. These conditions can last several months or years, depending on the amounts of particulates. Finally, as the atmosphere is cleared of dust, any increased CO_2 and water vapor helps hold in radiated heat from Earth's surface, leading to warming of the atmosphere.

Summary

The universe formed about 13.7 billion years ago and our Solar System and Earth began to develop about 4.6 billion years ago. Rotational forces of the Sun and its solar winds threw off material that cooled and created the planets, which consist of terrestrial-like inner planets and the much larger, gaseous giants. Some of this material formed a belt of asteroids that revolves around the Sun in the space between Mars and Jupiter. Changes in gravitational forces occasionally cause some asteroids to leave their orbit and head toward the Sun. As smaller objects move closer to Earth, Earth's gravitational forces pull them through the atmosphere, at which point they either burn up or collide with the surface. More than 170 larger bodies have impacted Earth and formed craters on the land, with a larger, undetected number of impacts lying in the oceans.

The periodic orbits of comets vary greatly with some passing near Earth every few years and others taking several hundred years to reappear. Halley's Comet returns every 75 or 76 years. Each year observers discover new comets. Spacecraft continually send back images of new discoveries.

On the far edge of the Solar System the Kuiper Belt is the source of comets, icy, rocky bodies that depart from their orbit and move toward the Sun. Other comets are generated in the Oort Cloud, an expansive, spherical zone located between 50,000 and 100,000 AU from the Sun.

Both comets and asteroids present a threat to Earth, as each type of body moves at high velocities through space. Geologic evidence exists for a major impact of an asteroid with Earth 65 million years ago. The Chicxulub impact site in Mexico is believed to be caused by a major impact that is tied to the death of the dinosaurs and other life on Earth at that time. The environmental impact of impacts includes changes in atmospheric conditions that both lower and then raise global temperatures, as well as initial increases in temperature due to the fireball associated with the impact.

References and Suggested Readings

Bevan, A. and J. de Laeter. 2002. *Meteorites: A Journey through Space and Time*. Washington, D.C.: Smithsonian Institution Press.

Lewis, J. S. 1996. *Rain of Iron and Ice: The Very Real Threat of Comet and Asteroid Bombardment*. New York: Perseus Publishing Group.

Moore, Sir Patrick. 2003. *Firefly Atlas of the University*. Buffalo: Firefly Books.

Shawl, S. J., K. M. Ashman, and B. Hufnagel. 2006. *Discovering Astronomy*, 5th ed. Dubuque, IA: Kendall/Hunt Publishing Company.

Smith, R., ed., 1998. *The Solar System*, 3 vols: Pasadena: Salem Press.

Woolfson, M. M. 2000. *The Origin and Evolution of the Solar System*. Bristol: Institute of Physics Publishing.

Web Sites for Further Reference

http://astrogeology.usgs.gov/About/People/GeneShoemaker/
http://impact.arc.nasa.gov/
http://www.jpl.nasa.gov/asteroidwatch/asteroids-comets.cfm
http://www.jpl.nasa.gov/solar_system/asteroids_comets/asteroids_comets_index.html
http://solarsystem.nasa.gov/deepimpact/index.cfm
http://www.nasa.gov
http://www.unb.ca/passc/ImpactDatabase/
http://www2.jpl.nasa.gov/comet/index.html

Questions for Thought

1. Explain the difference between a meteor and a comet.

2. How is a meteor different from a meteorite?

3. Distinguish between the asteroid belt and the Oort Cloud.

4. What is the basic concept associated with the Big Bang theory?

5. What is the relationship of the age of the Sun and its Solar System as measured against that of the Earth?

6. Give two major differences between the inner planets and the outer planets?

7. What can cause an asteroid to change its travel path?

8. Why are relatively few impact features on Earth well preserved?

9. Describe two types of tail structures associated with comets.

10. What was significant about Comet Shoemaker-Levy?

11. Explain the significance of the Chicxulub impact crater in Yucatan, Mexico.

Global Climate and Hazardous Weather

8

FEMA, photo by Greg Henshall.

An EF5 tornado destroyed 95 percent of the town of Greensburg, Kansas, in early May 2007. Twelve lives were lost as the twister touched down more than 75 times as it crossed the southwestern edge of Kansas.

Key Terms

air mass
anticyclone
blizzard
carbon sequestration
climate
climate change
Coriolis Effect
cyclone
Doppler radar
global warming
greenhouse effect
greenhouse gas
hail
heat wave
jet stream
latent heat of condensation
latent heat of fusion
latent heat of vaporization
lightning
nor'easter
precession

(Continued)

relative humidity	tornado	updraft
sublimation	tornado warning	weather
thunderstorm	tornado watch	

Earth's weather and climate are strongly influenced by the oceans, which cover 71 percent of Earth's surface. The oceans along with the atmosphere serve as key reservoirs for much of the water in the hydrosphere. The atmosphere gets the majority of its heat and moisture from the oceans, which control weather patterns and climate. The Sun provides the energy that heats the ocean and drives the hydrosphere.

One purpose the oceans serve is to regulate temperature in the lower part of the atmosphere. The atmosphere in turn provides energy through winds to create waves and to help move currents below the surface. Weather that is created by the interaction of the oceans and the atmosphere can become a problem, both on land and at sea. Severe weather, such as hurricanes, tornadoes, severe thunderstorms, and winter blizzards, generates violent winds and heavy precipitation that can produce loss of property and lives.

Weather and Climate

weather
The composite condition of the near earth atmosphere, which includes temperature, barometric pressure, wind, humidity, clouds, and precipitation. Weather variations over a long period create the Climate.

climate
The long-term average weather, usually taken over a period of years or decades, for a particular region and time period.

People often confuse the terms *weather* and *climate*. **Weather** is the short-term condition of the atmosphere. Often measured over periods of hours, days, or even weeks, these conditions are usually localized or confined to a small region of the country. Temperature, precipitation, wind speeds, and humidity are associated with the weather. The ever-changing, complex nature of the atmosphere makes it difficult to forecast weather conditions very far into the future.

Climate describes the atmospheric conditions over long periods of time, ranging from a few years to thousands or even millions of years. These persistent conditions can have a major effect on a region, a continent, or even the entire globe. In the geologic past there are numerous examples of long-lasting ice ages or warming trends that affected large areas of the Earth's surface. In relatively recent history, areas of Europe experienced a Little Ice Age, which lasted from about 1400 to 1800 (**Figure 8.1**). During this period, glaciers increased in size, and temperatures dropped, which affected farming and crop production.

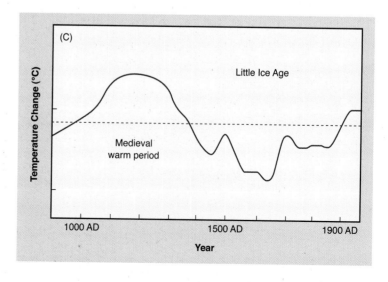

Figure 8.1 Changes in global temperatures between 1400 and 1800 produced the Little Ice Age, a time when temperatures were lower and glaciers advance. The dashed line shows the baseline average temperature.

Weather and climate are interrelated to the atmosphere and the hydrosphere in complex ways. Varying amounts of water in the oceans change the climate and thus affect weather. Changes in solar radiation patterns over long periods of time can increase or decrease global temperatures, which can cause changes in the amount of ice held in polar regions. Changes in atmospheric gases can also alter global temperatures.

The Atmosphere

Earth's atmosphere is composed of many gases. In the lower atmosphere, air consists primarily of nitrogen (78%) and oxygen (21%). Argon makes up slightly less than one percent of the atmosphere. Other gases present in very small amounts include carbon dioxide, methane, and nitrous oxides. Depending on atmospheric conditions, water vapor can make up as much as four percent of the atmosphere when the air is saturated. Water content varies with air temperature.

We measure the amount of water present in air in terms of the saturation level. **Relative humidity** (measured as a percent) indicates the amount of saturation present as measured against total saturation (100%) at a given temperature. Total saturation results in condensation of moisture as dew and precipitation (either rain or snow). A relative humidity of 50 percent tells us that the air is holding one half of the amount it would hold if totally saturated at a given temperature.

A cross section of the atmosphere shows that it is layered (**Figure 8.2**). Heat in the lowest portions of the troposphere helps drive the weather that occurs here. Air becomes less dense with increasing altitude, as there are fewer gas molecules present. There is also less gravitational attraction to hold the molecules close to the surface.

As altitude increases, there is less air pressure pushing down on the surface. Also with a thinner atmosphere, temperatures decrease rapidly. For example, the air outside an airplane flying at 10 km is about −42°C. In the troposphere, the layer nearest the surface, air temperatures decrease steadily to about −80°C at an altitude of 18 to 20 kilometers. The tropopause is the boundary separating the troposphere and the overlying stratosphere, in which the ozone layer is found. This layer, consisting of ozone (O_3), is the primary line of defense for harmful ultraviolet (UV) radiation that is produced by the sun. Above this layer is the stratopause, which lies below the mesosphere, the layer that extends to an altitude of about 50 to 55 kilometers. Although other layers of the atmosphere lie above the mesosphere, they do not have any significant effect on the climate and surface conditions. These upper layers do play a role in protecting Earth from extraterrestrial objects, as discussed in Chapter 7.

relative humidity
The percentage of moisture present in the air as measured against the amount it can hold at a given temperature and pressure to be saturated.

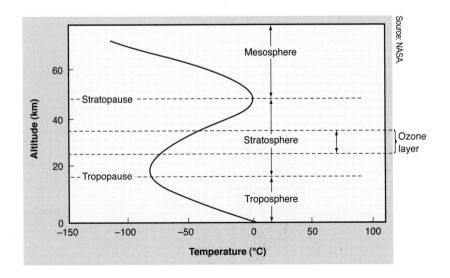

Figure 8.2 A simplified cross section of Earth's atmosphere up to 60 km. Human activity and weather are confined to the troposphere. The solid, curved line shows the change in temperature as altitude changes.

Solar Radiation and the Atmosphere

In Chapter 2 we saw that electromagnetic radiation from the sun has a wide range of wavelengths. Part of this spectrum of energy includes visible light. However, a much wider range of energy bombards Earth, including many wavelengths that are harmful. Certain gases in the atmosphere interact with selected wavelengths in different ways. As mentioned earlier, ozone shields the lower atmosphere and Earth's surface from UV radiation. Destruction of the ozone layer has occurred by the release of organic compounds, including chlorofluorocarbons (CFCs) into the atmosphere. Ozone reacts readily with the CFCs and is destroyed in the chemical reaction. Thus UV radiation can then penetrate the atmosphere.

Greenhouse Gases

Once solar radiation penetrates the atmosphere and strikes the surface, some of the energy is absorbed by the land and water, while some of it is reflected back into space. Gases and dust particles also reflect radiation back into the upper atmosphere, as well as ice.

When this energy is reflected, there is a change in the wavelengths of the radiation. Longer wavelength infrared waves are absorbed by moisture and gases in the middle and upper troposphere (**Figure 8.3**). This absorption of heat by these gases causes the temperature of the atmosphere to rise. Radiation is also re-reflected back toward the surface. This trapped energy causes a **global warming** effect, similar to that that takes place in a greenhouse.

Several gases contribute to this **greenhouse effect**. The primary gases are carbon dioxide, water vapor, methane, and ozone. Because so much water is present on Earth's surface, it is easily heated, and rises to become a significant contributor to the warming process. However, carbon dioxide, a gas that is generated by the oxidation (burning) of fossil fuels that contain carbon, is also a key greenhouse gas. Over the past 200 years, there has been a 35 percent increase in the presence of carbon dioxide in the atmosphere. This has contributed to the warming of the atmosphere.

Methane (CH_4) is the simplest of the hydrocarbons. It is a byproduct of the decay of organic material, and it also forms during the digestive process of organisms. Cattle are the leading producers of methane, as they belch methane. Estimates show that as much a 20 percent of all methane is produced by cattle. Methane is also released during the processing and transportation of petroleum products.

global warming
The gradual increase in global temperatures caused by the emission of gases that trap the sun's heat in the Earth's atmosphere (greenhouse effect). Gases that contribute to global warming include carbon dioxide, methane, nitrous oxides, chlorofluorocarbons (CFCs), and halocarbons (the replacements for CFCs). The carbon dioxide emissions are primarily caused by the use of fossil fuels for energy.

greenhouse effect
The heating that occurs when gases such as carbon dioxide trap heat escaping from the Earth and radiate it back to the surface.

greenhouse gases
Atmospheric gases, primarily carbon dioxide, methane, and nitrous oxide restricting some heat-energy from escaping directly back into space.

carbon sequestration
The storage or removal of carbon from the environment or the reducing or elimination of its presence.

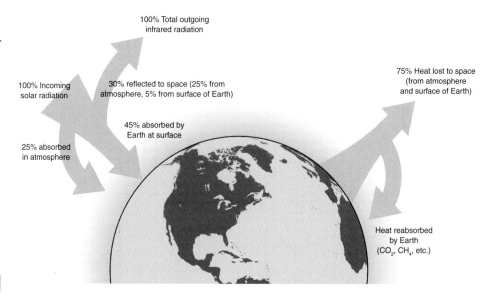

Figure 8.3 Energy from solar radiation is divided between reflected and reradiated energy. Original Source: Pruitt and Underwood, Bioinquiry: Making connections in Biology, 3rd ed., John Wiley & Sons, p. 555.

Effects of Volcanic Activity

The eruption of volcanoes often involves expulsion of gases and other pyroclastic material (**Figure 8.4**). Large amounts of carbon dioxide are added to the atmosphere which increases the concentration of **greenhouse gases**. Sulfur gases are able to reflect short wavelength solar radiation, thus preventing its entering the lower atmosphere. This process produces a cooling effect. The combination of sulfur gases and the pyroclastic dust can produce a significant drop in temperature. Recent examples of volcanic eruptions that altered the atmosphere and the amount of incoming radiation include the 1982 eruption of El Chichón in Chiapas, Mexico, and the 1991 eruption of Mount Pinatubo in the Philippine Islands. More than 20 million tons of sulfur dioxide were thrown into the atmosphere, causing global temperatures to drop 0.5°C for a two-year period following the Pinatubo event. In 1912 the eruption of Novarupta Volcano on the Katmai Peninsula of Alaska produced more than twice the pryoclastics and gases of Mount Pinatubo. When measured in the context of geologic time, major eruptions similar to these can occur rather frequently. One major eruption every 100 or 200 years adds great volumes of gases to the atmosphere, and water to the hydrosphere.

In the geologic past, several episodes of massive flood basalt eruptions sent countless millions of tons of gases into the atmosphere. The flood basalts of Siberia that formed about 250 million years ago, and the Cretaceous basalt eruptions in India and the Pacific Northwest had a disastrous effect on living organisms. The continual eruption of basalts associated with seafloor spreading adds gases directly into sea water, thereby changing the chemistry of the largest body of water on Earth.

United States Geological Survey.

Figure 8.4 Gases and other pyroclastic materials produced by volcanoes contribute significantly to affecting the amount of sunlight that can reach the surface.

Reducing the Presence of Carbon

The role that carbon dioxide plays in enhancing the greenhouse effect, and to air pollution, is evident to many scientists. If the amount of carbon can be reduced, the end result would be a better environment. The federal government, in conjunction with many private industries, is working on the process of **carbon sequestration**, which involves removing carbon from the environment or reducing or eliminating its presence. The increased use of hybrid vehicles decreases the demand for of petroleum fuels and reduces the amount of carbon dioxide put into the atmosphere. The rapidly increasing use of wind turbine technology to generate electricity reduces the need for coal-fired power plants (**Figure 8.5**) and reduces the production of carbon dioxide.

Courtesy of David M. Best.

Is Global Warming Occurring?

For centuries people have recorded temperatures in cities and other locations throughout the world. Obviously seasonal variations exist at a given site, but annual data for the past 125 years show that average global temperatures have been increasing significantly (**Figure 8.6**), with an increase of more than 1°C in the past century. Documented cases exist of glaciers having retreated several kilometers in the past 50 years. When viewed over the past 15,000 years, temperatures have risen more than 4°C.

Figure 8.5 This wind turbine farm in southern California is part of a larger plan nationwide to increase the production of electricity from clean, existing resources.

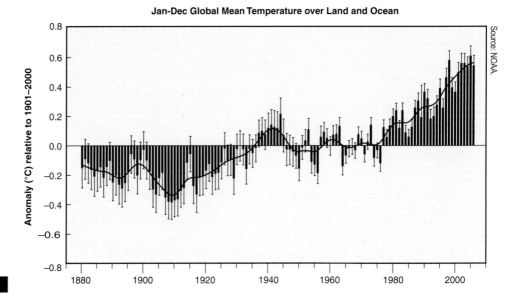

Figure 8.6 Annual global temperatures have been increasing over the past 125 years, although most of the increase has occurred since 1980.

Effects of Global Warming

Several changes occur on both regional and global scales in terms of temperature. Increased warming causes sea level to rise as glaciers and ice sheets melt. Because water has a high capacity to retain heat, the oceans absorb more heat, causing the molecules to expand, thereby raising sea level. Large bodies of water are slow to heat up, but they are also slow to cool down.

Warmer ocean waters have increased thermal energy, which leads to more tropical storms. Higher ocean temperatures allow cyclonic storms to intensify as they pass over these heat sources. Such was the case with Hurricane Katrina as it moved into the Gulf of Mexico before striking the Louisiana-Mississippi coastline in August 2005. Temperatures in the Gulf of Mexico were well above average, thus providing more thermal energy to intensify the hurricane.

Continental areas are affected by global warming as changes in the hydrologic cycle alter the distribution of precipitation, which can affect vegetation patterns. Droughts can destroy once productive agricultural areas, causing famine and the possible need to relocate people. These conditions are discussed at the end of this chapter.

Increased temperatures in high latitudes melt permafrost, surface material that is normally permanently frozen below a certain depth. Because these ecosystems contain large amounts of organic material, decay processes are accelerated, which release trapped methane into the atmosphere. This methane then becomes part of the greenhouse gases.

Long-Term Climate Changes

Variations in climate have occurred since the formation of the atmosphere several billion years ago. In addition to the role the oceans play in these variations, continents also contribute in less obvious ways. The positions of the continents have changed drastically through geologic time. The location of these large land masses has a direct effect on large-scale circulation patterns in the oceans. Today, many ocean currents flow near the edges of continents, moving massive amounts of water and heat across Earth's surface. Near the end of the Permian Period about 250 million years ago, one large land mass existed (called Pangaea). As it began to break up, the earliest shapes of the present-day continents began to form. This break-up altered oceanic circulation. Today's configuration of the continents allows water to move readily between the poles and the open oceans (**Figure** 8.7).

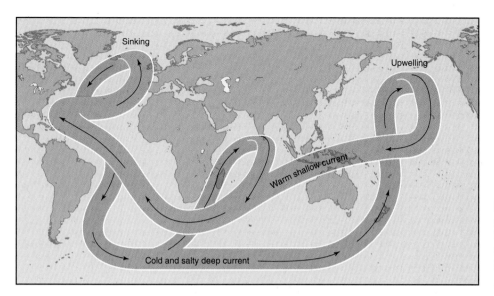

Figure 8.7 Movement of waters in the ocean can be thought of as a conveyor belt transporting deep, cold water to the surface where the warm, shallow water moves near the surface.

The rearrangement of the continents also puts the land masses in different locations. Positioned near the equator, an area would experience warmer temperatures and more precipitation than a land mass situated at a high latitude. Think of portions of central Africa and southern and central Alaska today. The warm equatorial regions of Africa have a wide range of life forms, while the harsh, colder climate of Alaska limits the varieties of plants and animals that can survive there (**Figure 8.8**).

Short-Term Climate Changes

The cause of short-term changes in Earth's climate is generally controlled by the amount of solar radiation received by Earth. Both atmospheric and astronomical factors contribute to these changes. Atmospheric factors include the amount of greenhouse gases and the amount of dust and other aerosols in the air. Researchers have examined ice cores from the polar regions and found that periods of increased glaciation correspond to increased particulates in the ice core. The additional atmospheric dust reflected solar energy back into space, thereby producing a drop in near-surface temperatures. As discussed previously, short-lived decreases in global temperatures can result from large volcanic eruptions throwing aerosols into the air.

Astronomical influences are mainly related to Earth's position in space in terms of the Sun. Variations in Earth's orbit around the Sun produce changes in the distance between the two bodies. Eccentricity is a measure of how elliptical a path is. As Earth revolves around the Sun, its path changes due to the gravitation attraction of the Sun, moon, and other more distance objects. When the orbit become more circular, the distance between the Earth and the Sun is less than when the path is very stretched (**Figure 8.9**). These variations occur over periods ranging from about 100,000 years to 400,000 years.

As the Earth rotates on its axis, there are times when the angle of tilt of the axis changes. Currently Earth is tilted 23.5° from the vertical. The period of change is roughly 41,000 years. As Earth rotates, it also tends to wobble on its axis, in the same way a spinning top begins to wobble as its rotation rate decreases. This wobble, also termed **precession**, has a period of about 23,000 years. The net effect of all these astronomic factors occurring together is that their period, together with their maximum influences, corresponds to times when active glaciations and lower glacial activity happened. Major changes occurred in the amount of solar radiation striking Earth.

National Park Service.

Figure 8.8 Terminus of Hubbard Glacier, Alaska.

precession
The wobble that occurs when a spinning object slows down.

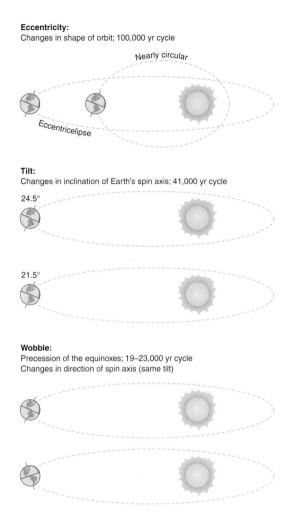

Figure 8.9 The eccentricity, tilt, and wobble of the Earth as it moves around the sun contribute to long-term changes in the climate.

The Coriolis Effect

Coriolis Effect
An imaginary force that appears to be exerted on an object moving within a rotating system. The apparent force is simply the acceleration of the object caused by the rotation. Along the equator, there will be no such rotation.

Earth rotates on its axis roughly once every 24 hours. This rotation produces an effect on the movement of fluids on or near the surface. The effect is termed the **Coriolis Effect**, after an early nineteenth-century French mathematician who proposed its existence. The effect is that any moving object in the northern hemisphere moves to the right (clockwise) and an object in the southern hemisphere moves to the left (counterclockwise) (**Figure 8.10**). Interestingly, the effect is negligible at the equator. The magnitude of the deflection increases toward the poles. Coriolis Effect helps explain why the paths of cyclonic storms tend to curve to the right in the northern hemisphere (refer to Chapter 11).

Atmospheric Circulation

air mass
A large body of air of considerable depth which are approximately homogeneous horizontally. At the same level, the air has nearly uniform physical properties, especially temperature and moisture.

The atmosphere is a fluid and thus moves readily. Several factors produce movement in the troposphere. Differential heating of the surface of the Earth and the overlying air generate warm and cool **air masses** that then move across the surface. The upward motion of warm air into the upper troposphere is caused by convection. The equatorial regions receive the greatest amount of radiation (**Figure 8.11**). Rising, warm air moves toward the poles in the upper portion of the troposphere. This air contains large amounts of water vapor derived from the evaporation of ocean waters near the equator. The area near the equator, where warm air is rising, is one of low pressure because of the suction effect of the rising air.

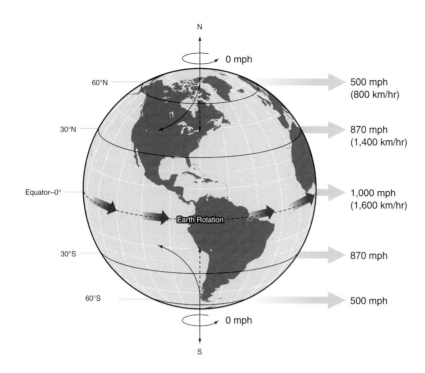

Figure 8.10 The heavy curved lines in each hemisphere show the Coriolis Effect experienced by a moving object in those regions.

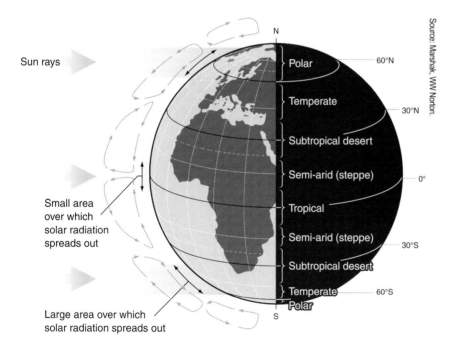

Source: Marshak, WW Norton.

Figure 8.11 Large amounts of solar radiation near the equator cause heating of the surface and air. Rising air then moves poleward where it descends in the subtropical regions near 30 degrees north or south of the equator.

Colder temperatures in the upper troposphere cause condensation of this water vapor, which returns to Earth as precipitation. The air in the upper troposphere is now cold and depleted of moisture, so it begins to sink in a region near 30° to 35° north or south of the equator. This region is called the subtropical high-pressure zone. High pressure exists in those areas where the cooler, dry air is descending back to the surface. As the air descends, it is compressed and begins to heat up, so the surface air is warm and dry. These conditions form many of the mid-latitude deserts on Earth, such as the deserts of North America, the Middle East, and Saudi Arabia in the northern hemisphere, and in Australia in the southern hemisphere. These descending winds are also deflected by the Coriolis Effect and produce the prevailing westerlies.

Air Masses

Large-scale movement of the atmosphere often involves expansive air masses. North America is affected by these masses, as they often move down from regions around the North Pole or off the Pacific Ocean. Polar air masses are typically cold and dry, while those coming off the Pacific and Gulf of Mexico are warm and moist. General movement is from west and northwest to east and southeast. These directions are driven by Coriolis forces and the prevailing west-to-east **jet streams** that traverse the country. Occasionally these two different air masses will collide; the result is often very unstable weather conditions.

Low-Pressure Conditions

In an area where less dense air rises due to heating, the upward force generates an area of low pressure (similar to a vacuum cleaner). This lower pressure caused the air to move from areas of high pressure into the lower pressure area. In the northern hemisphere this rising air rotates in a counterclockwise manner (**Figure 8.12**). Because of the Coriolis Effect, low pressure rotates in a clockwise manner in the southern hemisphere. Low pressure systems are termed **cyclones**. This rotational system is very evident when we examine hurricanes and other cyclonic storms (see Chapter 11).

High-Pressure Conditions

Whenever air has been cooled, it becomes denser and sinks to the surface, thus producing a higher degree of pressure on the surface. The winds generated by the collapsing air move outward from the center in a spiral fashion. These areas of high pressure rotate in a clockwise fashion in the northern hemisphere (counterclockwise in the southern hemisphere) and are termed **anticyclones** (**Figure 8.13**).

Role of Land Masses on Global Circulation

The average elevation of Earth is 4,000 feet above sea level. An idealized circulation pattern for Earth does not take into account land masses (**Figure 8.14**). On several continents there are many mountain ranges that disrupt the flow of air in the

jet stream
A high-speed, meandering wind current, generally moving from a westerly direction at speeds often exceeding 400 kilometers (250 miles) per hour at altitudes of 15 to 25 kilometers (10 to 15 miles).

cyclone
An area of low atmospheric pressure having is counterclockwise circulation in the northern hemisphere and clockwise motion in the southern hemisphere.

anticyclone
An area of high atmospheric pressure having clockwise circulation in the northern hemisphere and counterclockwise motion in the southern hemisphere.

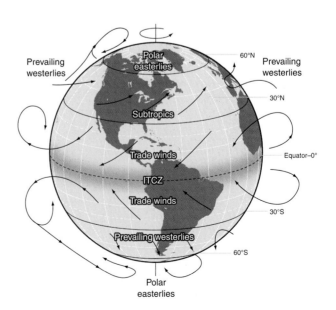

Figure 8.12 Circulation of Earth's atmosphere consists of a series of belts of air that produce the trade winds, westerlies, and polar easterlies.

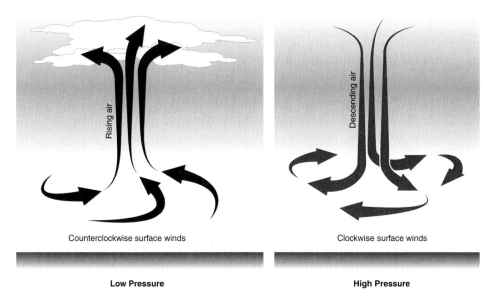

Figure 8.13 Rising air produces a low pressure condition on the surface in the northern hemisphere, producing a counterclockwise rotation; high pressure produced by colder, descending air generates clockwise winds at the surface.

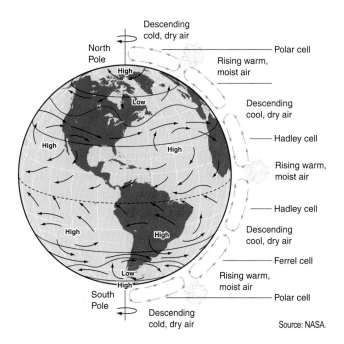

Figure 8.14 Hadley cells move warm, moist air from the equator to about 30° north and south latitudes, where it descends and produces high pressure. Ferrel cells lie between Hadley cells and Polar cells and move warm air to higher latitudes and shift cold air toward the subtropics.

atmosphere. The long, relatively linear stretches of mountains such the Rocky Mountains of North America and the Andes Mountains of South American stretch for thousands of miles in a north-south direction. These impedances cause the flow of air to be altered, and thus change the weather and climate associated with the theoretical flow patterns. The Himalaya Mountains of Asia are the highest on Earth, reaching 8,850 m above sea level. Air that strikes these peaks is driven higher into the upper troposphere where the moisture is concentrated and returned to Earth as snow. These features also disrupt the normal flow of air around the globe.

Role of Water in Global Climate

Water can exist in one of three states (or phases)—as a liquid, as a solid (ice), or as a gas (water vapor). When water changes from one state to another, heat is either absorbed or released, as seen in **Figure 8.15**. The amount of heat, measured in calories, needed to change 1 gram of water from one phase to another can be measured. For

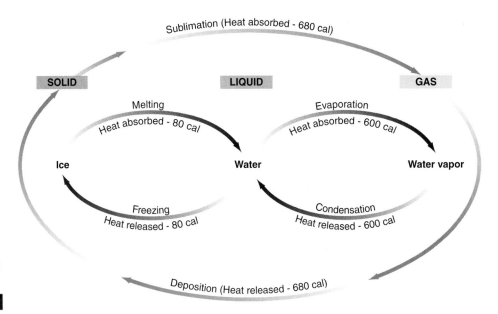

Figure 8.15 Heat is absorbed as water changes from a solid to a liquid and then to a gas. Heat is given off when it changes from a gas to a liquid and then a solid. The number of calories shown are those needed to change the state for 1 gram of water.

latent heat of vaporization
Heat stored in water vapor as as it changes states from a liquid to a vapor.

latent heat of fusion
Heat released when water freezes to form ice.

sublimation
The process that changes a solid into a gas, bypassing the liquid phase.

latent heat of condensation
Heat released when water vapor absorbs heat to be transformed to water.

water to be transformed from a liquid to a gas, the liquid must absorb 600 calories to evaporate. This is termed the **latent heat of vaporization**. When liquid water freezes, heat is released (80 cal per gram). This is the **latent heat of fusion**. In a case where the liquid phase is bypassed (the transformation of ice directly into water vapor), 680 calories are absorbed. In this instance, the process is termed **sublimation**. The **latent heat of condensation** (600 cal per gram) is associated with the change from water vapor to a liquid. This is a primary source of energy in hurricanes, as countless billions of grams of water in the hurricane, when water vapor at high altitudes condenses to form water (as rain). This heat adds to unstable atmospheric conditions and also helps fuel the storm (refer to Chapter 11).

Ocean water and that in lakes and other surface features cover about 75 percent of Earth's surface. Of all commonly occurring substances, water has one of the highest measures of heat capacity. This characteristic means that water requires a large amount of heat in order to increase its temperature . The equatorial regions of Earth receive the greatest amount of solar radiation, so this heat can be stored by the oceans. Evaporation is most effective at latitudes at or near the equator. These processes that absorb and release heat are key to driving the convection process in the atmosphere and the oceans.

Fronts and Mid-Latitude Cyclones

Whenever two different air masses collide with one another, they generally do not mix. A cold air mass with little moisture in it will not intermix with a warm, moisture-laden air mass. In North America cold air masses descend from the North Pole regions toward the equator. As these cold air masses pass through the mid-latitudes (30° to 45° north), they often collide with warm air masses that have moved northward from the Gulf of Mexico or off the western edge of the Atlantic Ocean. When a cold air mass collides with warm, moist air, the cold air pushes the warm air to higher altitudes. Condensation takes place as the moist air encounters colder temperature (**Figure 8.16a**). A line of high rising, vertical clouds results.

When warm air collides with cold air, the warm air rises above the denser, cold air and pushes out along a long surface. The warmer air is at a higher altitude so condensation occurs. This elongated string of clouds produces cloudy conditions that extend over hundreds of kilometers (**Figure 8.16b**).

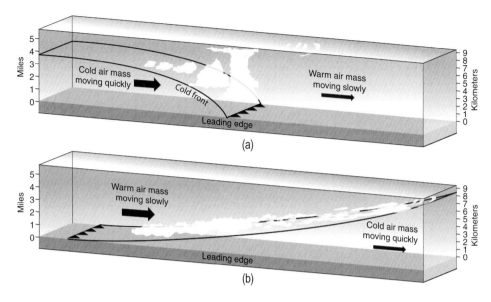

Figure 8.16 (a) The leading edge of a cold air mass force the warmer air upward, causing condensation to produce clouds, (b) when a warm air mass overrides a cold air mass, the warmer air is spread out along a long distance, producing far reaching clouds.

Elements of Hazardous Weather

The collision of air masses which have very different characteristics often creates hazardous or severe weather. These weather events are often relatively short-lived in duration, but can be very destructive to property and can cause a significant loss of life if populated areas are struck. Most violent weather begins as picturesque cumulus clouds that eventually are pushed higher in the atmosphere, where their dynamics change (**Figure 8.17**).

Thunderstorms

The conditions that must exist for a **thunderstorm** to develop include the heating and rising of air to higher altitudes where moisture will condense upon cooling, and the development of electrical charges that generate **lightning**. The generation of lightning through the atmosphere superheats the air and produces thunder, a tell-tale sign of such storms. One rule of thumb to remember is that once you see a bolt of lightning, for each five seconds you count, the lightning is one mile away (the sound is traveling at roughly five seconds per mile). Three stages are in the life cycle of a thunderstorm: the developing or towering cumulus stage, the mature stage, and the dissipating stage.

The developing stage involves the formation of cumulus clouds (puffy clouds that resemble large cotton balls or heads of cauliflower) that are pushed upward by an **updraft** of warm air (**Figure 8.18**). Little or no rain forms as the clouds are coalescing at elevations of five to seven kilometers.

A continuation of the updraft pushes moisture to higher altitudes where condensation takes place and precipitation begins to fall to the surface. This combination of updraft and downdraft generates shear within the thunderstorm. The storm now has reached the mature stage, when the storm becomes its most violent. Strong downdrafts can generate high winds; hail and heavy rain can fall, and intense lightning can develop. If sufficient rotation exists within the storm, tornadoes can be spawned.

When the amount of downdraft exceeds the rising updraft, the storm reaches the dissipation stage. Descending cold air intercepts warm air near the base and prevents the warm air from rising to fuel the storm. Precipitation is in its final stages, although lightning can still be a threat.

Figure 8.17 Cumulus clouds are the early stage of thunderstorm development. As more updraft moves moisture to high altitudes, the storm enters the mature stage.

thunderstorm
A local storm produced by a cumulonimbus cloud and accompanied by lightning and thunder.

lightning
A visible electrical discharge produced by a thunderstorm. The discharge may occur within or between clouds, between the cloud and air, between a cloud and the ground or between the ground and a cloud.

updraft
A small-scale current of rising air. If the air is sufficiently moist, then the moisture condenses to become a cumulus cloud or an individual tower of a towering cumulus.

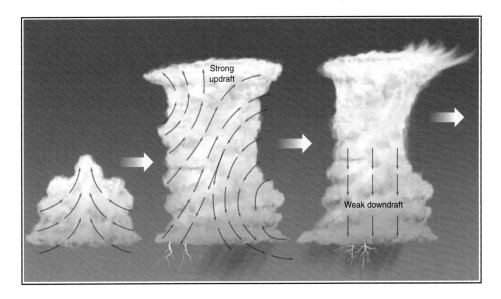

Figure 8.18 Stages of development of a thunderstorm. The developmental or towering cumulus stage begins as warm, moist air rises. The mature stage is characterized by updrafts and downdrafts. The dissipation stage occurs when the upward movement of air has ceased.

Thunderstorms can take on several different appearances. An isolated or single cell storm is often a short-lived event, lasting maybe 20 or 30 minutes. Their isolated nature generally prevents them from being very severe as they do not have sufficient energy to become very large.

The most common occurrence of thunderstorms is as a multicell cluster. Several storm cells move along as a unit, with each cell representing a different phase in the life cycle. This configuration typically has a mature cell near the center of the cluster with dissipating cells on the downwind edge of the cluster.

If a long line of storms develops having a continuous well-developed gust front at the leading edge, then a squall line forms. These lines are known for having strong downdrafts ahead of the line, with large hail and heavy rainfall accompanying the winds.

The most intense of thunderstorms is the supercell, which is highly organized. Although rare, they can be a major threat. Supercells are characterized by having extremely strong updraft (estimated wind velocities of 240 to 280 kph) and a significant amount of rotation. Hail can exceed 5 cm in diameter due to the extreme updrafts pushing the moisture repeatedly upward. Violent downdrafts are common, and strong to violent tornadoes form. The extreme updrafts prevent precipitation from falling through the center of the storm.

The National Oceanic and Atmospheric Administration reports that the typical thunderstorm is 30 kilometers in diameter, and lasts about 30 minutes. Worldwide, about 2,000 thunderstorms are occurring at any given moment. In the United States approximately 100,000 thunderstorms occur each year with roughly ten percent of them being classified as severe. Thunderstorms can occur during any month, but the largest number occur during the summer months, due to the increased heating of the ocean and atmosphere, particularly in the Gulf of Mexico. All states experience these storms but the largest number occurs in Florida, where thunderstorms take place an average of 80 to 100 days per year (**Figure 8.19**). Unfortunately Florida also leads the nation in the number of deaths related to these storms. The high number of thunderstorms that occur in Colorado and northern New Mexico are due to the collision of warm and cold air masses along the Front Range of the Rocky Mountains. Note that California has very few thunderstorms because there is very little cold air in the atmosphere.

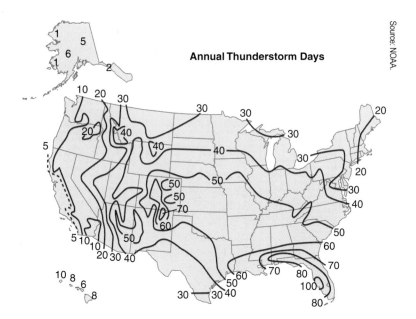

Source: NOAA.

Annual Thunderstorm Days

Figure 8.19 The average number of days that thunderstorms occur in the United States.

Hail

As thunderstorms move air vertically in the updrafts, moisture is being pushed to higher altitudes, where it freezes. Most hail takes on a spherical shape as the moisture droplets become larger as more water is frozen to its surface due to being repeatedly being moved to higher altitudes. At some point, the mass of the frozen particle is too much for the updraft to hold and the particle falls to Earth as **hail** (**Figure 8.20**). Although grapefruit-size hail has been reported in storms in the Midwest, hail seldom kills people (most people have sought refuge indoors). However, large hail can destroy agricultural crops and can damage roofs and automobiles.

What Causes Lightning?

Lightning has existed on Earth for billions of years. Early storms and volcanic eruptions produced lightning strikes. Once vegetation began to flourish about 400 million years ago, fires were ignited that burned unchecked until the fuel source was exhausted. Benjamin Franklin was fascinated by storms and lightning. In June 1752 he conducted experiments with his kite and discovered that lightning was a form of electricity. Soon after that he developed the lightning rod that was placed atop many buildings of the period. Lightning is a natural phenomenon that is present in large hurricanes, volcanic eruptions, extremely intense wildfires, heavy snowstorms, and (most commonly) thunderstorms.

The exact cause of lightning is not known because we cannot easily observe the processes that create the electrical charges within a storm cloud. Two schools of thought debate the formation of lightning. The so-called precipitation theorists suggest that as a cloud grows, two different sized particles form, and then collide, break apart, and separate. The smaller particles are driven higher in the cloud by updraft winds while gravity pulls the larger particles downward. The positively-charged smaller particles collect near the tops of the clouds, whereas the negatively-charged particles congregate at the cloud base (**Figure 8.21**).

© Jack Dagley Photography, 2010. Under license from Shutterstock, Inc.

Figure 8.20 Hailstones the size of golf balls can produce a great deal of damage, especially to agricultural crops.

hail
Solid, spherical ice precipitation that has resulted from repeated cycling through the freezing level within a cumulonimbus cloud.

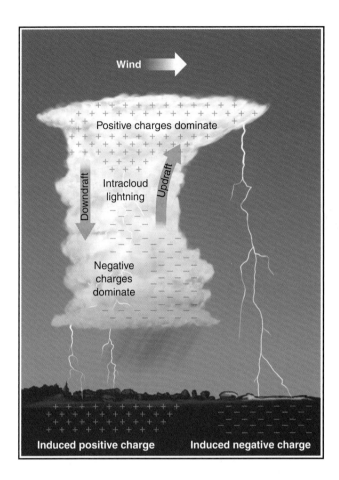

Figure 8.21 Electrical charges in a thundercloud are separated with positive charges in the top and negative charges toward the bottom. Once the energy is strong enough, the opposite charge attract and connect to create lightning.

Figure 8.22 Lightning is a common feature of thunderstorms.

A second group of theorists—the convectionists—explain the collection of positive particles in the upper cloud by convection having carried them upward. Down-drafts of colder air push the negatively charged particles to the lower portion of a cloud. Separation of these charges by several thousand or tens of thousands of feet produces an extremely large electrical potential within the cloud and between the cloud and the ground. Several million volts of electrical energy exist and at some instant the electrical resistance in the atmosphere is overcome as a flash signals the beginning of the bold of lightning (**Figure 8.22**).

On Earth's surface underneath a thunderstorm, positive charges collect and move along the surface with the storm. Eventually the electrical force between the negatively-charged cloud base and the positively-charged ground surface creates lightning. The forces within a storm tend to be much stronger, explaining why about 75 to 80 percent of lightning occurs within storm clouds.

Types of Lightning

Ground flashes involve the ground or something attached to the ground being hit. Natural flashes are triggered from the cloud, and strike the positively-charged surface of the Earth. Artificially-triggered lightning includes strikes to airplanes, rocks, and tall buildings and towers. The lightning moves from the ground upward to the cloud. Lightning bolts travel about 100,000 kilometers per second and have an estimated width of 1 to 2 cm.

Cloud flashes involve lightning that travels between clouds without striking the surface. Either cloud-to-air or sheet lightning illuminates the sky but does not do any damage to the surface.

Following a bold of lightning, thunder travels out at the speed of sound, roughly 1236 km per hour. The sound is the result of rapidly expanding gases that were superheated by the lightning. Air adjacent to a bolt is heated to 10,000°C. This superheated air expands and creates the rumbling sound that can be used to estimate the distance to the lightning strike. Thunder is generally heard within 25 km of a storm. Sound travels roughly 343 m per second. Begin counting when you see a lightning flash and count until you hear the thunder, then multiply your number of seconds by 350 m per second to get the approximate distance to the lightning.

Effects of Lightning on Humans

Lightning can be deadly! Each year an average of 60 people die in the United States and about 300 more are injured by lightning strikes. A study completed by the National Weather Service and the National Severe Storms Laboratory in the mid 1990s found that men accounted for 84 percent of the 3,239 deaths and 9,818 injuries caused by lightning between 1959 and 1994. Only flash floods and river floods caused more weather-related deaths during that period. A study carried out in France found that only 20 percent of lightning victims died immediately upon being struck.

Property damage caused by lightning strikes increased substantially over the 35 years of the NWS/NSSL study. Most of the increase was due to population increases and new construction. The report, entitled "Demographic of United States Lightning Casualties and Damages from 1959 to 1994," by Holle and Lopez, described almost 20,000 property damage reports due to lightning strikes. Pennsylvania had the highest number of damage reports, while the highest rates of damage reports corrected for population differences were in North Dakota and Oklahoma.

In terms of casualties, Florida has twice as many lightning deaths and injuries as any other state, and ranks first among the states that have lightning casualties (**Figure 8.23**). Nationwide the greatest number of deaths and injuries occurred between noon and 4 p.m. (local standard time). Sunday had 24 percent more deaths than any other day; Wednesday was second (these are very popular golfing and fishing days). The worst month was July, when thunderstorms are most common across much of the country.

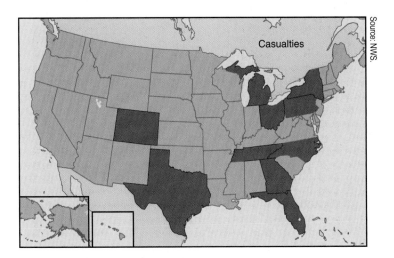

Figure 8.23 Of the ten states with the highest number of lightning casualties (deaths and injuries combined), Florida has the highest number.

In what location were most people struck?

- Walking in an open field
- Swimming
- Holding a metal object (golf club, fishing pole, umbrella)

Obviously lightning is a naturally occurring event to be taken seriously. If you are in the vicinity of a thunderstorm and lightning is being generated, seek shelter in order to reduce your exposure to the elements. Documented cases exist of lightning striking up to 15 km from a thunderstorm. If you can hear thunder, you are potentially vulnerable to being struck.

Tornadoes

tornado
A rotating column of air usually accompanied by a funnel-shaped downward extension of a cumulonimbus cloud and having a vortex several hundred yards in diameter whirling destructively at speeds of up to 600 kilometers per hour (350 miles per hour).

The intense atmospheric dynamics of a thunderstorm can produce a **tornado**, a rapidly rotating column of air that ranks as the most violent type of naturally occurring weather condition. These funnel-shaped clouds, which extend down from the base of severe thunderstorms, may or may not make contact with the ground (**Figure** 8.24). The appearance of a tornado ranges from light gray in color (containing mainly moisture droplets) to pitch black (one with a high degree of dust, dirt, and debris). The siphoning effect of the updrafts carries material high into the atmosphere and spreads it across the landscape as the tornado moves along.

How Do Tornadoes Form?

The most common cause of tornadoes is a supercell thunderstorm. These are highly organized, extremely intense storms that derive their energy from the strong updrafts that rotate and become tilted. Lasting more than one hour, these supercells contain

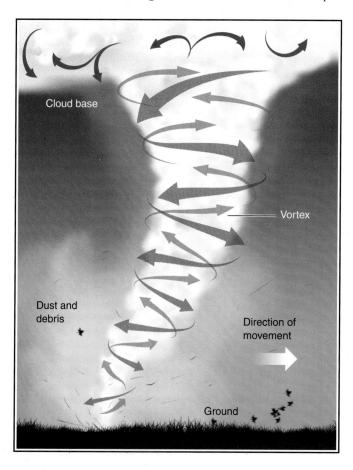

Figure 8.24 The rotational vortex of a tornado lifts dust and debris from the surface high into the funnel cloud.

vertical air currents that stretch 15 km in diameter and reach altitudes of 15,000 m. Rotation within the cloud system produces a mesocyclone which can be seen on Doppler radar. A tornado represents a very small extension of the larger roational cell. Because of the large size of a supercell, most large and violent tornadoes are formed from supercells.

Occasionally tornadoes can develop without a rotating updraft. Shearing forces can generate a whirl of debris or dust, sometimes along the front edge of a storm that has wind gusts on its edges. These tornadoes are often small in size and the winds are not very violent or long-lived.

Researchers do not fully understand how tornadoes form, in spite of years of lab research and observations in the field. A supercell must exist and have a well-defined rotating updraft. Winds traveling in two opposite directions at different altitudes can produce a shear force (similar to placing your flat palms together and sliding them in opposite directions). This shearing force produces a horizontal rotation that gets transformed into a vertical rotation by updrafts in the storm. Warm, moist air near the surface is lifted and cooled. The temperature differences in the air at altitude adds to the stresses at work.

Fewer than 20 percent of supercells spawn tornadoes. Scientists are undertaking large-scale studies to learn more about these phenomena. In 2009 and 2010 Project Vortex 2, a collaborative endeavor involving governmental agencies and universities, is examining the conditions that form tornadoes in the Midwest. The study will, in their terms, be "fully nomadic," as it will roam across several states in search of severe weather outbreaks.

Once a tornado forms, the rotational vortex spins in a counterclockwise direction (less than 1 percent rotates clockwise). Most tornadoes form at the trailing end of a thunderstorm, stretching down from the cloud base. Tornadoes move along with the thunderstorm at velocities ranging from almost nothing to 100 km per hour. The rotational speed of winds can reach more than 500 km per hour. As tornadoes stretch down they can lift themselves up so that they appear to skip along the surface.

Tornado Intensity

Wind velocity values are used to classify the intensity of tornadoes. These velocities are either measured directly or extrapolated from the damage the storms produce. Twenty-eight different measures go into the determination and assignment of the value for a given tornado. The Enhanced Fujita Scale is used to assign a value to a given tornado (Table 8.1). The scale was originally set up in 1971 by Professor Ted Fujita,

TABLE 8.1	The Enhanced Fujita Tornado Scale				
	Fujita Scale			Operational EF-Scale	
F Number	Fastest 1/4-mile (mph)	3 Second Gust (mph)	EF Number	3 Second Gust (mph)	
0	40–72	45–78	0	65–85	
1	73–112	79–117	1	86–110	
2	113–157	118–161	2	111–135	
3	158–207	162–209	3	136–165	
4	208–260	210–261	4	166–200	
5	261–318	262–317	5	Over 200	

Source: http://www.ncdc.noaa.gov/oa/satellite/satelliteseye/educational/fujita.html

a world-renowned researcher who worked at the University of Chicago. His research into severe storms and weather led to the discovery of phenomena such as microbursts and downbursts, sudden violent blast of air that produces damaging results.

Frequency of Tornado Occurrences

Tornadoes occur everywhere across the world. However, the highest number of tornadoes occur in the United States, where all fifty states have experienced a tornado at some point in time. Because tornadoes are directly related to thunderstorms, they are more common in those regions struck by thunderstorms. Because clashes of cold and warm, moist air generate unstable atmospheric conditions, the Midwest has the greatest number of tornadoes. Warm, moist air from the Gulf of Mexico drawn up into the Midwest collides with cold air from Canada. The result creates severe thunderstorm and tornado conditions.

Since 1950, the number of tornadoes that have occurred in the United States has increased about seven-fold (**Figure 8.25**). One possible reason for this dramatic increase is the introduction of new technology that can more readily identify tornadoes in the atmosphere, even though they might not be spotted by humans. Another reason for the increase could be that the population of the country has become more spread out in areas that once had few, if any, people. Also there could be an effect produced by **climate change** that has created more warm and cold air masses that collide with each other.

The highest number of tornadoes occur during the spring and summer months, when clashing air masses of different temperatures and moisture content collide. Long weather fronts often spawn swarms of tornadoes rather isolated events. A megaswarm of tornadoes occurred on April 3 and 4, 1974 (**Figure 8.26**). Plots of the paths of the tornadoes showed the general northeast movement of tornadoes. This is normal for tornadoes in the United States, as they are driven by the overall weather pattern and westerlies. One-hundred forty-eight tornadoes struck 13 states in a 16-hour period. Included were six F-5 twisters. At the end 307 people lost their lives and more than 6,000 were injured. Property damage amounted to more than $600 million (in 1974 dollars). In 2008 there were a record 1,691 tornadoes that struck the United States, including 461 in the month of May.

climate change
The long-term fluctuations in temperature, precipitation, wind, and other aspects of the Earth's climate.

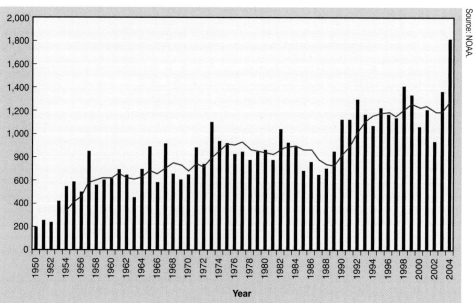

Figure 8.25 The number of tornadoes that have occurred in the United States has increased significantly since 1950.

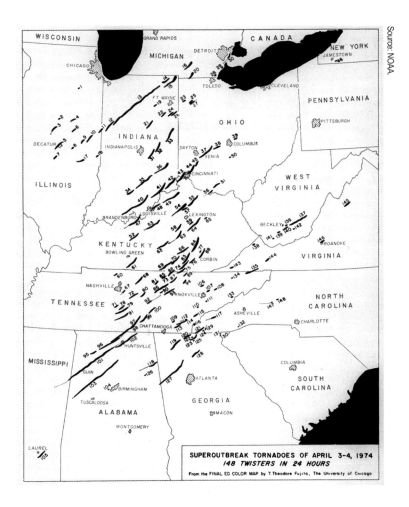

Figure 8.26 The super outbreak of tornadoes across the Mississippi Valley region on April 3 and 4, 1974, generated 148 tornadoes.

Tornado Damage

Tornado damage is caused by high wind velocities and a large difference in atmospheric pressure between the tornado and its surroundings. The rotating winds can destroy weaker structures, and the extremely low pressure inside the tornado generates strong pressure differences between the inside and outside of buildings. This pressure difference causes roofs to be lifted and removed. The high winds pick up smaller objects including small structures, animals, people, cars, and especially mobile homes, and can carry these objects up to several kilometers. The debris picked up by the winds becomes rapidly moving projectiles that can become lethal when hurled against a human body.

Homes in the Midwest have storm cellars that allow families to go underground to wait out the high winds. As seen in the opening image of this chapter, an EF5 tornado can turn a community into a flattened mass of rubble. Winds from a less violent EF3 storm can still have enough energy to strip trees of their leaves and throw building roofs into the cleared area (**Figure 8.27**).

Tornado Prediction and Warning

Tornadoes cannot be predicted with precision. However, when strong thunderstorm activity is detected, a **tornado watch** is generally issued for all areas that may fall in the path of the thunderstorm. **Doppler**

Figure 8.27 An EF3 tornado hit Mena, Arkansas on April 11, 2009. The tree was stripped of all its leaves and caught an aluminum roof from a nearby building.

tornado watch
This is issued by the National Weather Service when conditions are favorable for the development of tornadoes in and close to the watch area. Their size can vary depending on the weather situation.

Doppler radar
Radar that can measure radial velocity, the instantaneous component of motion parallel to the radar beam (i.e., toward or away from the radar antenna).

tornado warning
This is issued when a tornado is indicated by the WSR-88D radar or sighted by spotters; therefore, people in the affected area should seek safe shelter immediately.

radar can detect rotating motion within a thunderstorm and when this is detected, or a tornado is actually observed, a **tornado warning** is issued for all areas that may fall in the path of the thunderstorm. Tornado safety (from the Federal Emergency Management Agency, FEMA) includes the following.

If at home:

- Go at once to the basement, storm cellar, or the lowest level of the building.
- If there is no basement, go to an inner hallway or a smaller inner room without windows, such as a bathroom or closet.
- Get away from the windows.
- Go to the center of the room. Stay away from corners, because they tend to attract debris.
- Get under a piece of sturdy furniture such as a workbench or heavy table or desk and hold onto it. Use arms to protect head and neck.
- If in a mobile home, get out and find shelter elsewhere.

If at work or school:

- Go to the basement or to an inside hallway at the lowest level.
- Avoid places with wide-span roofs such as auditoriums, cafeterias, large hallways, or shopping malls.
- Get under a piece of sturdy furniture such as a workbench or heavy table or desk and hold onto it. Use arms to protect head and neck.

If outdoors:

- If possible, get inside a building.
- If shelter is not available or there is no time to get indoors, lie in a ditch or low-lying area or crouch near a strong building. Be aware of the potential for flooding. Use arms to protect head and neck.

If in a car:

- Never try to out-drive a tornado in a car or truck. Tornadoes can change direction quickly and can lift up a car or truck and toss it through the air.
- Get out of the car immediately and take shelter in a nearby building. If there is no time to get indoors, get out of the car and lie in a ditch or low-lying area, away from the vehicle. Be aware of the potential for flooding.

After the tornado:

- Help injured or trapped persons. Give first aid when appropriate. Don't try to move the seriously injured unless they are in immediate danger of further injury. Call for help.
- Turn on radio or television to get the latest emergency information.
- Stay out of damaged buildings. Return home only when authorities say it is safe.
- Use the telephone only for emergency calls.
- Clean up spilled medicines, bleaches, gasoline or other flammable liquids immediately.
- Leave the building if you smell gas or chemical fumes.
- Take pictures of the damage—both to the house and its contents—for insurance purposes.
- Remember to help your neighbors who may require special assistance—infants, the elderly, and people with disabilities.

Inspecting utilities in a damaged home:

- Check for gas leaks—If you smell gas or hear a blowing or hissing noise, open a window and quickly leave the building. Turn off the gas at the outside main valve if you can and call the gas company from a neighbor's home.
- If you turn off the gas for any reason, it must be turned back on by a professional.

Mitigation of Potential Tornado Damage

Tornadoes can occur wherever thunderstorms occur—basically anywhere. Therefore it is financially impossible to prepare every possible locality for a tornado. The best preparation is to ensure that people are aware of warning systems and have taken some precaution, such as building a storm cellar if they live in an area that is likely to experience tornadoes. Building codes can also require more sturdy construction. However, if a structure is hit directly by an EF 5 tornado it will certainly be severely damaged, if not totally destroyed. Warnings broadcast by governmental agencies must be timely, accurate, and heeded by those for whom they are sent.

Other Severe Weather Phenomena

Nor'easters

A **nor'easter** is an extratropical cyclonic storm that forms in a region away from the tropics. It has some of the characteristics of a tropical storm, winds that rotate in a counterclockwise direction, and a low pressure center. These storms that affect the east coast of the United States form off the south Atlantic coast. The counterclockwise winds come out of the northeast (hence the name). Because these usually occur in the fall and winter months, they are less intense than hurricanes because the colder ocean waters do not provide a large amount of thermal energy. However, damage can be extensive (**Figure 8.28**).

nor'easter
A strong low pressure system with winds from the northeast that affects the mid-Atlantic and New England states between September and April. These weather events are notorious for producing heavy snow, copious rainfall, and tremendous waves that crash onto Atlantic beaches, often causing beach erosion and structural damage.

Winter Blizzards and Severe Weather

Severe winter weather is often accompanied by high winds, significant snowfall, and cold temperatures. These conditions can produce a **blizzard**, which the National Weather Service Service defines as having sustained 35 mph (56 kph) winds that lead to blowing snow, and causes visibilities of ¼ mile or less, lasting for at least 3 hours (**Figure 8.29**). Although no specific temperatures are associated with a blizzard, the high winds often produce sub-zero wind chill conditions. These conditions form when a ridge of high-pressure interacts with a low-pressure system; this results in the horizontal movement of air from the high-pressure zone into the low pressure area.

blizzard
A severe winter storm that has the following conditions that are expected to prevail for a period of 3 hours or longer: sustained wind or frequent gusts to 35 miles an hour or greater and significant falling and/or blowing snow (i.e., reducing visibility frequently to less than ¼ mile).

Figure 8.28 A nor'easter hit New England in April 2009. Damage in Saco, Maine, was extensive along the shoreline.

Figure 8.29 A blizzard hits Denver, Colorado, in December 2006 with more than two feet of snow.

Other winter weather condition watches and warnings that are issued by the National Weather Service include:

- **Winter Storm Watch.** Conditions are favorable for hazardous winter weather conditions including heavy snow, blizzard conditions, or significant accumulations of freezing rain or sleet. These watches are issued by the Weather Service Forecast Office in Chicago and are usually issued 12 to 36 hours in advance of the event.
- **Winter Storm Warning.** Hazardous winter weather conditions that pose a threat to life and/or property are occurring, imminent, or likely. The generic term, winter storm warning, is used for a combination of two or more of the following winter weather events; heavy snow, freezing rain, sleet, and strong winds
- **Heavy Snow Warning.** Snowfall of 6 inches or more in 12 hours or less, or 8 inches or more in 24 hours or less.
- **Lake Effect Snow Warning.** Lake effect snowfall of 6 inches or more in 12 hours or less, or 8 inches or more in 24 hours or less. The source of moisture for these storms comes from large bodies of water, such as the Great Lakes. Michigan, northern Ohio and Pennsylvania, and western New York state are often hit with heavy snow that developed from coming from the Great Lakes.

Heat Waves

heat wave
A prolonged period of excessively hot weather and high humidity.

Several significant heat waves have occurred throughout the world over the past 30 years. A **heat wave** is a prolonged period of excessively hot weather, which may be accompanied by high humidity. There is no universal definition of a heat wave, as the term is relative to the usual weather in the area. Temperatures that people from a hotter climate consider normal can be termed a heat wave in a cooler area if they are outside the normal climate pattern for that area. Increased humidity adds to the effect of the elevated temperatures to create deadly conditions.

The term *heat wave* is applied both to routine weather variations and to extraordinary spells of heat which may occur only once a century. Severe, prolonged heat waves have caused catastrophic crop failures, thousands of deaths from hyperthermia, and widespread power outages due to increased use of air conditioning. NOAA reports that among the group of natural hazards that occur over continents, only the cold of winter—*not lightning, hurricanes, tornadoes, floods,* or *earthquakes*—takes a greater toll. In the 40-year period from 1936 through 1975, nearly 20,000 people were killed in the United States by the effects of heat and solar radiation. In the disastrous heat wave of 1980, more than 1,250 people died, mainly in the central and southern Plains.

In the summer of 2003 Europe was hit by a heat wave that killed more than 37,000. Much of the heat was concentrated in France, where nearly 15,000 people died. Elderly people were the most affected. In July 2006, the United States experienced a massive heat wave, and almost all parts of the country recorded temperatures above the average temperature for that time of year. Temperatures in some parts of South Dakota exceeded 115°F (46°C), causing many problems for the residents. Also, California experienced temperatures that were extraordinarily high, with records ranging from 100 to 130°F (38 to 54°C). On July 22, the County of Los Angeles recorded its highest temperature ever at 119°F (48.33°C).

When comparing the various types of weather-related disasters that have occurred in the United States in recent years, heat-related events have caused the largest loss of life (**Figure 8.30**). Over a longer time span, floods have produced the greatest loss of life. Heat takes its toll on people who might otherwise have pre-existing medical conditions that become worst with increased heat. People with weak heart and lung systems experience increased stress on their bodies, often leading to death.

Drought and Famine

Drought is an extended period lasting months or years when a region receives much less than normal precipitation, resulting in a significant shortage of water. Both surface water and groundwater supplies are drastically reduced. This shortage rapidly

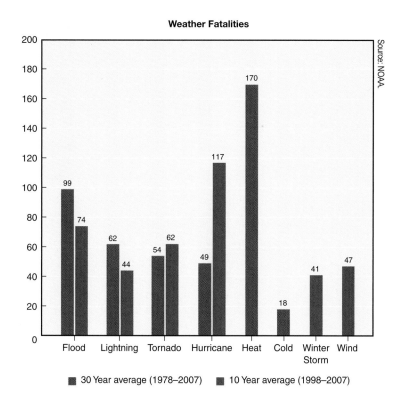

Figure 8.30 Deaths attributable to weather related events show a wide range of values.

reduces the ability of people to grow crops and to sustain life, and the ecosystem of the area is impacted to the point that it begins to deteriorate.

The onset of a drought is caused by changes in the upper-level air flow. In the United States it is associated with a prolonged ridge of high pressure that pushes dry air down to the surface. This sinking air is warmed by compression and begins to reduce the humidity even more. The dry air reduces moisture in the soil, reducing the ability to grow crops. In the heartland of the United States these conditions began to develop in the early 1930s and lasted for several years. The result was severe dust storms throughout the central part of the United States, giving the name "Dust Bowl" to the region. Widespread crop failures resulted in the malnutrition of hundreds of thousands of people, Numerous farms were abandoned and there was a mass migration to the West, especially California, where former farmers hoped to reestablish the way of life they had known previously. These events came on the heels of the Great Depression, making living conditions and survival very problematic for thousands of people.

Summary

Oceans cover 71 percent of Earth's surface and play a major role in the weather and climate. Shorter-term weather is an atmospheric condition, while climate is a longer-term condition that affects the hydrologic cycle of a large region on Earth or the entire globe. Solar radiation is the driving energy force for both weather and climate. Changes in the amount of solar radiation occur when there is an increase in greenhouse gases, which causes more heat to be held close to the surface. These gases result from the burning of fossil fuels, natural decay of organic substances, and volcanic activity.

Global warming can produce changes in vegetation patterns on Earth, more storm activity, changes in the

amount of sea ice and sea level, and alterations in the hydrologic cycle. Climatic changes are also controlled in the very long term by plate tectonics as the movements of plates change oceanic circulation patterns. Variations in Earth's orbit around the Sun and fluctuations in Earth's tilt on its axis also affect its climate.

Weather systems move in the troposphere, and include areas of low and high pressure. The movement of fronts and mid-latitude cyclones produces short-term weather conditions that are associated with the seasons. Hazardous weather conditions are generally short-lived. Thunderstorms and associated lightning generate conditions that can prove deadly. Tornadoes sometimes form when cold and warm air masses collide, a situation common to the central United States. This region is one that has experienced the greatest number of tornadoes on Earth and there has been a significant increase in activity during the past thirty years.

Other less frequent weather hazards include winter blizzards, heat waves, and drought-related famines. In terms of weather-related deaths, heat waves and flooding have killed the most people in the United States in the past 30 years.

References and Suggested Readings

Burt, C. C. 2004. *Extreme Weather: A guide and record book*. New York: W. W. Norton & Company.

Fagan, B. 2002.*The Great Warming*: New York : Bloomsbury Press.

Houghton, Sir J. 2004. *Global Warming: The Complete Briefing*. Cambridge, England: Cambridge University Press.

Mayewski, P. A. and F. White. 2002. *The Ice Chronicles*. Hanover, NH : University Press of New England.

Schmidt, G. and J. Wolfe. 2008. *Climate Change—Picturing the Science*. New York: W. W. Norton & Company.

Silver, J. 2008. *Global Warming and climate change demystified*. New York: McGraw Hill.

Uman, Martin. 1986. *All about Lightning*. New York: Dover Publications.

Williams, J. 1997. *USA Today: The Weather Book*: New York: Vintage Books.

Web Sites for Further Reference

http://www.climatecrisis.net/

http://www.ncdc.noaa.gov/oa/climate/severeweather/tornadoes.html#history

http:/www.ncdc.noaa.gov/paleo/milankovich.html

http://www.nssl.noaa.gov/users/brooks/public_html/tornado/

http://www.srh.noaa.gov/srh/jetstream/atmos/layers.htm

http://www.srh.noaa.gov/srh/ssd/html/heatwv.htm

http://www.stormfax.com/elnino.htm

http://www.wrh.noaa.gov/fgz/science/svrwx.php

Questions for Thought

1. Explain the difference between weather and climate.

2. What are the three layers of the atmosphere closest to Earth's surface, and what is one key characteristic of each?

3. What effects do volcanic eruptions have on the makeup of the atmosphere, and how can they alter solar radiation reaching Earth?

4. What is the main cause of long-term climate change, and explain how it changes the climate?

5. Distinguish between low pressure and high pressure conditions. What types of weather are associated with each?

6. Explain how changes in Earth's eccentricity affect global climate.

7. How does La Niña differ from El Niño?

8. How does hail form?

9. Why do most thunderstorms form in the central and southeast parts of the United States?

10. What are the general weather conditions associated with a drought?

Flooding and Streams

9

Heavy rains produce significant runoff. This road in the Midwest is under water from an intense summer thunderstorm.

© Tony Campbell, 2010. Under license from Shutterstock, Inc.

Key Terms

acre foot
discharge
drainage basin
floodplain
flood stage
hydrologic cycle
ice-jam flood
river
runoff
stream
tributary
watershed

All climates, especially arid ones, are prone to flooding. Every year, numerous floods affect millions of people throughout the world. Weather systems can stall over an area and generate copious amounts of rainfall, causing streams and rivers to overflow their banks when too much water enters a drainage basin. Rapid melting of snow and ice also create abnormal water discharge. Areas that are most affected tend to lie toward the lower portions of a river system, as the topography there is less steep and the increased downstream flow cannot readily move water away from the affected region.

Floods can also be the result of poor urban planning or dam construction that causes water to flow unexpectedly in places where it was not meant to go. Coastal regions become flooded when maritime storms land ashore. The strong winds of hurricanes, cyclones, and typhoons generate storm surges that bring massive amounts of water inland, flooding low-lying areas. Tsunami although relatively rare, rapidly push sea water past beach zones, inflicting severe damage to communities in the path of the waves. Floods can also result from volcanic eruptions that melt snow and ice atop a volcano, the condition that contributed to the lahars of Mount St. Helens (see Chapter 3).

The U.S. Geological Survey (USGS) reports that flooding in the United States annually causes between 140 and 160 deaths and an average of almost $4 billion in damages. In many low-lying areas, home and landowners cannot purchase flood insurance and the losses they might sustain are not covered. Often the lower socio-economic classes reside in these topographically lower locations and suffer immense losses in terms of property and human life. Although floods can be caused by storms, tsunamis, and volcanoes, this chapter will address flooding associated with streams.

Stream Processes

Earth's hydrosphere includes all water at or near the surface in addition to what is contained in the atmosphere. Water, which can be a liquid, a gas (water vapor), or a solid (ice), moves around in the **hydrologic cycle**—a continuous circulation of water around the globe. Solar energy drives this cycle; it stretches from the equator to the poles, where water movement is obviously slower but nevertheless part of the cycle.

Geologists define a **stream** as a body of water that flows within a confined channel on the surface. When several streams or **tributaries** join to produce a larger flowing body, the term **river** is then applied (**Figure 9.1**). Water that falls onto Earth's

hydrologic cycle
The cycle that moves water and water vapor among the oceans, land, and atmosphere through evaporation, condensation, precipitation, transpiration, and respiration.

stream
A body of water that flows downhill under the influence of gravity and lies within a defined channel.

tributary
A stream that flows into another larger stream.

river
A large stream.

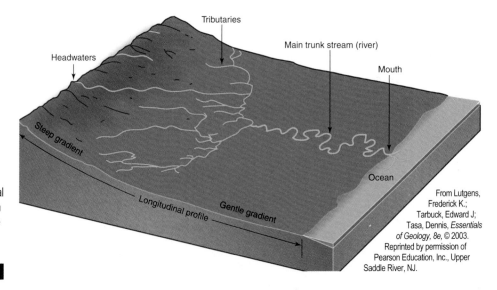

Figure 9.1 Streams flow from high to low elevations. In the upper portions, small streams form the tributary system to feed and create the trunk stream. The longitudinal profile shows the steep gradient, or slope, in the higher elevations becoming more gentle toward the mouth. Streams eventually feed into a sea or ocean.

From Lutgens, Frederick K.; Tarbuck, Edward J; Tasa, Dennis, *Essentials of Geology, 8e,* © 2003. Reprinted by permission of Pearson Education, Inc., Upper Saddle River, NJ.

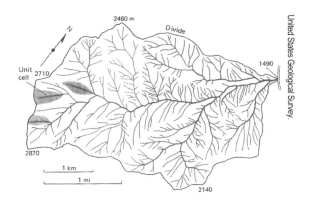

Figure 9.2 A watershed is outlined by its divide, which separates it from adjacent watersheds. Elevations are in meters. Notice how the stream patterns collect water and channel it into a larger stream downhill.

───────────────────■

surface can produce **runoff**, which travels along the surface and ends up in a channel, or it can infiltrate into the ground, where it temporarily resides in underground collection areas before eventually finding its way into a stream system. In arid climates, infiltrated water often evaporates and reenters the hydrologic cycle as water vapor.

runoff
Water that flows across a surface and into a stream or other body of water.

Drainage Basins

A **drainage basin** or **watershed** is an area on the surface that collects water that flows into a stream and forms a drainage pattern. Drainage divides, which are often ridges or a series of high points that divide downhill slopes, separate one watershed from its adjacent neighbors (**Figure 9.2**). The amount of water collected by a drainage system is dependent on the amount of precipitation, the size of the area, and the subsurface characteristics that control infiltration and runoff. As long as the amount of water flowing into streams and falling on the watershed is carried away by the existing system, water remains in the channels and does not present a problem.

Think of a stream configuration as the transportation system that is moving water to some end point, either a temporary one, such as a lake, or the ultimate end point—a sea or an ocean. A stream begins to cut a channel into the landscape, thereby creating the "highway" that is allowing water to move through an area. Each channel has a cross sectional view, in which we see the width and depth of the channel. This area, coupled with the length of a particular stream segment and its drop in elevation, defines the volume or how much water is contained (and moved) by the stream and at what velocity (**Figure 9.3**). **Discharge** is the volume of water that flows downstream past a given point in a given period of time. We measure this volume in cubic feet per second (cfs) or cubic meters per second (cms), and the amount can vary widely depending on weather, surface conditions, and stream characteristics. The world's largest rivers move enormous volumes of water each second (**Table 9.1**). All of these rivers have very large watersheds and many of the rivers flow through regions that have wet, temperature climates.

drainage basin
An area that drains water to a given point or feature, such as a lake.

watershed
See drainage basin.

discharge
The volume of water flowing through a stream channel in a given period of time, usually measured as cubic feet per second or cubic meters per second.

Floodplain

A stream channel is bordered by its banks and the area to either side, which is termed the **floodplain**. When water spills over the banks, it is no longer moving in its channel. As it spreads out, the velocity of the water decreases rapidly, causing any sediment carried by the stream to be deposited. This process is repeated every time the stream floods. The continual buildup of sediment along the banks creates natural levees that increase in height, deepening the stream channel. Repetition of this process permits the stream to carry more water than before, because its cross-sectional area has

floodplain
The flat area alongside a stream that becomes flooded when water exceeds the banks of the stream.

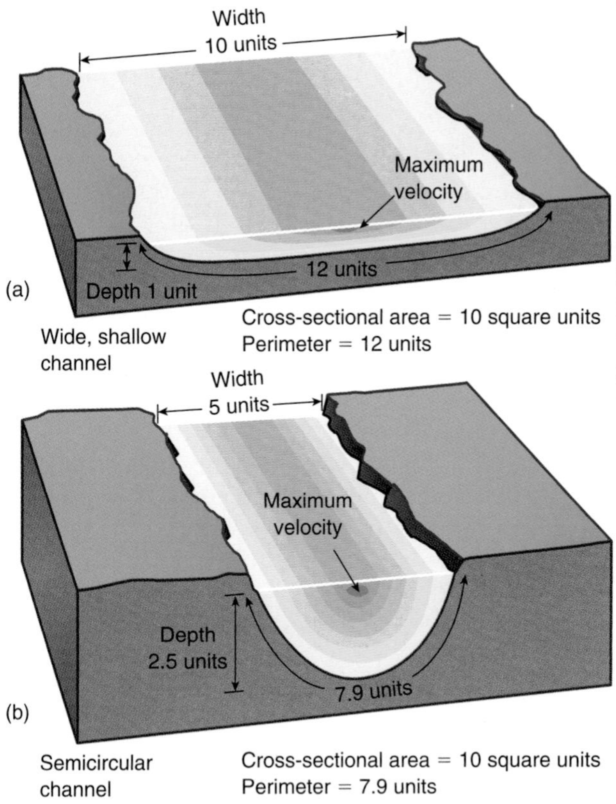

From Lutgens, Frederick K.; Tarbuck, Edward J.; Tasa, Dennis, *Essentials of Geology, 8e,* © 2003. Reprinted by permission of Pearson Education, Inc., Upper Saddle River, NJ.

(a)

Width
10 units

Maximum
velocity

12 units

Depth 1 unit

Wide, shallow
channel

Cross-sectional area = 10 square units
Perimeter = 12 units

Width
5 units

Maximum
velocity

(b)

Depth
2.5 units

7.9 units

Semicircular
channel

Cross-sectional area = 10 square units
Perimeter = 7.9 units

Figure 9.3 Notice how the shape of the channel dictates the velocity. (a) Water in a wide, shallow channel moves more slowly than in a narrow, deep one because of increased friction. (b) This stream has the same cross-sectional area as the one in (a) but there is less friction, hence a higher velocity.

TABLE 9.1	World's 10 Largest Rivers by Discharge	
River	Country	Average Discharge at Mouth (Cubic Feet per Second)
Amazon	Brazil	7,500,000
Congo	Congo	1,400,000
Yangtze	China	770,000
Brahmaputra	Bangladesh	700,000
Ganges	India	660,000
Yenisey	Russia	614,000
Mississippi	USA	611,000
Orinoco	Venezuela	600,000
Lena	Russia	547,000
Parana	Argentina	526,000

Source: http://www.waterencyclopedia.com/Re-St/Rivers-Major-World.html.

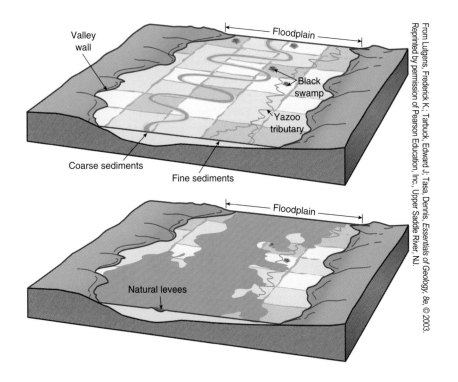

From Lutgens, Frederick K.; Tarbuck, Edward J.; Tasa, Dennis, Essentials of Geology, 8e. © 2003. Reprinted by permission of Pearson Education, Inc., Upper Saddle River, NJ.

Figure 9.4 The stream and its main channel lie within the floodplain, which is the area covered by water during flood stages.

FACT BOX	Equivalent Measures and Weights of Water at 4°C.
	1 gallon = 0.134 cubic feet = 8.35 lb (3.79 kg)
	1 cubic foot per second (cfs) = 7.48 gal/sec = 62.4 lbs/sec (28.3 kg/sec)
	= 449 gals/min

increased. When water spills over onto the floodplain, it does not tend to drain back into the main river channel because the natural levees now act as a dam. This water slowly moves downhill on the floodplain until it finds a site where it can flow back into the main stream or off the floodplain (**Figure 9.4**).

Streams provide water, transportation, food, irrigation, and soil to people living nearby. The finer silt carried by streams is deposited to create arable farmland, such as in the delta region of the Nile River in Egypt. Before the High Aswan Dam was built, these rich farmlands were frequently flooded, destroying crops, homes, and killing people, but also depositing valuable silt that enriched farming areas near the mouth of the river. Construction of dams lessened the occurrence of major floods, but the dams now hold back valuable silt, rendering many downstream regions unsuitable for growing crops.

Types of Floods
Regional River Floods

Flooding that is related to seasonal rains or snow melts, or a combination of rain falling on snow, produces a large volume of water that cannot be handled by existing stream systems. Such flooding is common in wetter climates that have generally larger, more established rivers. Floods can occur anywhere along the length of a stream. In the upstream regions, such floods are caused by intense rainfall or snow

(a)

Photo by James S. Best, by permission.

(b)

Photo by James S. Best, by permission.

Figure 9.5 The Reedy River in Greenville, South Carolina. (a) normal flow conditions in March; (b) following a June thunderstorm.

■

acre foot
The amount of water that cover one acre to a depth of one foot; equivalent to 325,851 gallons of water.

melt over a watershed that flows into smaller streams and tributaries. When several watersheds feed into a larger stream, the volume of water can be immense and produce widespread flooding.

These vast amounts of water can have an effect on the landscape. One foot of water covering one acre is termed an **acre foot**. This amounts to 325,851 gallons of water, so for every inch of rain falling on an acre, there are 27,154 gallons that can flow across the surface. If one acre foot of water moves into a river, it will contribute 43,560 cubic feet (1233 cu meters) of flow to the stream's volume. Generally, these intense rainfalls are so rapid that very little water percolates into the subsurface, producing stream flows that become rushing torrents (**Figure 9.5**). As the gradient of the stream flattens out and the downstream river channels become wider, the velocity drops and the water becomes calmer. However, the volume of water continues to increase, especially if more streams are present in the system.

Many watersheds consist of hundreds or thousands of acres, resulting in massive amounts of water moving downslope and downstream. The lateral and down cutting erosive power of the water causes dimensions of the stream channel to increase. Upstream regions generally have fairly steep slopes and the longitudinal profile of the stream shows a rapid drop in the elevation of the stream (refer to Figure 9.1). Therefore water will tend to move downslope quickly and generate large-scale flooding.

As higher elevation streams move water downstream, they join with other streams and increase the size of the trunk stream (in a fashion similar to the trunk of a deciduous tree having many branches that feed into the main trunk). If a widespread rain storm covers several different watersheds and feeds water into the trunk stream, the downstream region can experience flooding. Widespread saturation of the ground prevents water from percolating into soils, so the water must flow under the force of gravity to lower elevations. This downslope flow is not instantaneous, so there is often a lag that gives some warning to communities at the lower end of a drainage area. Downstream communities generally have some preventive measures in place in preparation for recurring floods. Increasing the height of levees and deepening the river channel help the river to handle a greater flow, thus preventing flooding. However, these measures are not always successful, particularly if there is a flow that greatly exceeds the normal amount. The result can be flooding so severe and widespread that the governor of the state requests presidential declaration of the region as a disaster area so that it can receive federal help (**Figure 9.6**). As we see from the map in Figure 9.6, very few regions in the United States are exempt from flooding over a period of several decades.

Other natural hazards play a role in changing the drainage regime of an area. The 1991 eruption of Mount Pinatubo, a volcano in the Philippine Islands, deposited massive amounts of ash across the countryside. Within a few months, copious rainfalls struck the region and moved the ash downslope as sediment. When flooding occurred in the streams, deposits were created along the banks and formed natural levees that increased the depth of the channels. After several episodes of this redistribution of the ash, streambeds were flowing at a level higher than the original surface. When the streams experienced later flooding, the water easily flowed into the lower lying areas adjoining the raised stream channels.

Flash Floods

Flash floods occur with little or no warning. They are usually caused by torrential rainfall that takes place over a very short time, often associated with severe, localized thunderstorms or from a series of storms continually soaking an area. Although

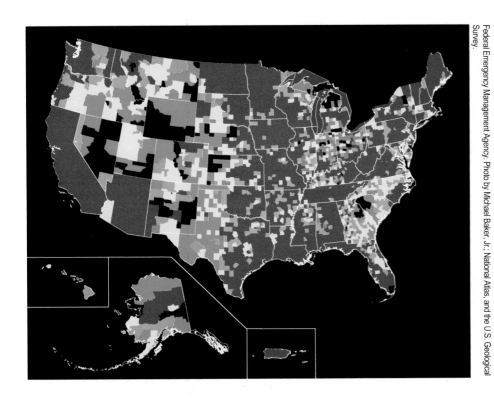

Federal Emergency Management Agency. Photo by Michael Baker, Jr.; National Atlas; and the U.S. Geological Survey.

Figure 9.6 Presidential disaster declarations related to flooding in the United States, shown by county: Green areas represent one declaration; yellow areas represent two declarations; orange areas represent three declarations; red areas represent four or more declarations between June 1, 1965, and June 1, 2003. Map not to scale.

flash floods can occur anywhere, they tend to be more devastating in two areas: (1) in mountainous areas, where steep slopes funnel water into narrow streams, and (2) in desert regions, where normally dry or low flow streambeds are quickly transformed into raging torrents. Rainfall does not have any chance of infiltrating into the subsurface and is often flowing in a rapid, sheet-wash manner across impervious surfaces to collect in dry stream beds. The intense nature of flash floods makes them capable of causing extensive damage and loss of life. When people drive their vehicles through normally dry washes and streambeds that are filled with fast-moving water, their vehicles begin to float, are pushed along by the flow, or are overturned, trapping the victims inside. The force with which fast-moving, sediment-laden water hits a surface is great (**Box 9.1**)

Winter Climate-Driven Floods

Ice-jam floods are a problem in regions where rivers freeze and then begin to thaw or receive surface water from rainfall or nearby melting. In the winter, some regions receive excessive snowfall, which rests on frozen ground. Rivers that normally drain the snowmelt freeze and cannot move any water downstream. Ice builds up whenever a small period of melting occurs and then refreezing takes place, thus making the drainage situation worse.

During the winter of 1996 and 1997, the watershed of the Red River of the North, which forms the state boundary of North Dakota and Minnesota, was besieged by excessive rainfall and a series of blizzards. Precipitation totals were more than three times normal and cold weather early in the winter froze the soil, preventing any percolation of water. In early April the region experienced record low temperatures along with another 10 to 12 inches (25 to 30 cm) of snow. Within a 10-day period, daytime temperatures swung from single digits to highs in the upper 50s (13–14°C). Rapid melting occurred that produced extensive flooding.

Because the Red River of the North flows north—one of few rivers in the United States to do so—the water was draining into an area where the river was still frozen.

ice-jam flood
A flood, usually in the spring, that results from broken pieces of river ice blocking the flow of a river, thereby flooding areas adjacent to the river.

| BOX 9.1 | How One State Deals with Not-So-Smart Drivers |

The State of Arizona has enacted the Stupid Motorist Law (Arizona Revised Statutes 28-910), which imposes a fine of up to $2,000 on drivers who have to be rescued from a flooded area. In spite of this law, people continue to drive around barricaded crossings and attempt to get through flooded roadways. A lack of adequate storm culverts causes water to flow across low points on streets and highways. The depth of water, even when it is flowing across a roadway that might be familiar to the driver, is very uncertain and the force of the water is much greater than one realizes.

DO NOT DRIVE THROUGH FLOODWATERS!

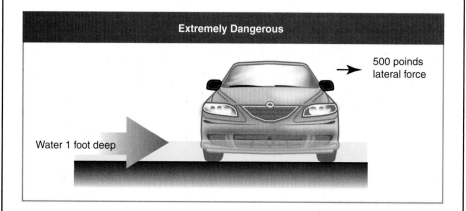

Extremely Dangerous

500 poinds lateral force

Water 1 foot deep

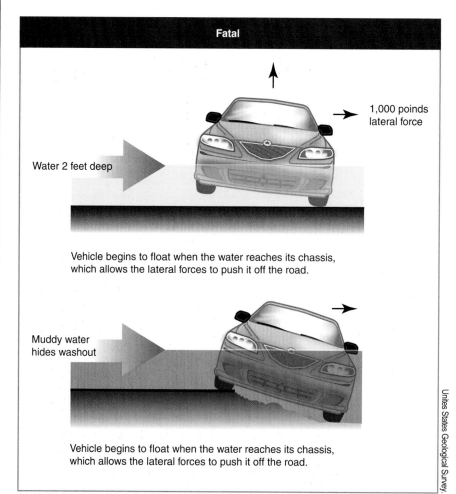

Fatal

1,000 poinds lateral force

Water 2 feet deep

Vehicle begins to float when the water reaches its chassis, which allows the lateral forces to push it off the road.

Muddy water hides washout

Vehicle begins to float when the water reaches its chassis, which allows the lateral forces to push it off the road.

Unites States Geological Survey.

The surrounding farmland had a very low gradient (less than a few inches per mile), so the water had nowhere to go. The result was that almost 4.5 million acres were covered in water. Grand Forks, a city of about 48,000 on the banks of the river, was flooded (**Figure 9.7**) and more than 24,000 homes were destroyed. The downtown area was under several meters of water. News photos recorded fires burning that destroyed several key buildings in the downtown area. Farms were under water, major crops were lost, and more than 120,000 cattle died. Losses amounted to several billion dollars. The damage continued into Canada, where more than $800 million worth of property was destroyed.

Dam-Related Floods

Buffalo Creek

Figure 9.7 Sorlie Bridge, which connects Grand Forks, North Dakota, with East Grand Forks, Minnesota, lies under water in the April 1997 floods. Losses exceeded $3.5 billion in these cities.

During a five-year period (1972 to 1977), several major dams failed in the United States. In 1972, a privately controlled, slag-heap dam on Buffalo Creek in West Virginia gave way following excessive rainfall, resulting in 125 deaths and more than $50 million in damage. After the main dam broke, water shot down the creek and destroyed two more dams downstream. In a matter of minutes more than 1,000 people were injured and 4,000 were left homeless. Interestingly, the U.S. Department of the Interior had warned officials in the state in 1967 that the potential for a disaster existed, but nothing was done to address the issues.

Teton Dam

On June 5, 1976, the Teton Dam on the Teton River in eastern Idaho, near the city of Rexburg, collapsed. The earthen dam was originally designed to serve as a multi-purpose structure, providing irrigation water to agricultural land in the area, hydroelectric capabilities, recreational opportunities, and flood control. The dam was the subject of lawsuits in federal court during its planning and early construction, but it was completed and put into service in January 1976. It failed on the first filling of the impounded reservoir. In the end, the federal government paid out almost $400 million in damage claims, and the dam was not rebuilt. The final toll was 11 lives lost and 13,000 head of livestock drowned. Flaws in the design of the dam were considered the source of the failure.

Vaiont Dam, Italy, 1963

Completed in 1961, the Vaiont Dam, the world's sixth highest dam, was built to generate hydroelectric power for northern Italy (**Figure 9.8**). The region behind the 859 foot (262 m) high dam included steep hillsides that were underlain by sedimentary rocks, including shales, in the Dolomite region, about 60 miles (100 km) north of Venice. In October 1963, after the reservoir filled naturally, abnormal rainfall caused a block of approximately 270 million cubic meters (9.4 billion cu ft) to detach itself from one wall and slide into the lake at velocities of up to 30 m/sec (100 ft/sec) or (110 kph or 65 mph). This generated a wave 100 m high (325 ft) that went over the top of the dam and blasted down the valley below. More than 50 million cubic meters (13.2×10^9 gallons) of water shot downstream. More than 2,500 people died in the disaster, which was very preventable. During planning and construction of the dam, geologists and engineers failed to note the potential of rock slippage. Fortunately the dam itself did not fail.

Figure 9.8 Vaiont Dam in northern Italy was topped by a massive wave generated by a landslide that went into the impounded reservoir. The resulting flood killed more than 2,500 people in villages along the downstream reach of the Vaiont River, in northern Italy.

Flood Severity

Some floods are accompanied by a slow rise in water level; others produce sudden raging torrents of water, silt, and other debris. Slow floods are generally forecast to happen whereas rapid ones provide little or no warning for people downstream. The degree of flooding is a function of the amount of water involved, the level it reaches, the expanse the water covers, and the slope of the land, which controls how rapidly an area might drain following an event. In addition to the damage caused by water, many floods deposit thick layers of silt and other debris. One of the major problems for homeowners is the pervasive nature of the mud that is left behind after a house is flooded.

Although flash floods are rapid and impart a great deal of damage, they are usually confined to relatively small areas. Large-scale flooding associated with major rivers or coastal areas affects huge areas and often is in localities that do not drain well, causing the water to linger for weeks. For anyone who has experienced a flood, the end results are personally devastating.

When flooding occurs in wet, humid regions, water takes a longer time to evaporate. Molds and mildew begin to grow and usually become a health hazard. Extensive flooding can also cause cholera and other water-related diseases to spread rapidly among the survivors of these disasters. More detail about these diseases is provided in Chapter 13, which discusses biological hazards.

Evidence from the Geologic Record

Flooding has occurred on Earth since water began falling from the sky. As the continents began to form several billion years ago, water was an important part of the erosion that began to wear these landmasses down. Eventually, stream patterns formed that allowed water to move more efficiently downhill.

Dubiel and others (1991) reported that exposed continental land lay astride the equator during the Triassic Period (about 225 million years ago). Paleoclimate models showed a maximum effect of monsoonal circulation in the atmosphere. Their study pointed to climatic conditions similar to present-day moisture and circulation patterns of Asia and the Indian Ocean. We could employ the principle of uniformitarianism ("the present is the key to the past") and conclude that the possibility existed for cyclonic storms to develop as they do now, thus leaving their mark on coastal and adjacent low-lying regions. Many of those sedimentary environments are now situated well above sea level as a result of later activity that uplifted much of the continents through plate tectonics.

Lessons from the Historic Record and the Human Toll

Floods have affected Earth's surface for millions of years, just like the other catastrophes we cover in this book. Historians have documented the occurrence of floods worldwide during recorded history. From their reports we learn that no part of the globe is exempt from flooding.

Yanosky and Jarrett (2002) have been able to analyze tree rings for past records of floods to determine the frequency and magnitudes with which they occurred. In some instances, they could determine the date of a flood within a few weeks. One technique they used was to notice where trees were injured by stream debris. By counting the rings, they were able to determine the year, and sometimes the season, when the major flooding took place. Such studies are helpful in determining the recurrence interval for various levels of floods in given areas.

Thousands of major floods have occurred throughout history. Some have killed many people, some have damaged or destroyed cities and surrounding areas. Sometimes the floods could have been prevented or at least the damage minimized. Several noteworthy floods have occurred in the past 120 years.

Johnstown Flood

On May 3, 1889, an earthen dam failed, flooding the town of Johnstown, Pennsylvania. The dam, which was built in 1853 to hold water for the Pennsylvania Canal, had not received maintenance for several decades. After the canal changed owners, the spillway was altered and a grating was installed to prevent fish from escaping from the lake. This led to a build up of debris that eventually clogged the outlet.

In the early spring of 1889 the region received significant snowfall that was later followed by heavy rains that melted the snowpack. With a watershed measuring 657 sq mi (1,725 sq km), several billion gallons of water flowed into the lake. By May 30 water was passing over the dam, and the following day the dam gave way with a flood crest of almost 40 ft (12 m). The estimated flow was approximately that of the water passing over Niagara Falls. Numerous debris flows resulted.

Figure 9.9 Debris flows were common in the region following the failure of the Johnstown dam in May 1889.

The town of Johnstown, located 15 miles (25 km) down the valley, was obliterated (**Figure 9.9**), and 2,209 people perished. Debates still continue about whether the townspeople were warned of the coming water or if they simply ignored the warning, having heard similar ones in the past when nothing happened. The event is memorialized at the Johnstown Flood National Memorial, part of the U.S. National Park system.

Florence, Italy

Rain began falling on Florence on November 2, 1966, and continued for two days. Early in the morning of November 4, the Arno River, which flows through Florence, burst its levees and the city began to be inundated with water. By midday the city was flooded, with water flowing through the lower floors of homes and buildings in the inner city. In sections of the city, water was more than 30 feet (10 m) deep. Numerous valuable objects of art and library volumes were damaged.

Within two days, the flood waters had receded and people began the restoration of millions of damaged items. Water was still almost five meters (16 ft) deep in places. With the assistance of many citizens from across Italy and other countries, life slowly began to return to normal. Volunteers came from many countries to help with the cleaning of the city and its valuable artifacts. Several countries contributed financial assistance and numerous world experts provided help with the restoration and repair of items damaged by water and mud.

Great Flood of 1993 in the U.S. Midwest

An unusual series of events produced the Great Flood of 1993, which resulted in the most catastrophic flooding the United States has ever experienced. At its peak, more than 15 percent of the contiguous United States was affected, more than 50,000 homes were damaged or destroyed, and water covered more than 400,000 square miles (1.04 million sq km) in Illinois, Iowa, Kansas, Minnesota, Nebraska, North Dakota, South Dakota, and Wisconsin. The infrastructure of America's heartland was disrupted and its economy was upset, affecting millions of people. For more than two months there was no barge traffic on the Mississippi and Missouri Rivers and railroad traffic stopped in the Midwest, one of the hubs of rail transportation in the United States. Because many highways and bridges were flooded, truck traffic came to a stop or had to be rerouted, delaying the delivery of many of the nation's goods.

Larson (1995) provides a good background for examining the conditions that led to the Great Flood of 1993. Flood forecasts are made by 12 National Weather Service River Forecast Centers located throughout the country. Components of the models used to forecast flood potential include temperatures, precipitation amounts, and the soil moisture content of various watersheds. Stream runoff is used to produce a flow model, which uses data from a unit hydrograph. The unit hydrograph represents one inch of runoff from a rain storm that evenly covers a defined headwater drainage basin over a given amount of time, usually taken as six hours. Data from different upstream drainage basins are combined to determine the amount of flow for a selected downstream location, allowing hydrologists to estimate the potential flooding severity.

The flooding that occurred in the Midwest in 1993 has been attributed by some researchers to El Niño, which is an oscillation of the ocean-atmosphere system located in the tropical Pacific Ocean (see **Box 9.2**). From January to June 1993, the upper reaches of the Mississippi River drainage basin received more than 1.5 times the normal rainfall (**Figure 9.10**). The continuing rains in July caused portions of North Dakota, Iowa, and Kansas to record more than four times their normal rainfall

---

BOX 9.2	**El Niño and Its Effect on Climate**

In normal conditions (non-El Niño years), warm surface waters in the western Pacific produce a convective rise of moisture that results in substantial rainfall in Indonesia and adjoining regions, while the trade winds blow to the west (Box Figure 9.2.1). Cooler, deep marine waters move eastward and their upwelling brings nutrient-rich water to the west coast of South America. This enriches the marine life and causes the region to have a major fishing trade.

During an El Niño year, trade winds in the western and central Pacific Ocean become weaker and a larger mass of warm water develops in the central Pacific. Atmospheric conditions change as low pressure exists in the eastern Pacific Ocean and high pressure dominates in the west. These conditions cause warm water to move toward South America. The colder waters at depth are unable to ascend to the surface. The increased atmospheric convection associated with the warm water along the eastern equatorial region produces more precipitation in the temperate latitudes of North and South America (Box Figure 9.2.2). Such was the case in early 1993 in the Midwest that led to the massive rainfalls that produced the Great Flood of 1993.

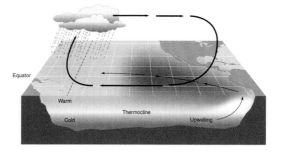

Box Figure 9.2.1 Normal conditions of atmospheric and oceanic conditions in the equatorial Pacific Ocean.

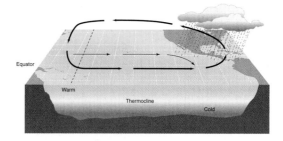

Box Figure 9.2.2 Atmospheric and oceanic conditions present during an El Niño event.

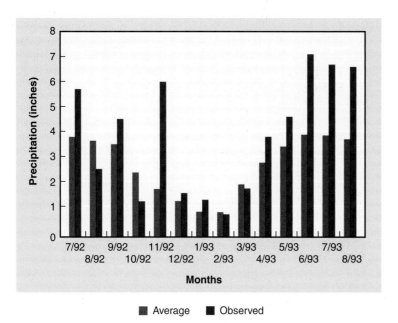

Figure 9.10 Comparison of observed and average monthly precipitation totals for the Upper Mississippi River Basin.

amounts. Rains fell often in very intense storms that passed over the region. Normally, floods only last a few days or perhaps a week or two at most. However, these prolonged rains produced supersaturated soils, causing subsequent rain to run off into the drainage basins and produce unprecedented flooding.

The Great Flood of 1993 lasted for months, with many locations experiencing flood stage conditions for five or six months. The duration of the regional flooding was caused by an extremely unusual set of weather patterns that kept the region very wet. Saturated soils could not absorb any rain, thereby causing it to run off into streams incapable of carrying the volume of water. Rainfall amounts set new all-time records for many recording stations, as the amounts recorded approximate those expected in a 75- to 300-year event (Larson, 1995).

With the continual rains and high humidity, very little evaporation occurred. The Mississippi River remained above flood stage in St. Louis, Missouri, for 146 days between April 1 and September 30, 1993 (**Figure 9.11**). In Cape Girardeau, a city

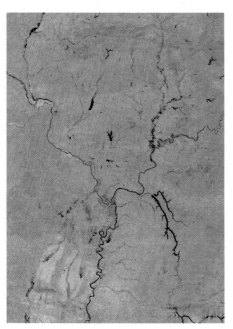
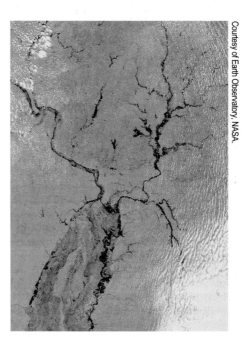

Courtesy of Earth Observatory, NASA.

Figure 9.11 A comparison of the river flows of the Illinois, Mississippi, and Missouri Rivers, August 1991 (normal conditions) and August 19, 1993 (full flood conditions).

(a)　　　　　　　　　　　　(b)

Figure 9.12 Levee along the Mississippi River at Cape Girardeau, Missouri. Arrows designate river level reached during the 1993 and 1973 floods.

---

flood stage
The point in time when a body of water, such as a river, rises to a level that causes damage to the adjacent areas.

located along the Mississippi River in southeastern Missouri, floodwaters crested 16.5 feet (5 meters) above **flood stage** (**Figure 9.12**). The river remained at flood stage in Cape Girardeau for 126 days. The floods of 1993 resulted in 50 deaths and damage estimates in the Midwest ranged between $15 and $20 billion.

Human Interactions and Flooding

Streams will naturally flood at some point during their existence. Whether flooding occurs regularly or on a very periodic time scale, water will flow over the banks and onto a floodplain. However, mankind has introduced new parameters into the flooding process that create flooding conditions more frequently than the natural processes.

Construction of Levees and Channels

Construction of natural levees, although meant to allow a stream to flood in a natural way less often, upsets the normal balance of a stream system. Increased height of levees causes more water to flow in the deeper channel, so when it does occasionally break through, major problems develop. Water cannot easily flow back into the main channel.

Construction of concrete channels to contain flow through urban areas only speeds up any excess water from heavy rainfall or snowmelt that enters the channels. The Los Angeles River has been channelized. More than 60 percent of its watershed is covered by impervious material (asphalt and concrete), so infiltration is greatly reduced. Water is sent to lower elevations in the channel system. There are times when too much water accumulates at the lower reaches of these channels, and flooding results in extensive, unexpected damage.

When flooding occurs, sandbags are often used to contain the flow of water. Bags are placed along the banks of a river as temporary, "quick-fix" levees that hold most of the water in a channel. Bags are also used in low-lying areas to slow the flow of water into buildings and their doorways or low windows. Sandbags work well in the short term for localized flooding but do not help with major overbank conditions. Sometimes large sheets of plastic are placed over the bags but these sheets are not continuous and will allow leaks. All of these techniques are insufficient in solving the larger problem of streams overflowing their banks.

Flood Control

When water is out of control, it creates major problems. Several techniques have been devised to help control and channel potential flood waters. Channelization can involve the clearing or dredging of an existing channel to speed up the flow. Another method is to construct artificial cutoffs that shorten the length of a stream, increasing its velocity and gradient. The result is that more water flows through an area with higher velocities. The goal is to reduce the chances of flooding, but channelization has produced mixed results.

Recall that natural levees form along rivers and deepen the channel. Piling extra earth atop natural levees or putting earth alongside a river creates artificial levees that deepen the channel and increase the volume of water in the stream channel. However, these are only temporary solutions that eventually prove unsuccessful in controlling floods.

Dams help to reduce flooding by storing water and releasing it in a controlled manner. In addition to the impounded water, however, the dams collect sediment that normally would have been transported downstream to enrich floodplain farmland or to create riparian habitats along the stream. Dams create an imbalance in the ecology of the river environment, a fact that was not recognized until the past 20 or so years. On the other hand, the thousands of dams in the United States also generate hydroelectric power, provide water for agricultural irrigation, and serve as recreation areas.

Predicting Floods

The occurrence of floods depends on many variables, most of which we have little control over. Weather-related floods can be forecast with some detail in the short term, but over longer periods of time we cannot foresee their arrival. However, researchers can analyze flood frequency over longer periods of time, provided adequate data exist.

The USGS has a system of gauging stations situated all over the country that provide real-time data. Stream flow is constantly monitored and the data are available for researchers and scientists to analyze. For the data to be useful, researchers attempt to determine some type of recurrence pattern or interval for flooding events. The term *100-year flood* is a standard time period used to describe the recurrence interval of floods. As Table 9.2 shows, the "100-year recurrence interval" means that a flood of that magnitude has a 1 percent chance of occurring in any given year. In other words, the chance that a river will flow as high as the 100-year flood stage in any given year is 1 in 100. Statistically, each year begins with the same 1 percent chance that a 100-year event will occur.

Even though a 100-year flood happens in one given year, its occurrence does not imply that another 100 years will pass before the water level would be that high again. Many variables are at work controlling the amount of moisture in an area so the event can indeed occur at any time, even twice within a given year. Should the 100-year flood begin to become more common, it will be necessary to reexamine all the data and redefine the 100-year flood, taking into account changes in the occurrence of such a flood level.

TABLE 9.2	Recurrence Intervals for Various Levels of Floods	
Recurrence Interval, in Years	Probability of Occurrence in Any Given Year	Percent Chance of Occurrence in Any Given Year
100	1 in 100	1
50	1 in 50	2
25	1 in 25	4
10	1 in 10	10
5	1 in 5	20
2	1 in 2	50

Summary

Floods occur every day somewhere on Earth. When a drainage system receives more water than it can handle, the excess water flows over river banks and across the landscape, flooding farmland and communities. Drainage systems found in watersheds serve to remove water that falls in the area. As streams flow downhill, they gather more water and have a higher flow in the lower elevations. If the amount of water exceeds the capacity of the stream, flooding occurs.

Regional floods result from prolonged rainfall or excessive snowmelt. Such floods usually last several days or weeks and create widespread damage to a region. Usually the loss of human life is small in these events. Flash floods are fast to form and move through an area rapidly. They provide very little warning and are the primary cause of loss of human life. Floods caused by ice dams or the breaking of artificial dams have widespread effects. Failure of artificial dams is often traced to engineering problems. The effects of historic floods throughout the world indicate that many regions are affected by the influx of water. Low-lying areas suffer the most because flood waters are slow to recede.

Humans have attempted to intervene in preventing floods but generally the results are mixed. Channelization of streams and overland flow moves water through an area more rapidly than normal but can produce unexpected effects if too much water ends up at the end of the drainage system. The construction of dams for flood control is often a temporary solution as the dams tend to fill with stream sediment that eventually reduces their capacity to impound the amount of water for which they were originally designed.

The ability of researchers to predict floods is helpful in forecasting potential problems in a given area. However, the fickle nature of weather systems and climate changes makes the prediction of major events very difficult. It is necessary to use recent information when establishing future patterns of floods and related events.

References and Suggested Readings

Benito, G., V. R. Baker, and K. J. Gregory, eds. 1998. *Palaeohydrology and Environmental Change.* Chichester: John Wiley.

Dubiel, Russell F., Judith Totman Parrish, J. Michael Parrish, and Steven C. Good. 1991. The Pangaean Megamonsoon—Evidence from the Upper Triassic Chinle Formation, Colorado Plateau. *Palaios* 6: 347–370.

House, P. Kyle, Robert H. Webb, Victor R. Baker, and Daniel R. Levish, eds. 2002. *Ancient Floods, Modern Hazards—Principles and Applications of Paleoflood Hydrology.* Washington, DC: American Geophysical Union.

Larson, Lee W. 1995. The Great USA Flood of 1993. http://www.nwrfc.noaa.gov/floods/papers/oh_2/great.htm.

Lutgens, F. and E. Tarbuck. *Essentials of Geology.* 8th ed. 2003, Upper Saddle River, NJ: Pearson Prentice Hall.

Mayer, L. and D. Nash, eds. 1987. *Catastrophic Flooding:* Boston, MA: Allen and Unwin.

Yanosky, Thomas M. and Robert D. Jarrett. 2002. Dendrochronologic evidence for the frequency and magnitude of paleofloods. In *Ancient Floods, Modern Hazards—Principles and Applications of Paleoflood Hydrology,* ed. P. Kyle House, Robert H. Webb, Victor R. Baker, and Daniel R. Levish. Washington, DC: American Geophysical Union.

Web Sites for Further Reference

http://pubs.usgs.gov.circ/2003/circ1245
http://water.usgs.gov/
http://water.usgs.gov/wid/index-hazards.html
http://waterdata.usgs.gov/nwis/rt
http://www.photolib.noaa.gov.historic.nws/index.html
http://www.usgs.gov.hazards/floods/
http://www.bt.cdc.gov/disasters/floods/index.asp
http://www.noaa.gov/floods.html
http://www.weather.gov/oh/hic

Questions for Thought

1. How does the size of a drainage basin affect the amount of discharge in a stream?

2. How does the cross-sectional area of a stream affect its velocity?

3. What factors lead to flash floods? What would you do in case of encountering a flash flood?

4. What are ice-jam floods and during what part of the year are they most likely to occur?

5. Give two examples of how poor planning resulted in disastrous failures of dams.

6. Explain the role that levees play in major rivers.

7. How predictable are floods?

8. What would be the effects of a two-foot flood in your hometown? Where would such a flood come from and what would be the long-term effects? Look at your hometown's website to see if any plans are in place for such as disaster.

Coastal Regions and Land Loss

10

Courtesy of Save the Light House, Inc.

Morris Island Lighthouse, near Charleston, South Carolina, was built about 400 meters inland in 1876. By 1938 changes in nearshore currents had eroded the land so that the lighthouse was at water's edge. Today it is about 300 meters from the shoreline.

Key Terms

backshore zone
backwash
barrier island
beach
beach drift
beach nourishment
berm
break
breaker
breakwater
circular orbital motion
coastal submergence
coastline
crest
ebb tide
emergent coastline
eustatic
fetch
flood tide
groin
headland
jetty

(*Continued*)

longshore current	seawall	trough
longshore drift	spilling breaker	wave
neap tide	spring tide	wave base
offshore zone	storm surge	wave height
overwash	submergent coastline	wave period
plunging breaker	surf zone	wave refraction
rip current	surging breaker	wave speed
rogue wave	swash	wave steepness
sea arch	swell	wave-cut platform
sea stack	tide	wavelength

Coastal regions, where the land meets the ocean or a large lake, are attractive places to live or vacation. Because large bodies of water moderate temperatures, coastal areas are cooler in summer and warmer in winter than farther inland. Humans have always been attracted to the coastal regions, to take advantage of the milder climate, abundant seafood, easy transportation, recreational opportunities, and commercial benefits. It is no wonder that the coastlines have become heavily urbanized and industrialized. Approximately 60 percent of the world's population lives within 100 kilometers of ocean coasts. In the United States alone, coastal population increased by 38 million between 1960 and 1990, and today 55 percent of the population live in coastal counties along the Atlantic and Pacific Oceans, the Gulf of Mexico, and the Great Lakes. This percentage is estimated to rise to over 70 percent by 2025.

Increasing coastal development is of major concern to geologists because they know that coastlines are among Earth's most geologically active environments. Water in oceans and lakes is constantly in motion due to winds, tides, currents, and, occasionally, tsunami. Therefore, coasts are dynamic and constantly changing from interactions between the energy in the water and the land. These coastal changes occur on two very different time scales. Short-term change (years to decades) is largely due to coastal erosion from waves, storms, and coastal flooding, while long-term changes (hundreds to thousands of years) are due to slower sea-level rise that causes a landward shift in the coastline. These natural processes posed no problem until people began to live along coasts.

Hurricanes and coastal storms are major hazards affecting most coasts in the United States, due to the high energy waves they bring to these areas. As these coasts become increasingly developed, they are highly vulnerable to these natural hazards, and storm damage continues to rise dramatically. Many coasts are affected by multiple hazards such as landslides caused by cliffs continually being undermined by large waves. Today, coastal erosion affects businesses, homes, public facilities, beaches, cliffs, and bluffs (cliffs along lakes) built close to the water's edge. It is estimated that within the next 60 years, coastal erosion may claim one out of four structures within 150 meters of the coastline of the United States.

Even though the coasts are dynamic environments, predicting future coastal change is often difficult, because of the many variables inherent in world climate, weather, nature of the coastline, and human activity. However, scientific studies show evidence that sea level will continue rising, and that storms will become more common and powerful in the coming years. If our present patterns and rates of development along coasts continue, then we are on a collision course with more disasters and catastrophes. It is therefore important to understand the nature of coastal processes and their inherent natural hazards as future development is planned along the coastal zone.

Coastal Processes

Coastal Basics

The **coastline** is a unique boundary where the geosphere, atmosphere, and hydrosphere meet and the systems interact (**Figure 10.1**). At this boundary, dynamic processes of erosion and deposition are constantly at work shaping and reshaping the landscape. Coastal processes active along the coasts are the result of interactions within the climate system and the solar system. Coastal surf and storms result from interactions between the atmosphere and the hydrosphere, with the Sun as the ultimate source of energy driving them. Wave activity that derives from blowing winds is the most important process acting along lake and marine coastlines. Gravity is also an important source of energy in producing rising and falling tides and currents that mostly affect oceanic coastlines. Tides are produced by gravitational interactions between Earth, the Sun, and Moon. Thus waves and tides are important processes in bringing energy to the coasts for erosion of the land and the transport and deposition of sediment on beaches.

Photo courtesy of Barbara H. Murphy.

Figure 10.1 The shoreline serves as a boundary between the hydrosphere, geosphere, and atmosphere, where all three meet and interact.

Waves

Wave Generation

A **wave** is simply energy in motion that is the result of some disturbance. The energy that causes waves in water to form is called a disturbing force. For example, a rock thrown into a still lake will create waves that radiate in all directions from the disturbance. Mass movement into the ocean, such as coastal landslides and calving glaciers (creating icebergs) produce waves commonly known as splash waves. Sea floor movements change the shape of the ocean floor and release tremendous amounts of energy to the entire water column and create very large waves. Examples include underwater avalanches, volcanic eruptions, and fault movement, all of which can generate a tsunami (Chapter 6). Human activities can also generate waves, such as when ships travel across a body of water and leave behind a wake, which is a wave. Wind blowing across a body of water disturbs surface waters and generates most waves that we commonly see on the surface of the oceans or large lakes. Wind-generated waves represent a direct transfer of kinetic energy from the atmosphere to the water surface. In all these cases, some type of energy release creates waves; however, wind-generated waves provide most of the energy that reaches land and shapes and modifies the coastlines.

Wave Characteristics

When wind blows unobstructed across the water, it deforms the surface into a series of wave oscillations. The highest part of the wave is the **crest**, and the lowest part between crests is the **trough** (**Figure 10.2**). The vertical distance between the crest and the adjacent trough is the **wave height**. The horizontal distance between any two similar points of the wave, such as two crests or two troughs, is the **wavelength**. The **wave period** is the time it takes for one full wavelength to pass a given point. The **wave speed** is the rate at which the wave travels and is equal to the wavelength divided by the period. **Wave steepness** is the ratio of wave height to wavelength. If the wave steepness exceeds 1/7, the wave becomes too steep to support itself and it **breaks**, or spills forward, releasing energy and forming whitecaps, often observed in choppy waters or the surf area along a beach. A wave can break anytime the 1:7 ratio is exceeded, either in the open ocean or along the shoreline.

coastline
Unique boundary where the geosphere, atmosphere, and hydrosphere meet and the systems interact.

wave
Energy in motion that is the result of some disturbance that moves over or through a medium with speeds determined by the properties of the medium. Ocean waves are usually generated by wind blowing across the water surface.

crest
The highest point of a wave.

trough
The low spot between two successive waves.

wave height
The vertical distance between the crest and adjacent trough of a wave.

wavelength
The distance between two successive wave crests or troughs.

wave period
The time it takes for one full wavelength to pass a given point.

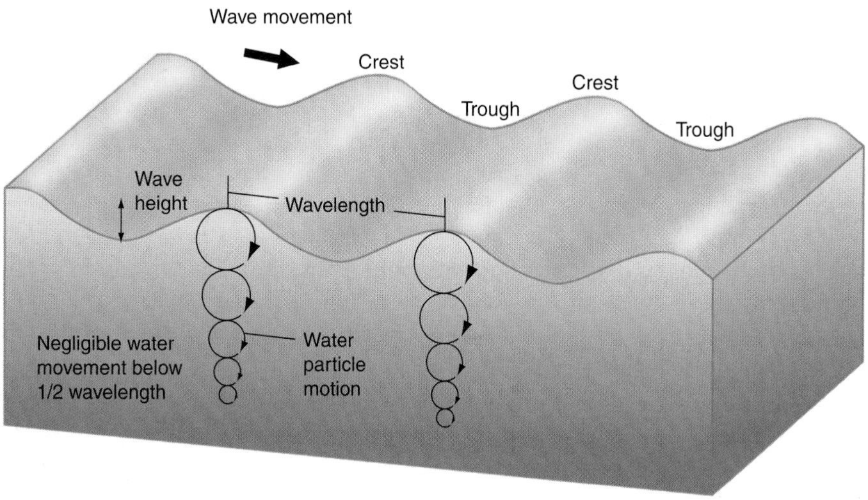

Figure 10.2 Wave form with characteristics. Crests and troughs alternate across the wave form. Note that water motion dies off at a depth of about one-half the wavelength.

wave speed
The velocity of propagation of a wave through a liquid, relative to the rate of movement of the liquid through which the disturbance is propagated.

wave steepness
The measured ratio of wave height to wavelength.

break
When a wave steepness exceeds 1/7, the wave becomes too steep to support itself and it breaks, or spills forward, releasing energy and forming whitecaps often observed in choppy waters or the surf area along a beach.

circular orbital motion
The movement of a particle by moving in a circle.

wave base
Depth equal to one-half the wavelength where there is no movement associated with surface waves.

Wave Motion

Waves are a mechanism by which energy is transferred along the surface of the water. Waves can travel many kilometers from their place of origin. Waves generated in Antarctica have been tracked as they traveled over 10,000 kilometers through the Pacific Ocean before finally expending their energy a week later on the shores of Alaska. But it is important to note that the water itself does not travel this great distance. The water is merely the medium for the waveform (the energy) to travel through, similar to earthquake seismic waves as they travel through solid rock.

In the open water, water particles pass the energy along by moving in a circle (**Figure 10.3**). This is known as **circular orbital motion**. This motion can be observed easily by observing a floating object as it bobs up and down and sways back and forth as the wave passes, but the object itself does not travel along with the wave. From the side, the object can be viewed moving in a circular orbit with a diameter at the surface equal to the wave's height. Beneath the surface, the orbital motion of the water particles diminishes downward with depth. At a depth equal to approximately one-half the wavelength, there is no movement associated with surface waves. This bottom depth of orbital motion is known as the **wave base**. Submarines can avoid large ocean waves by submerging below the wave base. Even seasick scuba divers can find relief by submerging into the calm water below the wave base. If the water depth is greater than the wave base, the waves are called deep-water waves and do not contact the ocean floor. The motion of water in waves is therefore distinctly different from the motion of water in currents, in which water travels in a given direction and does not return to its original position.

Figure 10.3 Circular motion of a wave form in open water. Individual particles of water rotate in a circle, producing an overall lateral movement of a wave.

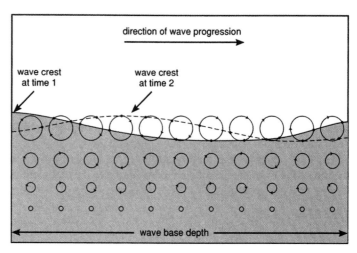

Wave Energy

As wind blows over the surface of a body of water, some of its energy is transferred to the water. The mechanism of energy transfer is related to frictional drag resulting from one fluid (the air) moving over another fluid (the water). Waves that crash along the coast in the absence of local winds are generated by offshore winds and storm events, sometimes thousands of kilometers away. The energy of a wave depends on its height and length. The higher the wave's height, the greater the size of the orbit in which the water moves. The height, length, and period of a wave depend on the combination of three factors: (1) wind speed (the stronger the wind speed, the larger the waves), (2) wind duration (the longer time the wind blows, the more time the wind can transfer energy to the water, and the larger the waves), and (3) **fetch**, the distance over which the wind blows (a longer fetch allows more energy to be transferred and form larger waves). If we compare waves on a lake and on the ocean, when wind speed and duration are the same, the waves will be higher on the ocean because the fetch is far greater than on a lake. When the maximum fetch and duration have been reached for a given wind speed, waves will be fully developed and will grow no further. This is because they are losing as much energy breaking as whitecaps as they are receiving from the wind.

In areas where storm waves are generated, waves will have different lengths, heights, and periods. When the wind stops, or changes direction, waves will separate into waves of uniform length called **swells**. Swells moving away from the storm center can travel great distances before the energy of the wave is released by breaking and crashing onto the coast.

fetch
The distance over which the wind blows across open water.

swell
Wave of uniform wavelength moving away from a storm center. They can travel great distances before the energy of the wave is released by breaking and crashing onto the coast.

Interference Patterns and Hazardous Rogue Waves

When swells of deep water waves from different storms run together, the waves interfere with one another to produce different interference patterns (**Figure 10.4**). The interference pattern produced when two wave systems collide is the sum of the disturbance that each would have created individually. Constructive interference occurs when two waves having the same wavelength come together in phase (meaning crest to crest and trough to trough). The wave height will be the sum of the two, and if it

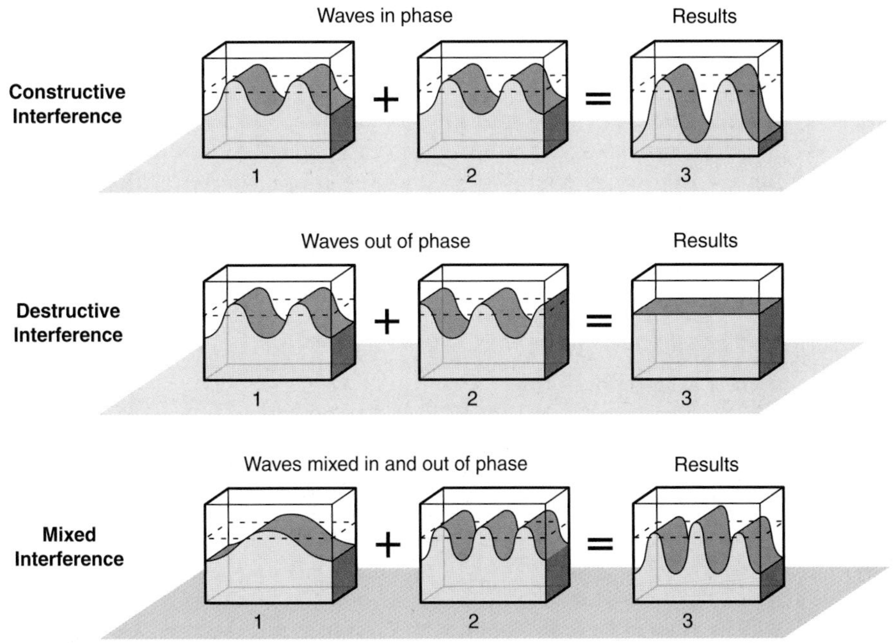

Figure 10.4 Interference patterns in water. Constructive interference creates larger waves, destructive interference reduces the waves to a flat surface, and mixed interference generates a mixture of small and large waves.

becomes too steep, the wave may break forming whitecaps. Destructive interference occurs when waves having the same wavelength come together out of phase (meaning the crest of one will coincide with the trough of the other). If the wave heights are equal, the energies will cancel each other and the water surface will be flat. If the waves are traveling in opposite directions, the waves will return to their normal heights once they travel through the interference area.

It is common for mid-ocean storm waves to reach 7 meters in height, and in extreme conditions such waves can reach heights of 15 meters. However, solitary waves called **rogue waves** can reach enormous heights and can occur when normal ocean waves are not unusually high. The word *rogue* means unusual, and in this case the waves are unusually large—monsters up to 30 meters in height (approximately the height of a 10-story building)—that can appear without warning. Rogue waves appear to be caused by an extraordinary case of constructive interference that can be very destructive and have been popularized in movies such as *The Poseiden Adventure* and *The Perfect Storm*.

In 1942 during World War II, the RMS *Queen Mary* was carrying 15,000 American troops near Scotland during a gale and was broadsided by a 28 meter high rogue wave and nearly capsized. The ship listed briefly about 52 degrees before the ship slowly righted herself.

Waves Reach the Shore: Shallow-Water Waves and Breakers

When deep-water waves approach shore, the water depth decreases and the wave base starts to intersect the seafloor. At this point the wave comes in contact with the bottom and the character of the wave starts to change (**Figure 10.5**). In this zone of shoaling (shallowing) waves, they grow taller and less symmetrical. Because of friction at the bottom, the wave speed decreases, but its period remains the same, and thus, the wavelength will decrease. The circular loops of water motion also change to elliptical shapes, as loops are deformed by the bottom. As the wave moves farther shoreward, the wavelength shortens considerably and the wave height increases. The increase in wave height, combined with the decrease in wavelength, causes an increase in wave steepness. With continued forward motion at the top of the wave and friction at the bottom, the front portion of the wave cannot support the water as the rear part moves over, and the wave breaks as surf. Here in the **surf zone**, actual forward movement of water itself occurs within the wave as all the water releases its energy as a wall of moving, turbulent surf known as a **breaker**.

There are three main types of breakers (**Figure 10.6**). **Spilling breakers** form on shorelines with gentle offshore slopes and are characterized by turbulent crests spilling down the front slope of the wave. **Plunging breakers** form on shorelines with steeper offshore slopes and have a curling crest that moves over an air pocket. Plunging breakers are prized waves for surfing. **Surging breakers** form when the offshore slopes abruptly and the wave energy is compressed into a shorter distance and the wave surges forward right at the shoreline.

rogue wave
Large solitary wave caused by constructive wave interference that usually occurs unexpectidly amid waves of smaller size.

surf zone
The nearshore zone of breaking waves.

breaker
A wave in which the water at the top and leading edge falls forward producing foam.

spilling breaker
Forms on shorelines with gentle offshore slopes and are characterized by turbulent crests spilling down the front slope of the wave.

plunging breaker
Forms on shorelines with more steep offshore slopes and have a curling crest that moves over an air pocket.

surging breaker
Forms when the offshore slopes abruptly and the wave energy is compressed into shorter distance and the wave surges forward right at the shoreline.

swash
A turbulent sheet of water that rushes up the slope of the beach following the breaking of a wave at shore.

backwash
The flow of water down the beach face toward the ocean from a previously broken wave.

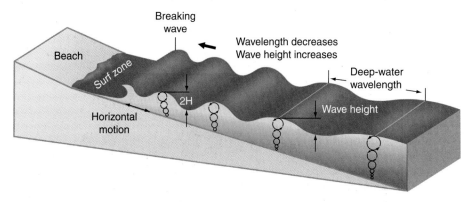

Figure 10.5 Wave hitting the shore with the bottom of the wave intersecting the sea floor. Waves form breakers and move toward the beach as the surface of the water moves faster than the water underneath.

Fundamentals: Breakers

There are three types of breakers:

- **Spilling breakers** break gradually over considerable distance.

- **Plunging breakers** tend to curl over and break with a single crash. The front face is concave, the rear face is convex.

- **Surging breakers** peak up, but surge onto the back without spilling or plunging. Even though they don't "break," surging waves are still classified as breakers.

Spilling

Plunging

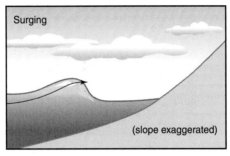

Surging

(slope exaggerated)

U.S. Army.

Figure 10.6 Spilling, plunging, and surging breakers. Sea floor topography plays a key role in the type of breaker that forms.

After the breaker collapses, a turbulent sheet of water, the **swash**, rushes up the slope of the beach (called the swash zone). The swash is a powerful surge that causes landward movement of sediment (sand and gravel) on the beach. When the energy of the swash is dissipated, the water flows by gravity back down the beach toward the surf zone as **backwash**. Therefore, as a wave approaches the shore, it breaks, and the stored energy in the wave is expended in the surf and swash zones, causing erosion, transport, and deposition of sediment along the coast (**Figure 10.7**).

Wave Refraction

As waves approach an irregular shoreline, or at an angle to the shore, the wave base will initially encounter shallower water areas first and begin to slow down before the rest of the wave does resulting in **wave refraction**, or bending of the wave. Refraction of waves approaching coastlines concentrates wave energy on protruding **headland** areas and dissipates energy in the bays (**Figure 10.8**). The concentrated energy on headlands erodes them into cliffs and causes deposition of sediment in the bays; thus, headlands erode faster than bays due to stronger wave energy. The result of wave refraction is to erode headlands and smooth out the coastline. The eroded sediments are deposited offshore, on beaches, or in bays.

© Dmitry Naumov, 2010. Under license from Shutterstock, Inc.

Figure 10.7 Swash forms as a breaking wave moves up the flat surface along the shore. Receding water produces the backwash.

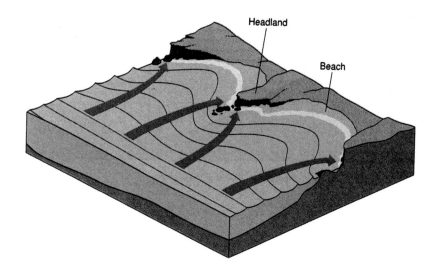

Figure 10.8 Refraction of wave energy around a headland. Incoming wave energy is bent or refracted toward the headland due to changes in the sea floor topography that cause the wave to change speed. This produced a zone of high energy that erodes the protruding headland.

wave refraction
The process by which the part of a wave in shallow water is slowed down, causing it to bend and approach nearly parallel to shore.

headland
A steep-faced irregularity of the coast that extends out into the ocean.

tide
The periodic rising and falling of the water that results from the gravitational attraction of the moon and sun acting on the rotating earth.

flood tide
The incoming or rising tide; the period between low water and the succeeding high water.

ebb tide
That period of tide between a high water and the succeeding low water; falling tide.

spring tide
The highest high and the lowest low tide during the lunar month. The exceptionally high and low tides that occur at the time of the new moon or the full moon when the sun, moon, and earth are approximately aligned. Contrast with Neap Tide.

neap tide
A tide that occurs when the difference between high and low tide is least; the lowest level of high tide. Neap tide comes twice a month, in the first and third quarters of the moon. Contrast with spring tide.

Tides

Tides produce short-term fluctuations in sea level on a daily bases. **Tides** are the rising and falling of Earth's ocean surface caused by the gravitational attraction of moon and the Sun on the Earth. Because the moon is closer to the Earth than the Sun, it has a greater effect and causes the Earth's water to bulge toward it, while at the same time a bulge occurs on the opposite side of the Earth due to inertial forces. These different bulges remain stationary while Earth rotates and the tidal bulges result in a rhythmic rise and fall of ocean surface, which is not noticeable in the open ocean but is magnified along the coasts. The changing tide produced at a given location is the result of the changing positions of the moon and Sun relative to the Earth, coupled with the effects of Earth rotation and the local shape of the sea floor.

The regular fluctuations in the ocean surface result in most coastline areas having two daily high tides and two low tides as sea level rises and falls onto the shore. A complete tidal cycle includes a **flood tide** that progresses upward on the shore until high tide is reached, followed by an **ebb tide** falling off the shore until low tide is reached and exposing the land once again. Tidal ranges between high and low tides along most coasts range about 2 meters. However, in narrow inlets tidal currents can be strong and cause variations in sea level up to 16 meters. High and low tides do not occur at the same time each day but instead are delayed about 53 minutes every 24 hours. This is because the Earth makes a complete axis rotation in 24 hours, but at the same time, the moon is orbiting the Earth in the same direction, which means the Earth must spin an additional 53 minutes for the same point on Earth to be directly beneath the moon again (and thus in the bulge at its highest). This explains why the moon rises in the sky about 53 minutes later each day, and in the same manner, why the tides are also about 53 minutes later each day.

Because the Sun also exerts a gravitational attraction on the Earth, there are also monthly tidal cycles that are controlled by the relative position of the Sun and moon to the Earth. Although the sun's gravitational pull on the oceans is smaller than the moon's, it does have an effect on tidal ranges (the difference in elevation between high and low tides). The largest variation between high and low tides, called **spring tides**, occurs when the sun and the moon are aligned on the same side of the Earth (new moon) or on opposite sides of the Earth (full moon) (**Figure 10.9**). Here the gravitational attractions of the moon and sun amplify each other and produce higher and lower tides. The lowest variation between high and low tides, called **neap tides**, occur when the moon is at right angles relative to the Earth and Sun (quarter moons).

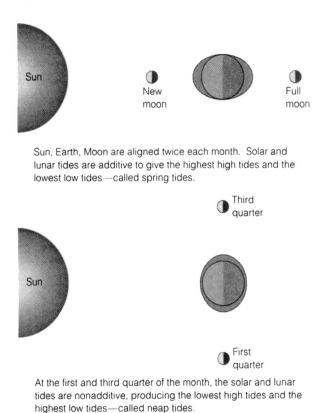

Sun, Earth, Moon are aligned twice each month. Solar and lunar tides are additive to give the highest high tides and the lowest low tides—called spring tides.

At the first and third quarter of the month, the solar and lunar tides are nonadditive, producing the lowest high tides and the highest low tides—called neap tides.

Figure 10.9 The tidal ranges are affected by the position of the moon with respect to Earth. New and full moons produce spring tides, while first and third quarter moons produce neap tides.

The timing when hazards, such as storms or tsunami events, strike the shoreline, especially at spring tides phases, is very important. The intensity of a disaster is magnified when rising water from a storm surge or tsunami arrives at the same time as the highest high tides. The combination of these events often sends the destructive power of the water farther inland, producing much more catastrophic results.

Coastal Erosion

The erosion of the coastlines is due to the constant battering of waves which causes the land to retreat. Most coastal erosion occurs during intense storms when waves and storm surges are more energetic. The rate of wave erosion varies greatly along coasts of different compositions but is typically rapid along sandy coasts and slower along rocky coasts. Water weathers and erodes coastlines by processes of hydraulic action, abrasion, and corrosion.

The force of water alone, called hydraulic action, is an effective erosional process. Breaking waves exert a tremendous force on the shores by direct impact of the water and are very effective on cliffs composed of sediment or fractured rocks. A large wave 10-meter high storm wave striking a 10-meter high cliff produces four times the thrust energy of the space shuttle's three main orbiter engines. A wave striking a cliff drives water into cracks or other openings in the rock and compresses air inside. As this happens, the water and air combine to create hydraulic forces on the surrounding rock that is large enough to dislodge rock fragments or large boulders. Repeated countless times, hydraulic action wedges out rock fragments from cliff faces which fall to the bottom of the cliff or sea bed. The debris can be picked up and used for another erosive wave action–abrasion. Loose sand is easily moved by wave action and by currents that run parallel to the shoreline.

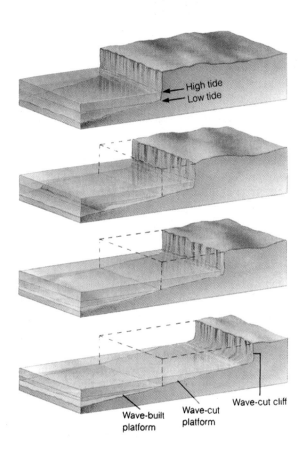

Figure 10.10 Wave-cut platforms and wave-cut cliffs are formed by active erosion produced by incoming tides.

wave-cut platform
A gently sloping surface produced by wave erosion, extending far into the sea or lake from the base of the wave cut cliff.

sea arch
An opening through a headland caused by erosion.

sea stack
An isolated rock island that is detached from a headland by wave action.

Landforms of Erosional Coasts

Along coasts where erosion dominates, rocky coastlines are common. Since erosion occurs at sea level, abrasion and hydraulic action undercut exposed bedrock forming a wave-cut cliff. As the cliff continues to erode, it leaves behind a flat or gently sloping **wave-cut platform** (**Figure 10.10**). Farther offshore a wave-built platform can be formed by transported sediments that are deposited as the water moves seaward. Locally, refraction of waves onto a narrow headland can cut a cave into the rock which may eventually erode all the way through the headland forming a **sea arch**. When the sea arch collapses, a small prtion of the headland may be isolated from the retreating sea cliff and remain as a **sea stack** (**Figure 10.11**). As waves continue to batter the rocks, eventually the sea stacks crumble. The overall effect is to produce a rugged shoreline that is constantly being hit by incoming waves (**Figure 10.12**).

Courtesy of David M. Best.

Figure 10.11 Arches represent the remains of a once-protruding headland that reached out into the ocean. The sea stack in the distance seen through the arch is the remnants of a collapsed arch.

Courtesy of David M. Best.

Figure 10.12 Waves strike a rugged shoreline in the Channel Islands National Park, California.

Coastal Sediment Transport

Sediment that is created by the abrasive and hydraulic action of waves, or sediment brought to the coast by streams, is picked up by the waves and transported. One of the most important processes of sediment transport within the shoreline area is longshore drift. Sediment is also transported by rip currents and tidal currents.

Longshore Drift and Currents

Longshore drift is the net movement of sediment parallel to the shore. The process starts when waves approach the shore obliquely, even after refraction. Waves striking the shore at an angle, as opposed to straight on, will cause the wave swash to move up the beach at an angle. The swash moves the sediment particles (typically sand) up the beach at this angle, while the backwash brings them directly down the beach slope, under the influence of gravity. This has the net effect of gradual movement of the sediment along the shore by the swash and backwash. The swash of the incoming wave moves sand up the beach in a direction perpendicular to the incoming wave crests and the backwash moves the sand down the beach perpendicular to the shoreline. Thus, with successive waves, the sand will move along a zigzag path along the beach parallel to shore. This process is known as **beach drift** (**Figure 10.13**).

A similar process, known as a **longshore current**, develops in the surf zone and a little farther out to sea (**Figure 10.14**). The movement of swash and backwash in and out from the shore at an angle creates turbulent water in the surf that transports sediments along the shallow bottom in the same direction as the beach drift. Substantially

longshore drift
The net movement of sediment parallel to the shore.

beach drift
The movement of sand along a zigzag path along the beach parallel to shore due to successive waves on the beach.

longshore current
A current that flows parallel to the shore just inside the surf zone. It is also called the littoral current.

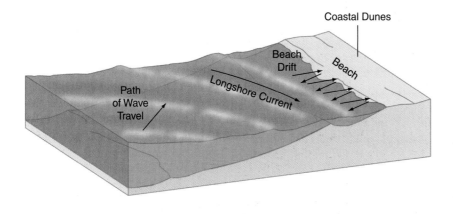

Figure 10.13 Beach drift is created by incoming waves hitting the beach at an angle moving sand that returns to sea in a direction perpendicular to the beach. Longshore current is the overall movement of water parallel to the beach but in the same general direction as the incoming wave direction.

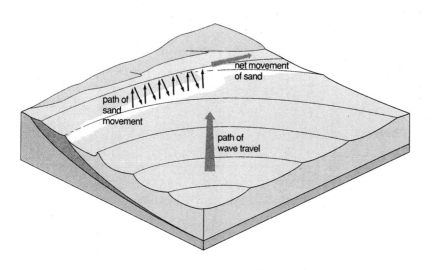

Figure 10.14 Longshore drift creates a net movement of sand along the beach and shallow water zone.

more sediment is transported along many beaches as a result of longshore currents than beach drift. Thus longshore currents and beach drift work together as longshore drift to transport huge amounts of sediment along a coast (Figure 10.14). At Sandy Hook, New Jersey, approximately 2000 tons of sediment per day moves past any point on the beach. No wonder beaches are often referred to by coastal geologists as rivers of sand.

Hazardous Rip Currents

rip current
Movement of water back into ocean through narrow zones through the surf zone.

There are times when waves can pile volumes of water on the beach that are much greater than normal. The only way this water can return to the ocean is to ebb back in a channel through the surf zone. This creates a **rip current** that typically flows perpendicular to the shore and the strong surface flow can have sufficient force to be a hazard to swimmers (**Figure 10.15**). It is often incorrectly called a "rip tide" or "riptide," because the occurrence is not related to tides, and also as an "undertow," although they do not drag people under water. Rates of return flow can range from 0.5 meters per second to as much as 2.5 meters per second. The position of rip currents can shift along the beach during the day as differing amounts of water are pushed up onto the shore.

Often two characteristics are present that allow us to identify a rip current. As the water is receding toward the ocean, its force counteracts the incoming force of the waves, thereby canceling out the waves. Thus a relatively smooth surface will be flanked by incoming waves. Also the receding water can carry along large amounts of sand and silt, which will discolor the water. It is advisable to look for the existence of a rip current before heading into the water. Such currents can be extremely dangerous, dragging swimmers away from the beach and leading to death by drowning when they attempt to fight the current and become exhausted. The United States Lifesaving Association reports that rip currents cause approximately 100 deaths annually in the United States, mainly on unguarded beaches. Over 80 percent of rescues by beach lifeguards are due to rip currents, totaling more than 37,000 lifeguard rescues in 2008.

If a swimmer is caught in a rip current, one should not try to swim directly back to shore but rather swim parallel to the shoreline in order to get out of the current. Rip currents can be between 15 and 45 meters wide. If you see a person caught in a rip current, yell at them to swim parallel to the shore and you should move along the shoreline in a direction that leads them out of the current.

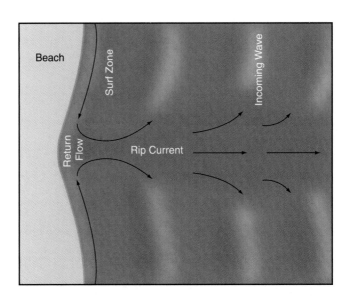

Figure 10.15 Rip currents form when too much water is pushed up onto the beach. Its return to the sea causes a rapid movement of water seaward.

Chapter 10 Coastal Regions and Land Loss   217

Coastal Sediment Deposition and Landforms

Sediment transported along the shore is deposited in areas of lower wave energy and produces a variety of landforms. Common landforms include *beaches*, *spits*, *tombolos*, and *barrier islands*. Erosion of headlands and sea cliffs is the source of some sediment, but probably no more than 5 to 10 percent of the total. The primary source of sediment is that transported to the coast by rivers that drain the continents and then redistributed along the shore by longshore drift.

Beaches

A **beach** is an accumulation of unconsolidated sediment along part of the coastline that is exposed to wave action. Beaches are formed from the wave-washed sediment along a coast, and represent interconnected zones of onshore and offshore sediment accumulation. Most beaches can be described in terms of three geomorphic zones (**Figure 10.16**). The **offshore zone** is the portion of beach that extends seaward from the low tide level. Strong backwash currents usually transport some sediment off exposed portions of the beach and deposit it offshore as submerged offshore bars. The foreshore zone represents the area between low and high tide levels. The **backshore zone**, which is commonly separated from the foreshore by a distinct ridge, called a **berm**, is the part of the beach extending landward from the high tide level to the area reached only during storms. Sediments in this zone are frequently redistributed by wind to form sand dunes. Behind the backshore may be a zone of cliffs, marshes, or additional sand dunes.

Even though beaches are areas of deposition, they are in a constant state of change, and dynamically responding to variations in the energy of waves and currents. The effect that waves have depends on their strength. Strong, also called destructive waves, occur on high energy beaches and are typical of winter storms (**Figure 10.17**). They

beach
An aggregation of unconsolidated sediment, usually sand, that covers the shore.

offshore zone
The portion of beach that extends seaward from the low tide level.

backshore zone
The part of the beach extending landward from the high tide level to the area reached only during storms.

berm
A low, incipient, nearly horizontal or landward-sloping area, or the landward side of a beach, usually composed of sand deposited by wave action.

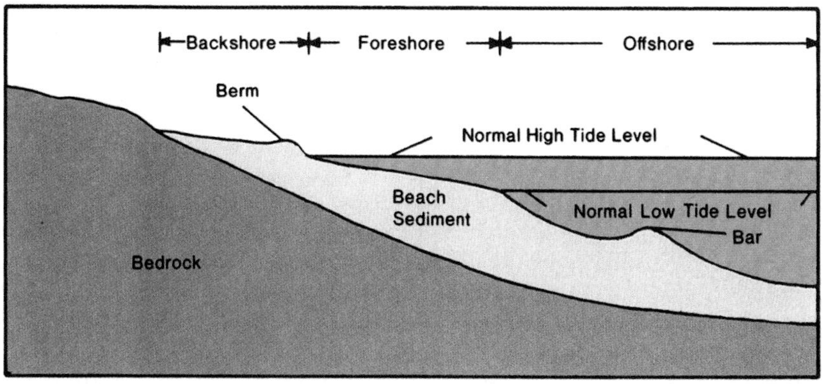

Figure 10.16 The topographic features of a beach area show the different "shore" zones that extend from the land out to sea.

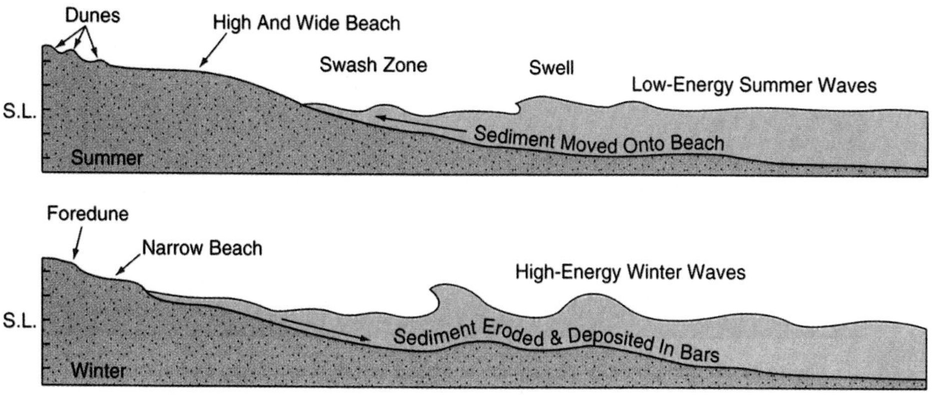

Figure 10.17 Beach profiles during the summer and winter differ in that the lower energy of the summer allows the beach to become enriched with sand. Winter storms erode most of the summer buildup and reduce the size of the beach.

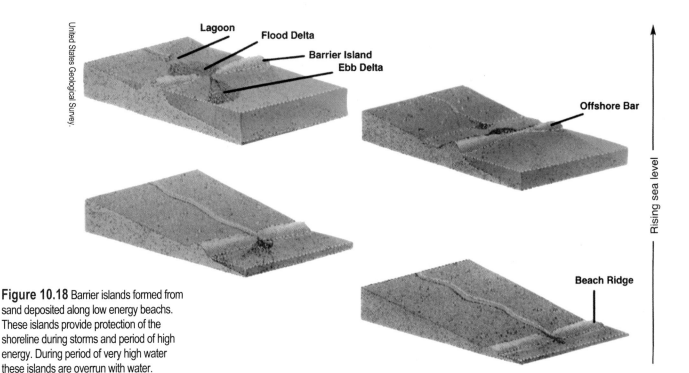

United States Geological Survey.

Figure 10.18 Barrier islands formed from sand deposited along low energy beachs. These islands provide protection of the shoreline during storms and period of high energy. During period of very high water these islands are overrun with water.

barrier island
A long, usually narrow accumulation of sand, that is separated from the mainland by open water (lagoons, bays, and estuaries) or by salt marshes.

James S. Best by permission.

Figure 10.19 Sand dunes serve as a line of defense against storm water that can overrun the beach zone. Salt-resistant sea grasses are often planted to stabilize the dunes.

reduce the quantity of sediment present on the beach by carrying it out to offshore bars under the sea. Constructive, weak waves are typical of low energy beaches, and occur usually during summer months. These are the opposite of destructive waves because they increase the size of the beach by removing sand from the offshore bars and piling it up onto the berm. This strong and weak wave activity alternates seasonally at most beaches. The weak wave activity produces a high and wide sandy beach at the expense of the offshore bars. The strong wave activity produces a narrow beach during the winter months and builds prominent offshore bars.

Barrier Islands

A **barrier island** is a long narrow offshore island of sediment running parallel to the coast and separated from it by a lagoon (**Figure 10.18**). Spits can form at the ends of barrier islands and as sand is transported into inlets by longshore and tidal currents.

Barrier islands are mostly between 15 and 30 kilometers long and from 1 to 5 kilometers wide. The tallest features are wind-blown sand dunes that reach heights of 5 to 10 meters (**Figure 10.19**). Barrier islands are common features along the Atlantic and Gulf coasts of the United States which forms the longest chain if barrier islands in the world. However, barrier islands are dynamic coastal features as they grow parallel to the coast by longshore drift and are often eroded by storm surges that often cut them into smaller islands. Despite their transient nature, many barrier islands heavily populated with homes and resorts. Even several major cities occupy barrier islands, including Miami Beach, Atlantic City, and Galveston.

Emergent and Submergent Coastlines

While sea level fluctuates daily because of tides, long-term changes in sea level have also occurred. Such changes in sea level result from uplift or subsidence along a coastline. Many coastal geologists classify coasts based on changes that have occurred in the past with respect to sea level. This commonly used classification divides coasts into two categories: *emergent* and *submergent*. An **emergent coastline** is a coastline that has experienced a fall in sea level, because of global sea level change, local land uplift, or isostatic rebound. Emergent coastlines are identifiable by the coastal landforms which are now above the high tide mark, such as raised beaches or raised wave cut benches (marine terraces) (**Figure 10.20**). Alternatively, a **submergent coastline** is one that has experienced a rise in sea level, due to a global sea level change or local land subsidence. Submergent coastlines are identifiable by their submerged, or "drowned" landforms, such as drowned valleys and fjords.

The type of coast produced is controlled mainly by tectonic forces and meteorological conditions (climate and weather). Tectonic processes can cause a coastline to rise or sink while lithospheric isostatic adjustment can depress or elevate sections of a continent. The tectonics at active plate margins can produce uplift or subsidence of a coast. In the northwestern United States, the coastline is slowly rising due to subduction of the Juan de Fuca plate beneath the North American plate. During the 1964 Alaska earthquake, parts of the coast rose and other parts subsided beneath sea level from a single event. Different isostatic adjustments to the land do not have to occur along plate margins.

During glacial periods, large continental ice sheets can displace the lithosphere into the plastic asthenosphere. About 18,000 years ago, a huge continental glacier covered most of Scandinavia, causing the land to sink isostatically. As the lithosphere sank, the displaced asthenosphere flowed southward, causing the Netherlands to rise. After the ice melted, the process was reversed and the asthenosphere flowed back north from below the Netherlands to Scandinavia today, Scandinavia is rebounding and the Netherlands is sinking. During the same glacial episode in North America,

emergent coastline
Is a coastline which has experienced a fall in sea level, because of global sea level change, local land uplift, or isostatic rebound.

submergent coastline
Is a coastline which has experienced a rise in sea level, due to a global sea level change or local land subsidence.

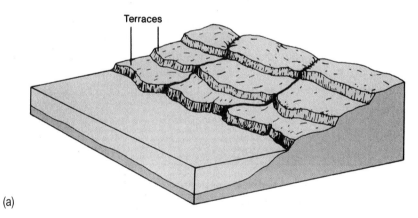

(a)

(b)

From *Physical Geology* by Dallmeyer. Copyright © 2000. Reprinted by permission of Kendall Hunt Publishing Co.

Figure 10.20 (a) An emergent coastline often has wave-cut terraces that represent odler period of higher sea level. (b) This wave-cut terrace is part of an emergent shoreline at Baker's Point, California.

Canada was depressed by ice, and asthenosphere rock flowed southward. The asthenosphere is flowing back north and much of Canada is rebounding while much of the United States is now sinking.

Global changes in sea level can also occur. Such global sea level changes are called **eustatic** changes and can occur by three mechanisms: the growth or melting of glaciers, changes in water temperature, and changes in the volume of the mid-ocean ridges. During glacial periods large amounts of water evaporated from the oceans becomes stored on the continents as glacial ice and causes sea level to become lower, resulting in global land emergence. Similarly, when glaciers melt, water flows back into the oceans and sea level rises globally, causing land submergence.

Changes in the volume of mid-ocean ridges can also affect sea level. Growth of a mid-ocean ridge displaces seawater upward. If lithospheric plates spread slowly from the ridge they create a narrow mountain ridge system that displaces relatively small amounts of seawater, resulting in lower sea level. In contrast, rapidly spreading plates produce a high-volume ridge system that displaces more water upwards, resulting in higher sea level. At times during Earth's history, sea floor spreading has been relatively rapid, and as a result, global sea level has been higher.

eustatic
Global changes in sea level.

Coastal Hazards: Living with Coastal Change

From our discussion of our coastal areas it is apparent that diverse and complex processes are at work continually changing the coastal landscapes. Vast areas of coastal land have been lost since the mid 1800s as a result of natural processes and human activities (Table 10.1). The most important causes that have the greatest influence on coastal land loss are relative sea level rise, erosion from frequent storms, and reductions in sediment supply; whereas the most important human activities are sediment excavation, river modification, and coastal construction. Any one of these causes may be responsible for most of the land loss at a coast, or the land loss may be the result of several of these factors acting at the same time.

From a hazard point of view, coastal erosion is the most widespread and continuous process affecting the world's coastlines and contributing to land loss destruction. Global warming and sea level rise are slow-onset hazards that greatly contribute to the erosional process. However, catastrophic, rapid-onset events play a very significant role both for coastal erosion and human suffering. These include erosion and destruction from storms, landslides, and tsunami.

Coastal Land Loss by Global Sea-Level Rise and Subsidence

A significant amount of coastal erosion presently plaguing today's coastlines is the result of gradual but sustained global sea level rise (**Figure 10.21**). Most of this rise is from the melting of polar continental ice sheets, coupled with expansion of the water itself as global temperatures increase. The sea level rise is currently estimated at about 0.3 meter per century. Although this amount does not sound very threatening, additional factors increase the risk. First, the slope of many coastal areas is very gentle so that a small rise in sea level results in a far larger inland advance of the coast than steeper sloping areas. The vulnerability of coastal regions along the eastern seaboard of the United States shows a wide range of potential risks (**Figure 10.22**). Estimates of the amount of coastline retreat in the United States from a sea level rise of 0.3 meters would be 15 to 30 meters in the northeast, 65 to 130 meters in California, and up to 300 meters in Florida. Second, the documented rise of atmospheric carbon-dioxide

TABLE 10.1	**Common Physical and Anthropogenic Causes of Coastal Land Loss**

Natural Processes

Agent	Examples
Erosion	waves and currents storms landslides
Sediment Reduction	climate change stream avulsion source depletion
Submergence	land subsidence sea-level rise
Wetland Deterioration	herbivory freezes fires saltwater intrusion

Human Activities

Agent	Examples
Transportation	boat wakes, altered water circulation
Coastal Construction	sediment deprivation (bluff retention) coastal structures (jetties, groins, seawalls)
River Modification	control and diversion (dams, levees)
Fluid Extraction	water, oil, gas, sulfur
Climate Alteration	global warming and ocean expansion increased frequency and intensity of storms
Excavation	dredging (canal, pipelines, drainage) mineral extraction (sand, shell, heavy mins.)
Wetland Destruction	pollutant discharge traffic failed reclamation burning

Source: United States Geological Survey.

levels (discussed in Chapter 8) suggests that global warming from the increased greenhouse-effect will melt the glaciers more rapidly, as well as warming the oceans more, thereby accelerating the rise of sea level. Some estimates put the anticipated rise in sea level at 1 meter by the year 2100. In addition to increased beach erosion, it would also flood 30 to 70 percent of coastal wetlands in the United States that protect shores from storm flooding events.

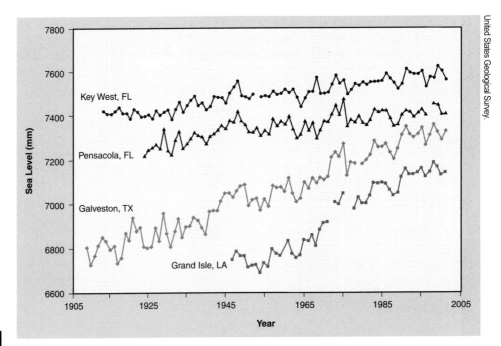

United States Geological Survey.

Figure 10.21 Long-term trends in average annual sea level at selected tide gauges in the Gulf of Mexico. Data are from the National Ocean Service.

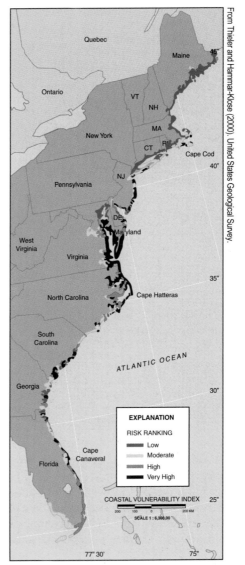

From Thieler and Hammar-Klose (2000), United States Geological Survey.

Figure 10.22 Map of the Coastal Vulnerability Index (CVI) for the U.S. East Coast showing the relative vulnerability of the coast to changes due to future rises in sea level. Areas along the coast are assigned a ranking from low to very high risk, based on the analysis of physical variables that contribute to coastal change.

The most widely assessed effects of future sea level rise are coastal inundation (submergence), erosion, and barrier migration. The USGS estimates the primary impacts of a sea level rise on the United States to be: (1) the cost of protecting ocean communities by pumping sand onto beaches and gradually raising barrier islands in place; (2) the cost of protecting developed areas along sheltered waters through the use of levees (dikes) and bulkheads; and (3) the loss of coastal wetlands and undeveloped lowlands. The total cost for a one meter rise is estimated to be $270–475 billion, ignoring future development.

Coastal submergence refers to permanent flooding of the coast caused by either a rise in global sea level or subsidence of the land, or both. At many coastal sites, submergence is the most important factor responsible for land loss and as sea level rises, or the land subsides, it will inundate present unprotected low-lying coastal areas and cities such as: Boston, New York, Charleston, Miami, and Los Angeles. Submergence also accelerates coastal beach erosion and landslides because it facilitates greater inland penetration of storm waves. In addition to accelerated land loss, coastal submergence causes intrusion of salt-water into coastal aquifers.

coastal submergence
The permanent flooding of coastal areas by global sea level rise or land subsidence.

Coastal Land Loss by Erosion: Impacts from Storms and Landslides

Superimposed on the slower sea level rise are shorter duration water fluctuations caused by storm events. The most damaging coastal storms for the United States are tropical cyclones (hurricanes) and extratropical cyclones (winter storms) that form around low-pressure cells (see Chapter 8 and 11 for discussions of these storms).

Hurricanes form in the tropics during summer to early fall and migrate northward and westward into temperate regions of the Atlantic and Gulf coasts. The extratropical storms occur mostly in winter (like nor'easters) and can cause erosion of the coastline at much higher rates than normal. Although each type of storm is unique, there are several factors common to all storm types which include strong winds, generation of large waves, and elevated water levels known as a storm surge.

Storm Impacts

Strong storms bring more energy to the coastline causing higher rates of erosion (**Figure 10.23**). Higher erosion rates are due to several factors:

- Wave velocities are higher during storms and thus larger particles can be carried in suspension causing sand on beaches to be picked up and moved offshore. This leaves behind coarser grained particles such as pebbles and cobbles, thus reducing the width of sandy beaches;
- Storm waves reach higher levels onto the coast and destroy and remove structures and sediment from areas not normally reached by normal waves;
- Wave heights increase during a storm and crash higher onto cliff faces and rocky coasts. Larger rock debris or debris from destroyed structures is flung against the rock causing rapid rates of erosion by abrasion;
- Hydraulic action increases as larger waves crash into rocks. Air and water occupying fractures in the rock becomes compressed and thus the pressure in the fractures is increased which causes further fracturing of the rock.

United States Geological Survey.

Figure 10.23 Upper figure shows erosion and property damage near Floridana Beach, Florida, caused by Hurricane Frances on September 4, 2004. The lower image shows the same area following the arrival of Hurricane Jeanne on September 25, 2004. Jeanne produced much greater beach erosion.

FEMA, photo by Marty Bahamonde.

Storm surge is responsible for about 90 percent of all human fatalities and damage during storms (**Figure 10.24**). A **storm surge** is an onshore flood of water created by a low pressure storm system. The surge is caused primarily by the strong winds of the storm blowing over the sea surface and causing the surface water to pile up above sea level. Low pressure at the center of the storm also elevates the surface water upward and enhances the height of the mound of water. As the storm nears land, the shallower sea floor prevents the piled-up water from collapsing and it floods inland as a deadly storm surge. Storm surges are at their highest and most damaging when they coincide with high tide (especially at the high tides during the spring tide cycles), combining the effects of the surge and the tide.

Figure 10.24 Property damage caused by a storm surge associated with the Patriot's Day nor'easter that struck New England in 2007.

storm surge
An onshore flood of water created by a low pressure storm system.

overwash
A deposit of marine-derived sediments landward of a barrier system, often formed during large storms; transport of sediment landward of the active beach by coastal flooding during a tsunami, hurricane, or other event with extreme wave action.

Dune and Beach Recession

High storm-generated waves erode large quantities of sediment from dune and beach areas. From March 5 through 8, 1962 a major coastal storm, known as the "Ash Wednesday" storm, moved northward and became stalled against the middle Atlantic coast through five high tides. The documented erosion that occurred at Virginia Beach, Virginia, showed that 30 percent of the beach and dune sand was removed. The crest of the dunes at Virginia Beach was reduced from an elevation of 4.9 to 3.4 meters thus enabling future storm surges to rise over the dunes and flood inland areas.

Dune and Beach Breeching and Overwash

Large amounts of sediment can be eroded from a beach and dunes during a major storm with some migrating along the shore by accelerated longshore currents and some moved offshore. In addition, storm waves may wash sediment through low areas between the dunes of islands and onto the back side it. This **overwash** is important because it maintains the barrier island's width as its front is eroded but can be devastating to homes and other structures). Overwash fans are lobe-shaped deposits eroded from the ocean side of a shore and deposited in the bays and lagoons behind barrier islands.

Barrier Island Breeching

Severe storms can cut through a barrier island to produce a tidal inlet and tidal delta. A storm-produced tidal inlet often is short-lived as it closes naturally in a few weeks by longshore drift deposits that cross the inlet and fills it. The filling of a tidal inlet produces a flat area that invites future housing development. However, this area will be the first to flood in a future storm and time and time again, barrier islands have been breached at the same location where they were cut through many years before.

El Niño Events

Along the Pacific coast, winter storms and unusual oceanographic conditions such as El Niño cause the most erosion and land loss. Approximately every four to five years, El Niño conditions cause warm surface water of the Pacific Ocean to flow eastward piling up water along the west coast of North and South America (refer to Chapter 8). The elevated water levels and the unusually strong storms during El Niño events cause extensive flooding and erosion beaches and cliffs. In 1983, an unusually strong El Niño caused torrential rainfall, rapid beach erosion, and massive landslides along the Pacific coast of the United States. Land loss was concentrated along the southern

California coast where numerous expensive homes built on cliffs were damaged or destroyed.

Two more major El Niño storms hit California in October 1997 and six months later in April 1998. Coastal areas were heavily eroded by these two events (**Figure 10.25**).

Landslides and Cliff Retreat

Coastal landslides occur where unstable slopes fail and land is both displaced down slope (**Figure 10.26**). Some of the fundamental causes of slope failures that lead to land loss are: (1) slope over-steepening (2) slope overloading, (3) shocks and vibrations, (4) water saturation, and (5) removal of natural vegetation. Sea level rise can elevate waves so they can erode and undercut cliffs at higher elevations, initiating mass movements. Cliffs may stay relatively stable and then retreat several meters in a single storm event which makes building structures near the edge of cliffs an especially risky during times of sea level rise. More discussion of landslides and their mechanisms is found in Chapter 5.

Coastal Land Loss by Human Activities

There is increasing evidence that recent land losses in many coastal regions are largely anthropogenic and attributable to human alteration of the coastal environment. Land losses indirectly related to human activities are difficult to quantify because they promote alterations and imbalances in the primary factors causing land loss such as sediment budget, coastal processes, and relative sea level changes. Human activities causing land loss are: transportation networks that tend to increase erosion, coastal construction projects that typically increase deficits in the sediment budget, subsurface fluid extraction and climate alterations that accelerate submergence and excavation projects that cause direct losses of land.

There are countless examples of human interference with coastal processes. The beach at Miami Beach must be restored periodically by sand pumped from offshore. In Louisiana the land is subsiding as sea level rises causing loss of natural coastal wetlands. Many California beaches are eroding due to the damming of rivers for irrigation and flood control. The river-supplied sediment that normally replenishes the beaches is being trapped in reservoirs behind dams. Since this sediment is not being supplied to the ocean, longshore currents cannot resupply the beaches with sediment. Instead, longshore currents carry the existing sediment in the downdrift direction, resulting in significant erosion of the beaches. Eliminating wetlands for development and agriculture removes the natural flood protection and storm-swollen estuaries now flood barrier islands from the bay side as storms move inland. Where dunes are removed, the most effective barrier to storm waters has been lost overwashing is more common. Groins built to trap sediment and widen updrift beaches have caused serious erosion problems downdrift. Over the last 10,000 years, most of the state of Louisiana has formed from the deposition of sediments by the flooding of the Mississippi River. Humans, however, have prevented the river from flooding by building levee systems that extend to the mouth of the river. As previously deposited sediments become compacted they tend to subside. Since no new sediment is being supplied by Mississippi river flooding, the subsidence results in a relative rise in sea level. This, coupled with a current rise in eustatic sea level, is causing coastal Louisiana to erode at an incredible rate.

United States Geological Survey, Center for Coastal and Watershed Studies.

Figure 10.25 Coastal region near Ventura, California. Upper image shows the coast following the El Niño storm of October 1997; lower image shows a definite change in the coastal morphology following the April 1998 El Niño storm.

Courtesy of David M. Best.

Figure 10.26 The small community of La Conchita, California, lies along the coast of the Pacific Ocean. Bluffs that overlie the town have collapsed twice since 1995 due to heavy rainfall that saturated the ground.

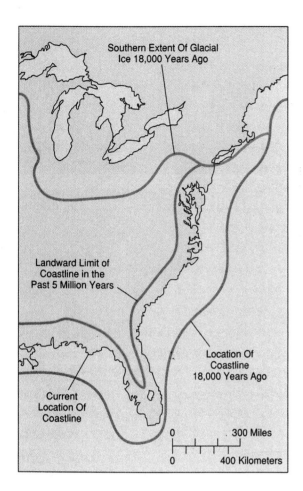

Southern Extent Of Glacial Ice 18,000 Years Ago

Landward Limit of Coastline in the Past 5 Million Years

Location Of Coastline 18,000 Years Ago

Current Location Of Coastline

0   300 Miles

0   400 Kilometers

Figure 10.27 Coastlines have fluctuated about 400 km over the past five million years. Notice that sea level was much lower in the past 18,000 years.

Tsunami

A tsunami is a shallow water sea wave generated by earthquakes, volcanic eruptions, meteorite impacts in the ocean, or landslides. Tsunami can cause coastal flooding and catastrophic destruction thousands of kilometers from where they where generated. Such waves can have wave heights up to 30 meters, and have great potential to wipe out large coastal cities. Tsunami are discussed separately in Chapter 6.

Lessons from the Geologic Record

Since the oceans formed over four billion years ago, they have influenced the coastlines of the newly forming continents. Sea level has risen and fallen repeatedly in the geologic past, and its coastlines have subsequently submerges and emerged throughout Earth's history. The rock record shows countless examples of seas transgressing and regressing over the continents. Marine fossils found thousands of kilometers inland from the present coast attests to this fluctuation in past sea levels.

At times during Earth's long history, tectonic movements arranged the continents into very different configurations from those of today. When there were large amounts of land near the poles, the rock record shows unusually low sea levels during past ice ages due to large ice sheets forming on the continents. During times when the land masses clustered around the equator, ice ages had much less effect on sea level. However, over most of geologic time, long-term sea level has been higher than today. Only at the Permian-Triassic boundary about 250 million years ago was long-term sea level lower than today.

The world's present coastlines are not the result of present-day processes but were affected by the rise of sea level caused by the melting of the Pleistocene glaciers beginning between 15,000 and 20,000 years ago (**Figure 10.27**). The rising sea flooded large parts of the low coastal areas, which are now part of the continental shelf, and moved the coastlines inland. Sea-level has risen about 130 meters since the peak of the last ice age about 18,000 years ago. It was during this time of very low sea level that there was a dry land connection between Asia and Alaska over which humans migrated to North America over the Bering Land Bridge. However, for the past 6,000 years (a few centuries before the first known written records), the world's sea level has been gradually approaching the level we see today.

Lessons from the Historic Record and the Human Toll

The increase in coastal populations together with rising sea level and intense storms combine to make coastal erosion and flooding very costly and life-threatening. Since the early 1900's, property damages in the United States have been on the rise. However, the death tolls have generally decreased (at least until Hurricane Katrina) because of advanced warnings and evacuations to the weather-related storm events. Unfortunately, people continue to build more structures along migrating shorelines and are unaware of the dynamic balance between erosion and deposition (refer to Box 10.1). Nearly all human intervention with coastal processes interrupts natural processes and thus can have an adverse effect on coastlines.

Coastal environments are a delicate setting in which conflicts arise when humans and natural processes attempt to coexist. With about 58 percent of people in the United States living within 150 km of a shoreline, it is inevitable that problems will

BOX 10.1   One Case Study Made By the United States Geological Survey

Ocean City, Maryland: An Urbanized Barrier Island

Relative recent changes in the shoreline have occurred in populated regions of the United States. USGS Circular 1075 provides a good summary of the events that have involved the shoreline near Ocean City, Maryland. For more than a century Ocean City, Maryland, has been a popular beach resort for vacationers from the Northeast and Mid-Atlantic States (Box Figure 10.1.1). During the Roaring 1920s, several large hotels and a boardwalk were built to accommodate visitors and development continued slowly until the early 1950s. Then a period of rapid construction began that lasted almost 30 years. Concerns about the coastal environment were raised in the late 1970s and led to Federal and State laws to limit dredging and filling of wetlands. The resort is built on the southern end of Fenwick Island, one of the chain of barrier islands stretching along the east coast (Box Figure 10.1.2). The Great Hurricane of 1933 (before names were assigned to hurricanes) opened the Ocean City Inlet by storm-surge overwash from the bay side. To maintain the inlet as a navigation channel, two large stone jetties were constructed by the U.S. Army Corps of Engineers. These jetties helped stabilized the inlet, but they have drastically altered the sand-transport processes near the inlet. The net longshore drift at Ocean City is southerly; it has produced a wide beach at north of the jetty, but Assateague Island, south of the inlet, has been starved of sediment. The result is a westerly offset of more than 500 meters in the once-straight barrier island.

The most damaging storm to hit Ocean City within historic times was the Five-High or Ash Wednesday northeaster of early March 1962 which caused severe erosion and flooding along much of the middle Atlantic Coast. For two days, over five high tide cycles, all of Fenwick Island, except the highest dune areas was repeatedly washed over by storm waves superimposed on a storm surge measuring 2-meters high. Property damage in Ocean City was estimated at $7.5 million. Given the dense development of the island over the last 20 years, damage from a similar storm would today be hundreds of millions of dollars.

Box Figure 10.1.1 The beaches at Ocean City, Maryland, are a popular vacation spot for many people who live in the heavily populated Atlantic Seaboard of eastern United States. Large amounts of construction are at risk along the beaches of the United States form cyclonic storms.

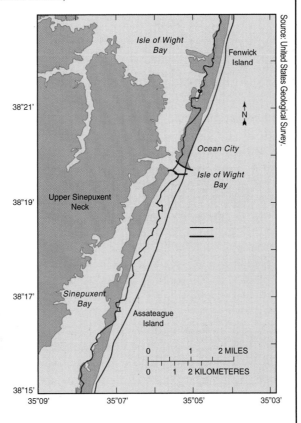

Box Figure 10.1.2 The natural sediment transport along the Fenwick Island-Assateague Island region has been altered by the construction of two large jetties at Ocean City Inlet. The landward shift of the southern barrier island was caused by a change in the longshore currents in the area.

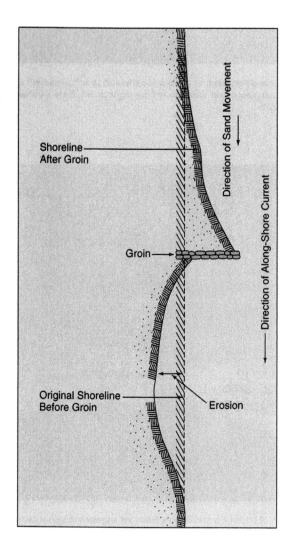

Figure 10.28 The effects of constructing a groin on a beach are that portions of the beach undergo erosion while others down drift experience erosion.

Courtesy of David M. Best.

Figure 10.29 This groin along the English Channel is constructed along a beach that consists of pebbles and cobbles rather than sand.

develop. These problems are all too common when cyclonic storms hit a coastline. The rising waters and wind cause large amounts of damage to property, and unfortunately take people's lives. Examples are well known to those who live in these areas, and others who live elsewhere become aware of the issues when large storms hit. The case of Hurricane Katrina striking the Gulf Coast region of the United States in late August 2005 brought these problems to everyone's attention.

Mitigation: Adapting to Coastal Erosion

Barrier islands and beaches, since they are made of unconsolidated sediment, and sea cliffs, since they are susceptible to landslides due to undercutting, are difficult to protect from the erosive action of the waves. Human construction methods can attempt to prevent erosion, but cannot always protect against abnormal conditions. In addition, other problems are sometimes caused by these engineering structures.

Protection of the Shoreline

Shoreline protection can be divided into two categories: hard stabilization in which solid structures are built to reduce wave action, and soft stabilization which mainly refers to adding sediment back to a beach as it erodes.

Hard Stabilization

Two types of hard stabilization are often used. The first type interrupts the flow of sediment along the beach. These structures include **groins** (**Figures 10.28** and **29**) and **jetties** (**Figure 10.30**), built at right angles to the beach to trap sand and widen the beach. The second type interrupts the force of the waves. **Seawalls** are built parallel to the coastline to protect structures on the beach (**Figures 10.31**) by allowing waves to crash against it and preventing them from running up the beach. **Breakwaters** serve a similar purpose, but are built offshore parallel to the beach (**Figure 10.32**), again preventing the force of the waves from reaching the beach and any structures.

While hard stabilization usually works for its intended purpose, it does cause sediment to be redistributed along the coast. A breakwater, for example, causes wave refraction, and alters the flow of the longshore current. Sediment is trapped behind the breakwater, and the waves become focused on another part of the beach where they can cause significant erosion (Figure 10.32). Similarly, because groins and jetties trap sediment, areas in the downdrift direction are not resupplied with sediment by the longshore current, and beaches are eroded and become narrower in the downdrift direction.

Soft Stabilization

Soft stabilization is primarily accomplished by adding sediment to the coastline which is called **beach nourishment**. This is usually done by dredging sediment from offshore and pumping it onto the coastline.

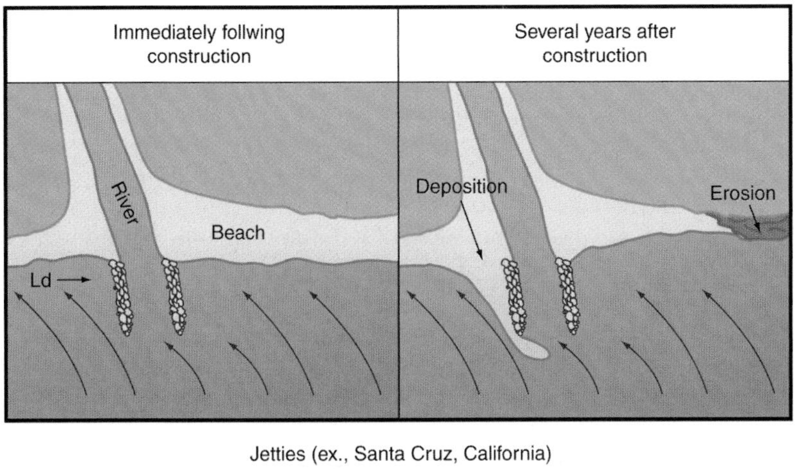

Immediately follwing construction
Several years after construction
River
Beach
Ld
Deposition
Erosion
Jetties (ex., Santa Cruz, California)

Figure 10.30 Jetties are extensions of pre-existing channels or river channels. Notice how depositional patterns change and sand begins to migrate around the jetty, eventually becoming a hazard to the entrance to the river.

From *Physical Geology* by Dallmeyer. Copyright © 2000. Reprinted by permission of Kendall Hunt Publishing Co.

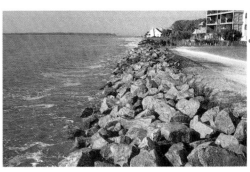

Figure 10.31 This rock sea wall is built to absorb the energy of storm waves coming off the ocean. Unfortunately this feature removes the possibility of having a recreational beach.

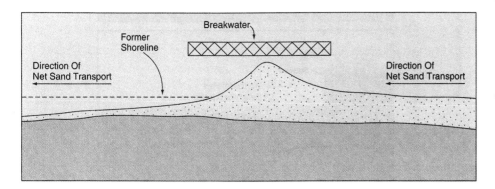

Breakwater
Former Shoreline
Direction Of Net Sand Transport
Direction Of Net Sand Transport

Figure 10.32 The construction of a breakwater off shore and parallel to the shoreline creates a buildup of sand that extends from the shoreline out to the breakwater.

groin
Solid structure built at an angle from a shore to reduce erosion from long shore currents, and tides.

jetty
A structure extending into the ocean to influence the current or tide in order to protect harbors, shores, and banks.

seawall
Massive structure built along the shore to prevent erosion and damage by wave action

breakwater
Structure built offshore and parallel to shore that protects a harbor or shore from the full impact of waves.

beach nourishment
A soft stabilization technique primarily accomplished by adding sediment to the coastline.

Adding sediment is necessary when erosion removes too much sediment, however, because the erosive forces are still operating, additional sediment will need to be periodically replenished at continuing expense (**Figure 10.33**). Less invasive approaches include construction of access walkways and steps to reduce erosion of fragile dunes, as well as planting and protection of well-rooted vegetation.

© Kochneva Tetyana, 2010. Under license from Shutterstock, Inc.

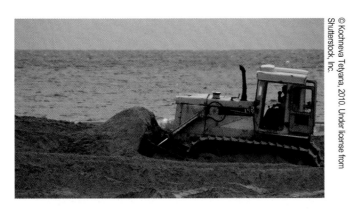

Figure 10.33 The replenishment of beach sands is key to maintaining the health of the shoreline. Each year millions of cubic yards of sand are added to coastal regions to maintain the balance between the ocean and the land.

Abandonment

Protecting ourselves from coastal hazards, such as beach erosion and coastal flooding, requires long-term strategic planning. Should we continue to attempt to defend the coasts from rising seas and ever larger storms? Or do we recognize awesome power of natural processes and strategically abandon the coast? Both methods are very costly and decisions will most likely require passage of governmental regulations. Presently, public policies are encouraging the development of hazardous areas by providing federal flood-insurance and disaster relief programs which encourages homeowners and businesses to rebuild after a disaster.

Many coastal geologists believe that instead of rebuilding our eroding beaches, we should return more of them to public use after they are devastated by a storm. Examples of such public coastal resources are the National Seashores, such as Cape Cod in Massachusetts, Padre Island in Texas, Hatteras in North Carolina, and Point Reyes in California. National Seashores are kept in as natural a condition as possible for all to enjoy. Storms will continue to damage their roads, parking lots, bathhouses, and concessions, but these can inexpensively be rebuilt, compared to the cost of restoring high-rise hotels, condominiums, and beach homes.

Summary

Coastlines are among Earth's most geologically active environments and are continually changing because of the dynamic interaction between the water and the land. Coastal changes occur on two very different time scales. Short-term change is largely due to seasonal severe storm events, while long-term changes are due to slow sea-level rise. Waves, longshore currents, and tidal currents interact with the geologic structures and plate tectonic processes of the coast to shape coastlines into a multitude of landforms.

Waves are formed by wind forces acting on the water surface and transfers energy in circular motion through the water. Motion in a wave ceases at a depth equal to one-half its wavelength. As a wave enters shallow water, its wave base intersects the bottom and the wave begins to break. Most of the energy carried within a wave is then released within the surf zone by breakers. Sediment is moved along the shoreline by longshore drift and into and out of tidal inlets by tidal currents. Beaches are accumulations of sediment along the shoreline.

Coastal storms and hurricanes are inherent natural hazards along most U.S. shorelines. As population and development continues to increase, storm damage has risen dramatically. Many coasts are affected by multiple hazards such as storm erosion, landslides, tsunami, and coastal flooding. Many scientists see evidence that sea level is rising, that storms and hurricanes will continue to occur as frequently as they do today, and that their magnitude may increase as global temperature rise.

Human interference with coastal processes and landform development is causing serious problems. Seawall construction protects structures against wave attack, but causes beach erosion. Groins cause the updrift beach to grow, but the downdrift beaches erode. Breakwaters and jetties prevent erosion of beaches in some areas, but their effect is to cause sand to move to other positions along the beach and waterfront. Hard and soft stabilization of beaches is only a short-term solution in the overall process.

References and Suggested Readings

Bush, D. M., O. H. Pilkey, Jr., and W. J. Neal. 1996. *Living by the Rules of the Sea.* Durham, NC: Duke University Press.

Coch, N. K., 1995, *Geohazards, Natural and Human.* Prentice Hall: Upper Saddle River, NJ.

Douglas, B. C., and W. R. Peltier, 2002. The Puzzle of Global Sea-Level Rise. *Physics Today* 55 (3): 35–41.

Komar, P. D. 1997. *The Pacific Northwest Coast:* Duke University Press: Durham, NC.

Morton, R. A. 2003. An overview of coastal land loss: with emphasis on the southeastern United States. U.S. Geological Survey, Open File Report 03-337.

Thompson, G.R. and Turk, J., 2007, *Earth Science and the Environment*. Thomson & Brooks/Cole: Belmont, CA.

Trujillo, A. P. and H. V. Thurmann. 2008. *Essentials of Oceanography*. Pearson Prentice Hall: Upper Saddle River, NJ.

Williams, S. J., K. Dodd, and K. K. Cohn, 1990. Coasts in Crisis. U.S. Geological Survey Circular 1075.

Web Sites for Further Reference

http://solidearth.jpl.nasa.gov/PAGES/sea01.html
http://coastal.er.usgs.gov/hurricanes/cch.html
http://coastalchange.ucsd.edu/index.html
http://pubs.usgs.gov/of/2003/of03-337/
http://marine.usgs.gov/kb/views/cch/index.php

Questions for Thought

1. How are waves that affect coastlines generated?
2. Explain longshore currents and how they cause sediment transport on beaches.
3. Describe what happens to waves as they shoal in shallow water.
4. What are the main causes of coastal erosion?
5. What are the main coastal hazards?
6. What are barrier islands, and how do they form?
7. Describe ways in which the relative elevation of land and sea may change.
8. What is the present trend in global sea level and what effect does it have on the coasts?
9. Describe the motion of a water particle as a wave passes.
10. Why are beaches often called rivers of sand?
11. What is wave refraction, and how does it affect coast erosion and deposition?
12. What influence do tides have on coastal processes and hazards?

Hurricanes, Cyclones, and Typhoons

11

Hurricane Fran approaches the southeastern United States, September 19, 1996.

NOAA

Key Terms

cyclone
cyclonic storm
eye
eye wall
hurricane
storm surge
thermal energy
tropical cyclone
tropical depression
tropical disturbance
tropical storm
typhoon

cyclonic storm
A generic terms that covers many types of weather disturbances that are typified by low atmospheric pressure and rotating, inwardly directed winds.

cyclone
A rotating mass of low pressure in the atmosphere that covers a large area; the warm air mass rotates counterclockwise in the northern hemisphere and clockwise in the southern hemisphere. The term is strictly applied to large low pressure storms in the Indian Ocean and South Asia.

typhoon
A cyclonic storm that forms in the central and western Pacific Ocean. Refer to cyclone and hurricane.

hurricane
Term applied to cyclonic storms that occur in the North Atlantic Ocean or eastern Pacific Ocean. Minimum wind velocity is 74 miles per hour. Refer to cyclone and typhoon.

Large-scale cyclonic storms occur in many areas of the globe, generated by warm, oceanic waters that develop rising air circulations that lift moisture and heat from the water surface into rotating storms. Few regions on Earth are spared from these weather phenomena. Both the geologic and human impacts can be long lasting when these storms encounter continental regions.

Hundreds of cyclonic storms have hit coastal and inland regions along the Pacific, Indian, and North Atlantic oceans, as well as the Caribbean Sea and Gulf of Mexico. Only a few specific storms are discussed in this chapter.

Areas of Cyclonic Storm Generation

Wind is the movement of air. It can be affected by changes in air pressure and the Earth's rotation. On a global scale the differences in heating near the equator and the cooling at the poles produce large circulation cells that drive large weather systems across Earth's surface. In some cases these are gentle breezes, but at other times winds can produce storms that become major natural disasters. In areas of low atmospheric pressure, which are associated with stormy conditions, rotational forces cause winds to create a swirling vortex. As the rotation intensifies, a cyclonic disturbance forms and moves through the Earth's atmosphere, with the potential of becoming a major cyclonic storm (**Figure 11.1**).

Three different terms are used to categorize these cyclonic storms, based on where they first develop. Cyclones are associated with the Indian Ocean and South Pacific Ocean and affect South Asia, Australia, and the east coast of Africa. Typhoons are generated in the western and southern Pacific, affecting southeast Asia, China, and Japan. Cyclonic storms in the western portions of the North Atlantic Ocean, the Caribbean Sea, and the eastern Pacific off Mexico and Central America are termed hurricanes (**Figure 11.2**). Unless we refer to a specific storm in a portion of the globe

National Geophysical Data Center, NOAA

Figure 11.1 A sequence of satellite images shows the position of Hurricane Andrew as it approaches and moves across Florida and the Gulf of Mexico. The storm image to the right was taken August 23, 1992; the center image is its position on August 24, and the image on the left was taken August 25, 1992. Notice the changes in the size of the eye of the storm.

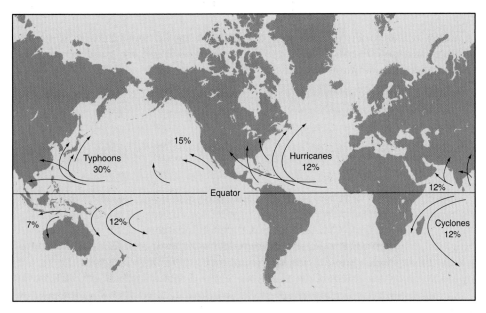

Figure 11.2 Global map showing where storms first form.

other than the North Atlantic and adjoining regions, we will refer to cyclonic storms as hurricanes.

Many variables are involved in the formation of these storms. They have a range of sizes and intensities as well as different forward velocities, so each storm is unique in terms of its characteristics and behavior. However, some traits are common to all cyclonic storms.

- They initially form in only a few areas over the eastern and southwest Pacific Ocean, eastern North Atlantic Ocean, and north Indian Ocean.
- Storms form between the latitudes of 5° and 20° N or S.
- They develop more frequently during the summer months of their respective hemispheres, but they can occur in any month.

The movement of cyclonic storms across Earth's surface is influenced by Earth's rotation. This rotation produces the Coriolis effect, which generates a force that causes storms in the northern hemisphere to curve to the right. The reverse movement is seen in the southern hemisphere. The physics of fluid dynamics also controls movement within the low-pressure zones. Areas at risk include the shorelines lying in the path of the storm in addition to the water surface over which the storm passes. Low-lying coastal areas are frequently subjected to major flooding produced by the landward movement of massive amounts of water being pushed ashore by the storm's winds. In general, east-facing coastal areas are affected in the Northern Hemisphere. An exception is the west coast of Mexico, which is often hit by storms coming off the eastern Pacific Ocean and the western side of India, which is affected by storms that form in the northern Indian Ocean.

Hurricane Formation and Movement Processes

In an average year, approximately 80 to 100 cyclonic storms develop in the tropical and subtropical regions on Earth, being born from the warm waters lying near the equator. Storms have a life span of several days up to a few weeks, during which time they can intensify from a small tropical depression into a major atmospheric event that can affect millions of people.

tropical disturbance
The beginning stage of a cyclonic storm lasting at least 24 hours, originating in the tropics or subtropics; clouds and moisture become organized and a vertically rotating wind mass creates atmospheric instability.

tropical depression
A slow-forming cyclonic storm with sustained surface winds of 38 miles per hour or less.

tropical storm
A cyclonic storm with sustained surface winds between 39 miles per hour to 73 miles per hour. At this level of activity the system is assigned a name to identify and track it.

thermal energy
The amount of energy in a system that is related to the temperature of its constituents; for hurricanes this is heat originally taken from the oceans.

Hurricane Conditions

For a **tropical disturbance** to evolve into a **tropical depression** and eventually a hurricane, several conditions must be met:

- Warm, ocean surface water (exceeding 24°C), along with high humidity, and unidirectional, constant winds
- Rotating winds produced by thermal heating and from the Coriolis effect in the lower latitudes
- Low atmospheric pressure near the equator that creates an atmospheric, tropical depression (**Figure 11.3**) that can develop into more intense storms.

Cyclones form in tropical and subtropical latitudes, generally between 5° and 20° north or south of the equator. This region on Earth receives the highest amount of solar radiation, which warms ocean waters that evaporate into the atmosphere (Figure 11.3). Sea water must be a minimum temperature of 24°C to be warm enough for heat and moisture to contribute to the formation of a storm. The resulting low-pressure conditions begin to rotate as the heat and moisture rise. Circulation is counterclockwise in the northern hemisphere and clockwise in the southern hemisphere.

The National Hurricane Center defines the progression of storm development as shown in Table 11.1. Over a period of days before the system reaches the threshold of a **tropical storm**, wind velocities increase and the storm becomes better defined; at this time it is assigned a name by the National Hurricane Center. Once hurricane conditions are reached, the high winds and torrential rainfall produce a significant threat to the environment and the communities the storm might impact. In the northern hemisphere, ocean conditions are at an optimum for storm development from August through October, when the largest number of tropical storms occur (Table 11.2). This period is toward the end of summer as ocean waters need the summer months to collect enough solar radiation and heat to generate storms.

Thermal energy—energy that comes from heat drawn from surface waters—is what drives hurricanes. As this energy rises into the atmosphere, it rotates and produces unstable conditions aloft. Further rotation concentrates energy toward a central column and the velocity increases (the spinning ice skater effect). The ocean surface becomes more agitated, which increases the heat and moisture in the atmosphere.

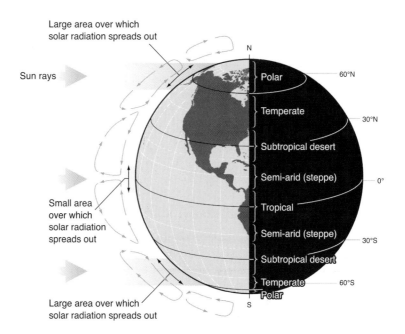

Figure 11.3 Global atmospheric circulation patterns. Note the formation of low pressure centered on the equator as heated air rises into the stratosphere and descends in the subtropical regions.

TABLE 11.1	Definitions of Tropical, Low-Pressure Weather Systems. Storms begin as disturbances and can become hurricanes.
System Designation	**Conditions**
Tropical disturbance	Organized mass of convectional air and thunderstorms with partial rotation present; generally 100 to 300 nautical miles in diameter; originate in the subtropics or tropics
Tropical depression	Closed circulation with sustained winds of 38 mph (33 kt or 62 kph) or less
Tropical storm	Sustained winds (1-minute measurement, 10 m above water) of 39 mph (63 kph or 34 kts) up to 64 kts
Hurricane	Sustained winds of 74 mph (119 kph or 64 kts) or more (see Box 11.1)

Source: National Hurricane Center.

TABLE 11.2 — Monthly Occurrence of Tropical Storms and Hurricanes in the North Atlantic Ocean, for the Period 1851 to 2004.

Total and Average Number of Tropical Storms by Month

Month	Tropical Storms† Total	Average	Hurricanes Total	Average	U.S. Landfalling Hurricanes Total	Average
January–April	5	*	1	*	0	*
May	18	0.1	4	*	0	*
June	76	0.5	28	0.2	19	0.12
July	94	0.6	47	0.3	23	0.15
August	336	2.2	214	1.4	74	0.48
September	448	2.9	309	2.0	102	0.67
October	273	1.8	154	1.0	50	0.33
November	58	0.4	38	0.2	5	0.03
December	8	0.1	4	*	0	*
Year	1316	8.5	799	5.2	273	1.78

* Less than 0.05.
† Includes subtropical storms after 1967.

Source: Chris Landsea, National Oceanic and Atmospheric Administration.

<table>
<tr><td colspan="2">BOX 11.1</td><td>Why Do Hurricane-Force Winds Start at 64 Knots?</td></tr>
</table>

Contributed by Neal Dorst

Beaufort Wind Scale	
Force 0	Calm
Force 1	Light Air
Force 2	Light Breeze
Force 3	Gentle Breeze
Force 4	Moderate Breeze
Force 5	Fresh Breeze
Force 6	Strong Breeze
Force 7	Near Gale
Force 8	Gale
Force 9	Strong Gale
Force 10	Storm
Force 11	Violent Storm
Force 12	Hurricane

In 1805–1806 Commander Francis Beaufort RN (later Admiral Sir Francis Beaufort) devised a descriptive wind scale in an effort to standardize wind reports in ship's logs. His scale divided wind speeds into 14 Forces (soon after pared down to thirteen) with each Force assigned a number, a common name, and a description of the effects such a wind would have on a sailing ship. And since the worst storm an Atlantic sailor was likely to run into was a hurricane, that name was applied to the top Force on the scale.

During the nineteenth century, with the manufacture of accurate anemometers, actual numerical values were assigned to each Force level, but it wasn't until 1926 (with revisions in 1939 and 1946) that the International Meteorological Committee (predecessor of the World Meteorological Organization, an agency of the United Nations) adopted a universal scale of wind-speed values. It was a progressive scale with the range of speed for Forces increasing as you go higher. Thus Force 1 is only 3 knots in range, while the Force 11 is eight knots in range. So Force 12 starts out at 64 knots (74 mph, 33 m/s).

There is nothing magical in this number, and since hurricane force winds are a rare experience, chances are the committee that decided on this number didn't do so because of any real observations during a hurricane. Indeed, the Smeaton-Rouse wind scale in 1759 pegged hurricane force at 70 knots (80 mph, 36 m/s). Just the same, when a **tropical cyclone** has maximum winds of approximately these speeds, we see the mature structure (eye, eyewall, spiral rainbands) begin to form, so there is some utility with setting hurricane force in this neighborhood.

Source: http://www.aoml.noaa.gov/hrd/tcfaq/.

tropical cyclone
A low-pressure system having a warm center that developed over tropical (sometimes subtropical) water and has an organized circulation pattern. The magnitude of its winds defines it as a disturbance, depression, storm or hurricane/typhoon.

eye
The central core of a cyclonic storms, normally relatively small and lacking clouds, moisture, and wind.

eye wall
The boundary between the eye of a cyclonic storms and the inner most band of clouds.

Hurricane Eye

Eventually an eye forms at the center of the rotating vortex. The eye is a region of calm air and relatively clear conditions characterized by a column of descending cool, dry air (**Figure 11.4**). The column is bordered by the **eye wall**, a thick mass of

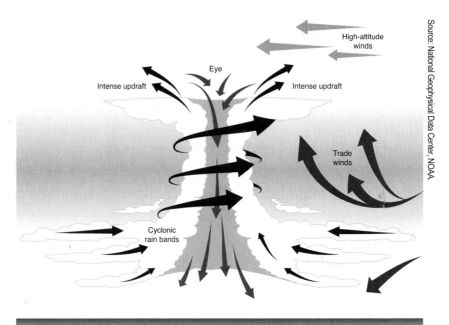

Figure 11.4 Generation of the eye wall and circulation pattern of a hurricane.

Source: National Geophysical Data Center, NOAA.

moisture-laden clouds spinning at a high velocity, and represents the most violent portion of a hurricane. Upward rotational movement in the eye wall lifts warm, moist water to higher altitudes, where condensation occurs and heat is given off.

Rotational forces throw most of the air and moisture outward, but some air flow is directed toward the center of the eye, where it is heated by compression. This warmer air can now absorb additional moisture, so most water vapor in the eye will be absorbed into the air. Any clouds in the eye dissipate, producing a zone devoid of clouds. The rising air creates lower pressure along the storm-ocean water boundary, which causes more moisture and heat to be drawn upward and the process continues. In the early stages of storm development, the eye is not well-defined but has better definition as overall rotation of the entire storm system increases (**Figure 11.5**). Once formed, the eye ranges in diameter from 5 to 40 miles (8 to 65 km), while the entire storm can stretch across 250 to 400 or more miles (400 to >650 km).

Figure 11.5 Photo taken in the eye of Hurricane Katrina, showing the eye wall. Note the clear conditions aloft and the rotating cloud pattern.

Measuring the Size of a Hurricane

Hurricanes have a huge amount of thermal energy inside them. The National Oceanic and Atmospheric Administration (NOAA) reports that an "average" hurricane has at least 1.3×10^{17} Joules per day or 1.5×10^{12} watts of energy, which is equivalent to about half the world's entire electrical generating capacity. Because of the extensive size of a hurricane and its slow forward speed, winds blow for many hours in a given location. At first winds are blowing from one direction and then as the storm passes directly over an area, the winds diminish to almost none at all as the eye passes overhead. For the uninformed, this might signal the end of the storm. After the eye passes, winds increase again—but from the opposite direction—and will continue to blow until the storm moves out of the area.

Causes of Damage

Wind Action

Forward motion of the entire mass of a hurricane is created by global circulation patterns prevalent in the area of the storm. In the North Atlantic Ocean, the prevailing westerlies push storms from their regions of origin in the east toward the west. The science of fluid mechanics also helps explain much of the movement of these storms. As a mass of fluid (in this case, water-laden, rotating air) moves from the equator toward the poles, it experiences increased rotation from a state of no rotation at the equator to a maximum value at the poles. In the northern hemisphere, there is a maximum rotation on the west side of a storm and a minimum value on the east side. This difference in the rotational strength, along with the Coriolis force, generates a deflection of the storm toward the north. Thus, hurricanes in the North Atlantic Ocean curve northward as they progress to the west (**Figure 11.6**). In its earliest stages, the forward motion of a hurricane ranges from almost 0 mph to more than 10 or 12 mph (0–20 kph). When storms move out of the latitudes of the trade winds (about 30° north and south of the equator), they increase their velocity to about 15 to 20 mph (25 to 35 kph). This forward motion, when coupled with the counterclockwise circulation of winds makes the northeast side of a storm the most devastating, as the two wind velocities are added together (**Figure 11.7**).

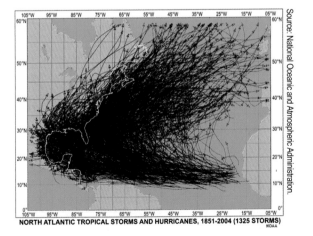

NORTH ATLANTIC TROPICAL STORMS AND HURRICANES, 1851-2004 (1325 STORMS)
NOAA

Figure 11.6 Paths of North Atlantic tropical storms and hurricanes, 1851–2004, showing a total of 1,325 storms. The east coast and Gulf coast areas of the United States are the primary targets for North Atlantic cyclonic storms.

Courtesy of The National Oceanic and Atmospheric Administration.

Source: National Oceanic and Atmospheric Administration.

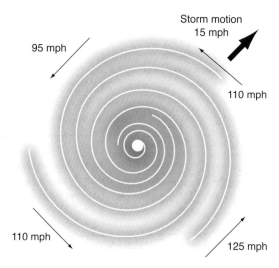

Storm motion
15 mph

95 mph

110 mph

110 mph

125 mph

Figure 11.7 Wind velocities of a moving hurricane are determined by the quadrant. In the northern hemisphere the highest velocities are in the northeast quadrant.

storm surge

An abnormal rise in sea level associated with intense storms, such as cyclones, typhoons, and hurricanes.

As storms get closer to the coast of the United States, they often encounter continental weather systems that have moved across the Midwest. These large weather fronts push the dissipating cyclonic low to the northeast and sometimes entrain the low pressure into its primary wind pattern, dragging it along and breaking it up. Tropical storms (which were previously hurricanes) have been recorded to move as fast as 50 mph (80 kph) across the mid-Atlantic and New England states northeastward into the North Atlantic, where colder waters fail to provide the energy needed to sustain the system.

All hurricanes are dangerous, but some are much worse than others. Wind, **storm surge**, and other factors combine to determine the destructive power of a hurricane. To make comparisons easier and to make the predicted hazards of approaching hurricanes clearer to emergency managers, National Oceanic and Atmospheric Administration hurricane forecasters use a scale that assigns storms to one of five categories (Table 11.3). This provides an estimate of the potential property damage and flooding expected along the coast with a hurricane.

Wind damage is perhaps the most obvious product of a hurricane. Photos and video images show buildings being torn apart, roofs flying through the air, and trees and limbs being bent. In many coastal areas that lie in the paths of hurricanes, building

Figure 11.8 Winds from Hurricane Dennis in September 1999 hit the Nags Head area of coastal North Carolina, leaving their mark.

Figure 11.9 Hurricane Fran hit the southeastern United States in September 1999 and its winds toppled mature pine trees, crushing this home.

TABLE 11.3	The Saffir-Simpson Scale Used to Categorize Intensities of Hurricanes. The scale was formulated in 1969 by Herbert Saffir, a consulting engineer, and Bob Simpson, director of the National Hurricane Center.

Category	Winds	Effects
1	74–95 mph	No real damage to building structures. Damage primarily to unanchored mobile homes, shrubbery, and trees. Also, some coastal road flooding and minor pier damage.
2	96–110 mph	Some roofing material, door, and window damage to buildings. Considerable damage to vegetation, mobile homes, and piers. Coastal and low-lying escape routes flood two to four hours before arrival of center. Small craft in unprotected anchorages break moorings.
3	111–130 mph	Some structural damage to small residences and utility buildings, with a minor amount of curtain wall failures. Mobile homes are destroyed. Flooding near the coast destroys smaller structures with larger structures damaged by floating debris. Terrain continuously lower than 5 feet above sea level may be flooded inland 8 miles or more.
4	131–155 mph	More extensive curtain wall failures, with some complete roof structure failure on small residences. Major erosion of beach. Major damage to lower floors of structures near the shore. Terrain continuously lower than 10 feet above sea level may be flooded requiring massive evacuation of residential areas inland as far as 6 miles.
5	Greater than 155 mph	Complete roof failure on many residences and industrial buildings. Some complete building failures with small utility buildings blown over or away. Major damage to lower floors of all structures located less than 15 feet above sea level and within 500 yards of the shoreline. Massive evacuation of residential areas on low ground within 5 to 10 miles of the shoreline may be required.

Examples			
Category	Sustained Winds (MPH)	Description	Examples
1	74–95	Minimal	Gaston 2004 SC
2	96–110	Moderate	Isabel 2003 NC
3	111–130	Extensive	Ivan 2004 FL
4	131–155	Extreme	Katrina 2005 LA
5	>155	Catastrophic	Labor Day Hurricane 1935 FL Keys | Camille 1969 MS | Andrew 1992 FL

Source: National Hurricane Center, formulated in 1969 by Herbert Saffir, a consulting engineer, and Dr. Bob Simpson, director of the National Hurricane Center.

standards have improved to lessen the effect of wind damage, but major storms are capable of producing severe wind damage. Roofs are very susceptible as winds catch under shingles or corrugated metal and rip these off. Roofs on older buildings are frequently blown off as a single unit as they are often not attached securely to the sides and internal walls of buildings. Isolated or free-standing structures are especially affected if they lack walls (**Figure 11.8**). Roofs on new construction in coastal communities are now required to be tie-bracketed to interior and exterior walls to help maintain some structural integrity for the building when strong winds strike. Tall trees are often toppled because the ground holding their root systems becomes saturated and can't hold the trees in place (**Figure 11.9**). Occasionally trees are snapped off, but that is mainly the result of sudden gusts of higher, velocity winds. Although the effects of wind are readily seen in an area hit by a hurricane, there is a more costly and devastating agent of destruction at work.

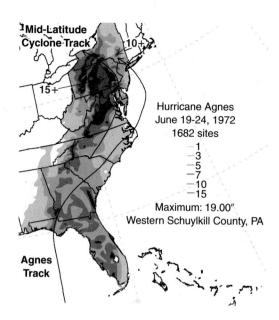

Figure 11.10 Rainfall map for tropical storm Agnes.
Source: NOAA.

Figure 11.11 Onshore winds push water over low-lying areas and produce beach erosion.

Figure 11.12 Large wind-driven waves produce significant erosion.

Storm Surge and Flood Hazards

Once a hurricane strikes land, it begins to lose its strength as it is no longer over the warm waters that feed it. Within a short period (usually less than one day), the hurricane is downgraded to a tropical depression. However, these reduced winds, along with intense rainfall, will continue to affect areas in the storm's path. Extensive flooding often results from this stage of the storm. In June 1972, Hurricane Agnes developed off the Yucatan Peninsula of Mexico, gained strength as it passed over the warm waters of the Gulf and Mexico, and hit land in the panhandle of Florida. After it made landfall, it was downgraded to a tropical depression and moved across Georgia and the Carolinas. It then joined another cyclonic low in the westerlies and regained strength as it went over the ocean to the east of Virginia. As it swung back inland, it was picking up moisture from the Atlantic Ocean and proceeded to drench the Mid-Atlantic states with intense rains (**Figure 11.10**). Portions of Maryland, northeastern Pennsylvania, upstate New York, and northern Virginia received more than 15 inches of rain. In addition to killing 122 people in the United States, Agnes caused more than $2 billion ($11.6 billion in 2005 dollars) in damage throughout the region.

Water damage from storm surges produces the most extensive damage of a hurricane, especially when the surge strikes at high tide. As a hurricane approaches land, it is pushing massive amounts of sea water toward the coastline. The momentum of this water is difficult to recognize because the wind and driving rain catch everyone's attention. Within hours, rising waters can appear. The storm surge, which can be 10 to 15 ft (3 to 5 m) or higher, inundates the shore and adjacent inland areas. The surge rises as the hurricane makes landfall. The rise in water level is usually gradual but relentless (**Figures 11.11** and **11.12**). The greatest recorded storm surge was associated with Tropical Cyclone Mahina that struck Australia in 1899. Reports stated a surge in excess of 13 m (42 ft). Dolphins were reported to be lying on the ground 45 ft above sea level!

Because coastal areas are relatively flat and often have low spots, flood waters do not recede very rapidly and the subsequent flooding can continue for several days or weeks. Flooding kills many more people than the wind, and freshwater flooding is often the silent killer.

Hurricane Camille in 1969 was one of only three category 5 hurricanes to hit the continental United States mainland (the other two were the unnamed Florida hurricane of 1935 and Hurricane Andrew in 1992, since upgraded by a review of its activity). Most of the 256 deaths associated with Camille were a result of flooding along inland portions of the Gulf coast states rather than along the seashore. From 1970 to 1994, 59 percent of the 589 deaths in the United States were by drowning in rain water that falls from an average hurricane at the rate of 2,300 cu meters per second. When measured against the average flow of the St. Lawrence River in New England and eastern Canada (a flow of 6,900 cu meters per second), an average hurricane produces that much water (in just three seconds). If rain falls in areas where streams funnel the runoff into developed areas, devastating conditions develop rapidly.

Hurricane Katrina, one of the strongest storms to hit the United States in the last 100 years, killed more than 1,800 people, most of whom drowned in flood waters that covered Alabama, Louisiana, and Mississippi. New Orleans, a city of almost 500,000 people before the onslaught of Katrina, is situated below sea level, surrounded by a system of levees that were constructed to hold back the Mississippi River and Lake Ponchartrain. Several major breaches in the levees resulted in rapid flooding that covered the city with more than 20 ft (7 m) of water (**Figure 11.13**).

Analysis of the region after Katrina passed led investigators to realize that the levees had not been sunk deep enough into the ground to withstand the enormous forces inflicted by the wind and water. Also, floodwaters undercut the bottom edges of the levees, thereby removing the material they rested on and they collapsed. Hurricane Katrina struck New Orleans as a category 3 storm, having lost some of its earlier punch when it was a category 5 storm over the central Gulf of Mexico. The relentless winds and storm surge battered the levees that were in place to protect the city. Once the levees broke, the entire city was flooded and remained in those conditions for several weeks until the water could be pumped out (**Figure 11.14**). Low-lying areas, such as the Ninth Ward, were inundated to depths exceeding 20 ft (6 m).

FEMA, photo by Jocelyn Augustino.

Figure 11.13 New Orleans experienced severe flooding due to a breeched levee system.

Courtesy of NASA.

Figure 11.14 New Orleans on August 31, 2005, two days after Hurricane Katrina hit the city. Extensive flooding is evident everywhere.

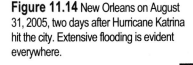

Figure courtesy of the United States Geological Survey.

Photo by Federal Emergency Management Administration, Mark Wolfe.

Figure 11.15 A storm surge and beach erosion caused by Hurricane Jeanne in October 2004 destroyed the seawall in New Smyrna Beach, Florida.

Figure 11.16 This house once sat on sand dunes along the beach in the Outer Banks of North Carolina. Severe beach erosion associated with Hurricane Dennis in August 1999 removed large amounts of sand and isolated many structures along the shoreline.

Any time there is a storm surge, there can be removal of sand from the beaches (**Figures 11.15** and **11.16**). This material is carried inland by the water and deposited once the surge abates. Thick sand deposits cover streets and yards and become a hazard (**Figure 11.17**). Beaches must be rebuilt to maintain the natural balance between the ocean and streams that normally supply the sand.

Photo courtesy of Mark Wolfe, Federal Emergency Management Agency.

Figure 11.17 Storm surge from Hurricane Frances in September 2004 pushed sand several hundred feet inland at Fort Pierce Beach, Florida.

Lessons from the Geologic Record

Hurricanes have existed on Earth for millions of years. In the geologic past, when continents were arranged differently, these storms affected a set of shorelines that were different from those of today. When there was much more expansive, open ocean, these storms undoubtedly moved about without affecting land masses in any significant way (**Figure 11.18**).

Whenever storms hit a land mass, they would produce the same effects we see today—beach erosion, flooding, storm surges, and winds. The geologic record has recorded the results of storm surges in the form of large sand deposits intermixed with soft clay deposits. However, it is hard for geologists to identify these preserved records definitively. Beach erosion and wave-cut terraces result from major storms. It is a challenge to distinguish a wave-cut terrace that formed over a long period from a geologically sudden cut terrace caused by a storm.

Figure 11.18 Configuration of the continents at 260 million years ago, showing that land masses being hit by cyclonic storms were in a variety of localities.

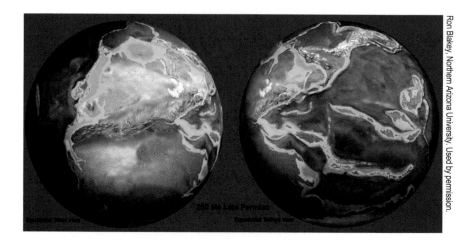

Ron Blakey, Northern Arizona University. Used by permission.

Lessons from the Historic Record and the Human Toll

Hurricanes have been reported by sailors since the first ships sailed the Pacific and Atlantic Oceans. Prior to that, the indigenous peoples of the Pacific Ocean and other regions all had stories of storms that were passed along through their oral histories. As countries took to the seas, sailors experienced the forces of nature when ships sailed through and around storms. Records of storms that have affected the United States have been kept for more than a century and detail some of the most destructive natural disasters to hit the U.S. mainland.

Often conflicting reports are given in terms of the number of deaths and the amount of property damaged by tropical storms and hurricanes. These data vary because of different reporting techniques that include a wide range of interpretations. We have attempted to provide reliable data that represent reports provided by U.S. government offices or by reputable research agencies, but differences will be encountered if you use a variety of sources.

The United States, certainly a country with a well-developed economic base and a means to deal with the ravages of hurricanes, is not exempt from the destruction that results when major storms hit the coastline. The National Oceanic and Atmospheric Administration (NOAA) reports that the average annual damage from hurricanes in the mainland United States is $4.9 billion. Damage increases exponentially with rising winds, so a category 4 hurricane can produce up to 250 times the damage of a category 1 storm. NOAA reported that in 2005 the country experienced a record 27 named storms, of which 15 reached hurricane status. Four hurricanes reached category 5, and five named storms developed in July—a record on both counts. This all-time record in terms of number and intensity of storms fortunately did not carry over into 2006, as some forecasters had feared. **Figure 11.19** shows that no hurricanes made landfall in 2006.

Table 11.4 shows the most costly storms to hit the United States since 1900; *60 percent have occurred since 1990*. Numerous coastal regions undergo a resurgence

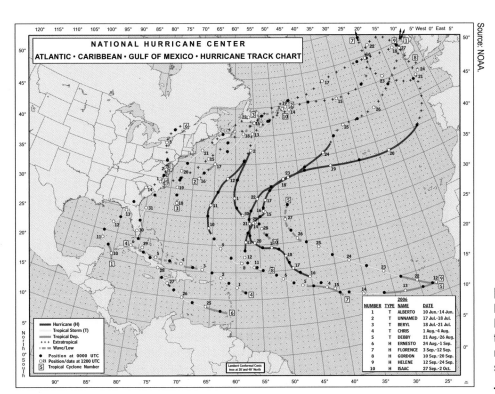

Figure 11.19 Tropical storm tracks in the North Atlantic Ocean for the 2006 season. No hurricanes hit the United States and the season had only 10 named storms, a marked contrast to 2005, when 27 named storms formed.

TABLE 11.4	Unadjusted Costs of Hurricane Damage in the United States, 1900–2006. (Several storms only reached tropical storm intensity.)			
Rank	**Hurricane**	**Year**	**Category**	**Damage**
1	Katrina	2005	3	$81,000,000,000
2	Andrew (SE FL, SE LA)	1992	5	26,500,000,000
3	Wilma (S FL)	2005	3	20,600,000,000
4	Charley (SW FL)	2004	4	15,000,000,000
5	Ivan (AL/NW FL)	2004	3	14,200,000,000
6	Rita (SW LA, N TX)	2005	3	11,300,000,000
7	Frances (FL)	2004	2	8,900,000,000
8	Hugo (SC)	1989	4	7,000,000,000
9	Jeanne (FL)	2004	3	6,900,000,000
10	Allison (N TX)	2001	TS	5,000,000,000
11	Floyd (Mid-Atlantic & NE U.S.)	1999	2	4,500,000,000
12	Isabel (Mid-Atlantic)	2003	2	3,370,000,000
13	Fran (NC)	1996	3	3,200,000,000
14	Opal (NW FL, AL)	1995	3	3,000,000,000
15	Frederic (AL, MS)	1979	3	2,300,000,000
16	Dennis (NW FL)	2005	3	2,230,000,000
17	Agnes (FL, NE U.S.)	1972	1	2,100,000,000
18	Alicia (N TX)	1983	3	2,000,000,000
19	Bob (NC, NE U.S.)	1991	2	1,500,000,000
20	Juan (LA)	1985	1	1,500,000,000

(continued)

of construction and population growth following major storms as many people think another storm will not come along to inflict damage again, but it certainly happens.

The naming of storms began in 1950, and currently the World Meteorological Organization prepares lists of names to be used for potential storms each year (see Box 11.2). To date, more than 60 names have been retired, as explained in Box 11.3, as these names still evoke unfortunate memories for people affected by those storms' destructive power.

In the past century, hundreds of thousands of people have died worldwide in cyclones, typhoons, and hurricanes. The densely populated regions of South Asia and the Far East have been subjected to catastrophic storms that hit coastal areas where many people live because of their need to be near the sea for their livelihoods and the

TABLE 11.4	Unadjusted Costs of Hurricane Damage in the United States, 1900–2006. (continued)			
Rank	Hurricane	Year	Category	Damage
21	Camille (MS, SE LA, VA)	1969	5	1,420,700,000
22	Betsy (SE FL, SE LA)	1965	3	1,420,500,000
23	Elena (MS, AL, NW FL)	1985	3	1,250,000,000
24	Georges (FL Keys, MS, AL)	1998	2	1,155,000,000
25	Gloria (Eastern US)	1985	3	900,000,000
26	Lili (SC LA)	2002	1	860,000,000
27	Diane (NE U.S.)	1955	1	831,700,000
28	Bonnie (NC, VA)	1998	2	720,000,000
29	Erin (NW FL)	1995	2	700,000,000
30	Allison (N TX)	1989	TS	500,000,000
30	Alberto (NW FL, GA, AL)	1994	TS	500,000,000
30	Frances (TX)	1998	TS	500,000,000
30	Ernesto (FL,NC,VA)	2006	TS	500,000,000
Addendum (Rank is independent of other events in group)				
19	Georges (USVI, PR)	1998	3	1,800,000,000
19	Iniki (Kaua'i, HI)	1992	Unknown	1,800,000,000
19	Marilyn (USVI, PR)	1995	2	1,500,000,000
25	Hugo (USVI, PR)	1989	4	1,000,000,000
30	Hortense (PR)	1996	1	500,000,000

Source: Costliest U.S. Hurricanes 1900–2006 (unadjusted), NOAA Technical Memorandum NWS TPC-5.

availability of easy, cheap transportation (Table 11.5). Unfortunately, it is the very sea these people rely upon that takes its toll and often wipes out entire communities and villages in the course of a few hours.

Deaths in the United States attributed to hurricanes have not been on the scale of those in other parts of the world, where storms strike areas that have high population densities, poor communication, and few means to evacuate people out of harm's way. In the past century more than 20,000 people have died in hurricanes in the United States. More than 71 percent of those deaths occurred in the three most deadly hurricanes to hit the United States (Table 11.6). With the exception of Hurricane Katrina in 2005, it has been more than 35 years since a hurricane has killed more than 50 people in the United States.

BOX 11.2	Hurricane Names for North Atlantic Ocean from National Hurricane Center

Experience shows that the use of short, distinctive given names in written as well as spoken communications is quicker and less subject to error than the older more cumbersome latitude-longitude identification methods. These advantages are especially important in exchanging detailed storm information between hundreds of widely scattered stations, coastal bases, and ships at sea.

Since 1953, Atlantic tropical storms have been named from lists originated by the National Hurricane Center. They are now maintained and updated by an international committee of the World Meteorological Organization. The original name lists featured only women's names. In 1979, men's names were introduced and they alternate with the women's names. Six lists are used in rotation. Thus, the 2005 list will be used again in 2011. Refer to Box 11.3 for a list of retired hurricane names.

Several names have been changed since the lists were created. For example, on the 2004 list (which will be used again in 2010), Gaston has replaced Georges and Matthew has replaced Mitch. On the 2006 list, Kirk has replaced Keith.

In the event that more than 21 named tropical cyclones occur in the Atlantic basin in a season, additional storms will take names from the Greek alphabet: Alpha, Beta, Gamma, Delta, and so on. If a storm forms in the off-season, it will take the next name in the list based on the current calendar date. For example, if a tropical cyclone formed on December 28th, it would take the name from the previous season's list of names. If a storm formed in February, it would be named from the subsequent season's list of names.

2010	2011	2012	2013
Alex	Arlene	Alberto	Andrea
Bonnie	Bret	Beryl	Barry
Colin	Cindy	Chris	Chantal
Danielle	Don	Debby	Dorian
Earl	Emily	Ernesto	Erin
Fiona	Franklin	Florence	Fernand
Gaston	Gert	Gordon	Gabrielle
Hermine	Harvey	Helene	Humberto
Igor	Irene	Isaac	Ingrid
Julia	Jose	Joyce	Jerry
Karl	Katia	Kirk	Karen
Lisa	Lee	Leslie	Lorenzo
Matthew	Maria	Michael	Melissa
Nicole	Nate	Nadine	Nestor
Otto	Ophelia	Oscar	Olga
Paula	Philippe	Patty	Pablo
Richard	Rina	Rafael	Rebekah
Shary	Sean	Sandy	Sebastien
Tomas	Tammy	Tony	Tanya
Virginie	Vince	Valerie	Van
Walter	Whitney	William	Wendy

Source: NOAA.

Hurricane Prediction and Mitigation

Numerous variables control the path of a hurricane, which makes the task of predicting its behavior and intensity a challenge. In the past 20 years, researchers have developed sophisticated computer models that provide a relatively reasonable estimate of the short-term path of a storm. Predictions for Hurricane Katrina were extremely precise (Figure 11.20).

BOX 11.3	Retired Hurricane Names 1954–2009 from the National Hurricane Center

The NHC does not control the naming of tropical storms. Instead, a list of names has been established by an international committee of the World Meteorological Organization. For Atlantic hurricanes, there is actually one list for each of six years. In other words, one list is repeated every seventh year. The only time that there is a change is if a storm is so deadly or costly that the future use of its name on a different storm would be inappropriate for obvious reasons of sensitivity. If that occurs, then at an annual meeting by the committee (called primarily to discuss many other issues) the offending name is stricken from the list and another name is selected to replace it.

There is an exception to the retirement rule, however. Before 1979, when the first permanent six-year storm name list began, some storm names were simply not used anymore. For example, in 1966, "Fern" was substituted for "Frieda," and no reason was cited.

Below is a list of retired names for the Atlantic Ocean, Caribbean Sea, and the Gulf of Mexico. There are, however, a great number of destructive storms not included on this list because they occurred before the hurricane naming convention was established in 1950.

List of Retired Names by Year

				1954	1955	1956	1957	1958	1959
				Carol	Connie		Audrey		
				Hazel	Diane				
					Ione				
					Janet				

1960	1961	1962	1963	1964	1965	1966	1967	1968	1969
Donna	Carla		Flora	Cleo	Betsy	Inez	Beulah	Edna	Camille
	Hattie			Dora					
				Hilda					

1970	1971	1972	1973	1974	1975	1976	1977	1978	1979
Celia		Agnes		Carmen	Eloise		Anita		David
				Fifi					Frederic

1980	1981	1982	1983	1984	1985	1986	1987	1988	1989
Allen			Alicia		Elena			Gilbert	Hugo
					Gloria			Joan	

1990	1991	1992	1993	1994	1995	1996	1997	1998	1999
Diana	Bob	Andrew			Luis	Cesar		Georges	Floyd
Klaus					Marilyn	Fran		Mitch	Lenny
					Opal	Hortense			
					Roxanne				

2000	2001	2002	2003	2004	2005	2006	2007	2008	2009
Keith	Allison	Isidore	Fabian	Charley	Dennis		Dean	Gustav	
	Iris	Lili	Isabel	Frances	Katrina		Felix	Ike	
	Michelle		Juan	Ivan	Rita		Noel	Paloma	
				Jeanne	Stan				
					Wilma				

Source: National Hurricane Center.

TABLE 11.5	Major Loss of Life Caused by Cyclones in South Asia.		
Date	**Location**	**Number Killed**	**Notes**
October 1942	Bengal, India	40,000	
October 1960	East Pakistan	6,000	
May 1963	East Pakistan	22,000	
May/June 1965	East Pakistan	47,000	
December 1965	Karachi, Pakistan	10,000	
November 1970	East Pakistan	>200,000	100,000 missing
September 1971	Orissa state, India	10,000	
December 1974	Darwin, Australia	50	City destroyed
November 1977	Andhra Pradesh, India	20,000	
April 1991	SE Bangladesh	131,000	
October 1999	Orissa state, India	9,500	10 million homeless
October 2007	SE Bangladesh	3,000	1.4 million homes destroyed or damaged
May 2008	Myanmar (formerly Burma)	125,000 estimated	2 million homeless

Source: Infoplease Almanac.

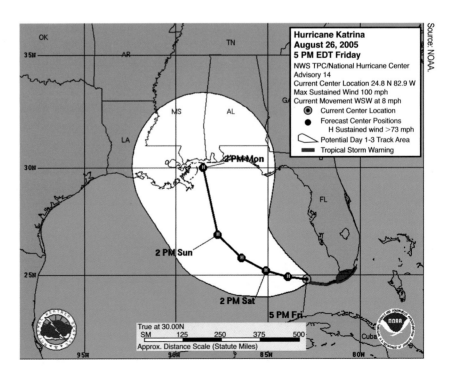

Figure 11.20 The projected storm track for Hurricane Katrina, published three days before landfall in the early hours of August 29, 2005. The forecast proved to be very precise in terms of where the storm came ashore.

TABLE 11.6	The Deadliest Hurricanes (60 or more deaths) in the United States, 1900–2005. The Top 20 Hurricanes for the U.S. Mainland			
Ranking	Hurricane	Year	Category	Deaths
1	TX (Galveston)	1900	4	8000*
2	FL (Lake Okeechobee)	1928	4	1836
3	LA/MS/AL (New Orleans)	2005	4	1810**
4	FL (Keys)/S. TX	1919	4	600†
5	NEW ENGLAND	1938	3‡	600
6	FL (Keys)	1935	5	408
7	AUDREY (SW LA/N TX)	1957	4	390
8	NE U.S.	1944	3‡	390††
9	LA (Grand Isle)	1909	4	350
10	LA (New Orleans)	1915	4	275
11	TX (Galveston)	1915	4	275
12	CAMILLE (MS/LA)	1969	5	256
13	FL (Miami)/MS/AL/Pensacola	1926	4	243
14	DIANE (NE U.S.)	1955	1	184
15	SE FL	1906	2	164
16	MS/AL/Pensacola	1906	3	134
17	AGNES (NE U.S.)	1972	1	122
18	HAZEL (SC/NC)	1954	4‡	95
19	BETSY (SE FL/SE LA)	1965	3	75
20	CAROL (NE U.S.)	1954	3‡	60
	ADDENDUM (Pre-1900 or not Atlantic/Gulf Coast):			

* May have been as high as 10,000 to 12,000.

† Over 500 of these lost on ships at sea; 600–900 estimated deaths.

‡ Forward speed was more than 30 miles per hour.

** Death count is still under revision as more data are analyzed

†† Some 344 of these lost on ships at sea.

Source: National Hurricane Center.

Models depend on adequate data to provide accurate forecasts, but data collection is sparse on the ocean surface. Most information is gained by upper-atmospheric reconnaissance flights by observational aircraft and satellite imagery. The National Hurricane Center in Miami, Florida, is the central clearing house for all observations in the North Atlantic, Gulf of Mexico, and the Caribbean regions. Reconnaissance flights by hurricane tracking aircraft ("hurricane hunters") provide real-time information about the wind speed, forward motion, location, and other key hurricane measurements. In addition, recording instruments are dropped from planes that penetrate the center of the storm. The instruments send back such information as wind velocity and moisture content readings that are linked to precise locations given by GPS coordinates within the storm.

Because researchers now understand hurricanes better than they did twenty years ago, they can provide relatively reliable computer models of a storm's behavior, fewer lives have been lost in recent years. The unfortunate exception is what occurred with Hurricane Katrina. Many people did not heed the warnings to evacuate and far too many people in New Orleans did not have the necessary resources to leave the city and its low-lying areas. As a result, more than 1,300 people died in nearby areas, mostly from drowning related to the storm surge and broken levees that were supposed to protect the city.

The Federal Emergency Management Agency (FEMA) has published a series of guidelines to assist in planning for natural disasters. One document, "Against the Wind—Protecting Your Home from Hurricane Wind Damage," outlines ways to strengthen a home by using reinforced construction techniques. Suggestions are given to build stronger roofs and to ensure that walls and gables are braced and tied down to lessen their chances of being blown down. However, many homes in regions affected by hurricanes have older homes and buildings that do not employ these newer concepts, so often these less-sturdy buildings are destroyed. Mobile homes, a very common housing option in storm-prone areas, offer no protection whatsoever against sustained hurricane-force winds and rising flood waters.

Attempts can be made, such as the ongoing effort by the Army Corps of Engineers in New Orleans, to improve our artificial defenses against Mother Nature. However, in spite of all our attempts to save our homes, communities, and other possessions, in the end these efforts will often prove costly and rather futile.

Summary

Cyclonic storms are generated in tropical and subtropical latitudes where warm ocean waters provide the thermal energy and moisture necessary for atmospheric circulation to expand into storms. Rotational forces in the atmosphere cause these disturbances to move across Earth's surface, gaining momentum. Once a well-defined circulation system has developed, the storm rotates around the eye, where vertically directed winds put more moisture and heat into the center of the storm. Once winds reach 119 km/hr (or 74 mph), the storm is classified as a hurricane. Hurricanes are classified in terms of wind strength—from a category 1 to a category 5.

In the United States the months of August, September, and October have the largest number of hurricanes. The East Coast and the Gulf Coast are the areas most frequently affected by hurricanes. Although we generally think of the wind damage associated with these storms, flooding caused by storm surges and torrential rainfall is their most damaging feature. In the past 20 years, coastal regions in the South and along the Gulf of Mexico have experienced devastating floods. The 12 most costly storms have occurred during that period, including Hurricane Katrina in 2005, which ranks as the worst natural disaster to occur in the United States in terms of property damage. Loss of life is still a major concern but has been decreased somewhat by employing better communication techniques to warn people of an impending disaster.

Cyclonic storms play a key role in generating destructive effects on the shoreline. Storms move massive amounts of sand and other material out to sea, and beaches must be rebuilt to maintain the equilibrium between the shoreline and the streams that bring sediment to the oceans.

Improvements in monitoring and modeling hurricanes allow forecasters and researchers to mitigate the impact of these storms by giving communities sufficient warning to prepare for the storms' arrival and by providing better information to engineers and builders so they can strengthen building codes and construct buildings that can withstand the hurricanes' destructive force. However, it is necessary that citizens heed these warnings and exercise good judgment when they are in harm's way.

References and Suggested Readings

Barnes, Jay. 1998. *North Carolina's Hurricane History.* Chapel Hill: University of North Carolina Press.

Burt, Christopher, C. 2004. *Extreme Weather—A Guide and Record Book.* New York: W. W. Norton.

Fitzpatrick, Patrick J. 1999. *Natural Disasters: Hurricanes.* Santa Barbara, CA: ABC-CLIO.

Gray, William M. 1968. Global View of the Origin of Tropical Disturbances and Storms. *Monthly Weather Review* 96 (October): 669–700.

Marshak, S. 2004. *Essentials of Geology.* New York: W. W. Norton.

Pielke, Roger A., Jr., and Roger A. Pielke, Sr. 1997. *Hurricanes—Their Nature and Impacts on Society.* New York: John Wiley.

Rosenfeld, Jeffrey. 1999. *Eye of the Storm—Inside the World's Deadliest Hurricanes, Tornadoes, and Blizzards.* New York: Plenum Trade.

Web Sites for Further Reference

http://www.fema.gov/hazard/hurricane/index.shtm
http://hurricanes.noaa.gov/
http://thehurricanearchive.com/Home.aspx
http://typhoon.atmos.colostate.edu/forecasts/
http://water.usgs.gov/wid/index_hazards.html
http://www.weather.gov/om/hurricane/
http://www.cnn.com/SPECIALS/2006/hurricanes/
http://www.nhc.noaa.gov/
http://www.ncdc.noaa.gov/oa/climate/severeweather/hurricanes.html

Questions for Thought

1. What are the three types of cyclonic storms and where does each most commonly occur?

2. What factors contribute to the formation of most cyclonic storms near the equator?

3. Describe the development of a cyclonic storm.

4. Describe the conditions that exist in and near the eye of a hurricane.

5. Why does water rather than wind cause the most damage in cyclonic storms?

6. Examine Table 11.4 that shows the amount of damage various hurricanes have generated in the United States. Why do we see that the majority of these entries are from the last 20 or so years?

7. What factors contribute to the much larger loss of life we see in Table 11.5 versus that in Table 11.6?

Wildfires

12

Raging wildfire in Alaska during July 2004. This was part of the Taylor Complex fire, which consumed more than 1,300,000 acres.

United States Forest Service.

Key Terms

backburn
brush fire
burnout
cellulose
conduction
controlled burn
convection
crown fire
firebreak
firestorm
fire triangle
forest fire
fuel
glucose
ground fire
heat
hydrophobic
hydrophobic layer
Incident Command System (ICS)
Incident Management Team (IMT)
land clearing
oxidation

(*Continued*)

oxygen	radiant heat	surface fire
prescribed burn	radiation	temperature of combustion
public information officer (PIO)	retardant	topography
pyrolysis	Santa Ana winds	wildfire

wildfire
An uncontrolled fire in a natural setting.

forest fire
A fire that occurs in a forest or stand of trees.

brush fire
A fire that burns primarily brush and material low to the ground.

pyrolysis
The process in which a series of reactions break down complex chemical compounds into simpler ones.

fire triangle
Consists of three parts that cause fire to occur; heat, fuel, and oxygen.

fuel
The part of the fire triangle that actually burns.

oxygen
A basic element that allows oxidation to take place; a necessary part of the fire triangle.

heat
The part of the fire triangle that provides the energy to create fire.

temperature of combustion
The temperature at which a specific type of material will ignite.

Fire Processes

The discovery of fire as a useful tool by the earliest humans allowed them to have heat and light, and they eventually learned that fire could be used to cook their food and improve their diets. Early man used fire to herd animals and to clear land. It was even used in warfare to drive combatants to their eventual death.

Today, we do similar things with fire but we also have the awareness that when fire gets out of control, it can become a destructive force. A **wildfire**, sometimes referred to as a **forest fire** or **brush fire**, is an unplanned and uncontrolled event involving the rapid oxidation of organic matter. These fires occur in either wildland or agricultural areas that are covered by grasses, brush, or trees. Fires are most common in those parts of the world where moist, warm climates permit vegetation to grow. When these areas experience periods of hot weather, coupled with low humidity, vegetation dries out and becomes fuel for wildfires.

The chemical decomposition of a substance by heat is defined as **pyrolysis**. The chemical products of this process are gases and solid matter that become flammable in the presence of sufficient heat. For a fire to occur, three components—fuel, oxygen, and heat—must be present (**Figure 12.1**). In this **fire triangle**, fuel must be available to burn, oxygen has to be present to allow the chemical reaction of oxidation to take place, and sufficient heat must exist to allow combustion to begin. Usually the heat source generating the fire comes from an open flame produced by a match, a spark, or a concentrated heat source, such as lightning.

Fires result from the accelerated combination of the elements in the fire triangle. Every flammable fuel has a **temperature of combustion**, which is a measure of the amount of heat necessary to cause the substance to ignite and burn in the presence of sufficient oxygen. The chemical process involves the formation of new bonds between oxygen and the carbon/hydrogen bonds present in the fuel.

Figure 12.1 The fire triangle. All three components (fuel, oxygen, and heat) must be present for fire to occur; remove any one and the fire goes out.

Cellulose

Wood, a common fuel source in nature, is composed mainly of **cellulose**, a complex carbohydrate containing hundreds of connected glucose molecules. **Glucose** is an organic substance with the chemical formula $C_6H_{12}O_6$. This organic molecule, when consumed in the presence of heat and oxygen, is broken down into carbon dioxide and water along with the release of heat:

$$C_6H_{12}O_6 + 6O_2 + heat \rightarrow 6CO_2 + 6H_2O + heat$$

Because other chemical compounds are found in complex living material, the combustion process forms many more chemical products besides carbon dioxide and water. Many different gases and solids, along with some liquids, are created as the fuel source is thermally decomposed.

Cellulose is also a major constituent of grasses, representing about 35 percent of their weight when they are dried out. Because grasses are not as compact as wood and can lose moisture much faster than wood in arid conditions, they are a prime candidate for wildfires. Grasses have a lower combustion temperature and can ignite much more rapidly than wood, producing fast-moving fires of higher intensity and shorter duration. Grasses also have more surface area per unit mass so they tend to ignite rapidly. Grass fires are beneficial, as they provide the occasional clearing-out process that removes the potential source to start larger fires.

cellulose
The primary substance that makes cell walls in plant tissue.

glucose
Is an organic substance with the chemical formula $C_6H_{12}O_6$.

Oxidation

The atmosphere consists of about 21 percent oxygen. However, only 16 percent oxygen is required for sustained combustion to take place. This combination of oxygen and fuel is the process of **oxidation**. In the preceding equation it is evident that removal of either component on the left-hand side (cellulose, the fuel or oxygen, or the lack of sufficient heat to produce combustion) will break the fire triangle and put out the fire. The products of the combustion process, carbon dioxide and water, along with heat, reenter the atmosphere and can later be used to form more cellulose.

Water cools the temperature below the ignition point and it temporarily removes oxygen from the process. Retardants and foam remove oxygen along with lowering the temperature below the ignition point. Almost all wildland fire suppression involves removing the fuel leg of the fire triangle.

oxidation
A process in which oxygen combines with a substance.

Heat Transfer

Heat can be transferred by means of **convection**, **radiation**, or **conduction**. Convection involves the rising of heat as the surrounding air becomes less dense. The updraft then draws in surface air located at the edges of the rising column, which then fans the flames. The convective movement is a primary cause of fires in canyons that expand rapidly. Rising, warmer air is funneled up the canyon sides toward the top. This draws in cooler air near the canyon base, thus fanning flames that then produce additional heated air, and the process continues in a continuous cycle.

Radiant heat is similar to what we encounter when we place our hand over a heating element on a stove that has been turned off. This process is always present in fires and serves to dry out fuel which then can become ignited more easily.

The process of conduction is the direct transfer of heat through a conducting medium, such as the handle on a cast iron skillet. Conduction is not effective in wildfires as wood is a poor conductor of heat.

convection
Transfer of heat by movement in the atmosphere caused by density differences produced by heating below.

radiation
The process in which heat energy is sent through space in the form of rays.

conduction
Transmission of heat through a material.

radiant heat
Produces an increase in temperature caused by radiation.

ground fire
Fire that moves along the ground and rises several feet above the surface.

hydrophobic
Incapable of absorbing water.

surface fire
Fire that moves along the ground and consumes ground litter.

Santa Ana winds
Seasonal winds that originate from high-pressure over the Great Basin of Nevada, pushing dry air from east to west across southern California.

TABLE 12.1	Causes of Wildfires in the United States in 2000			
	Number of Fires	Acres Burned	Sq mi	Sq km
Human caused (85%)	104,410	3,595,594	5,518	14,545
Arson (26%)				
Equipment (10%)				
Juveniles (4%)				
Smoking (4%)				
Campfires (3%)				
Railroads (3%)				
Other/unknown (50%)				
Naturally caused (15%)	18,417	4,826,643	7,542	19,525

Source: National Interagency Fire Center.

United States Forest Service.

Figure 12.2a Ground fire.

United States Forest Service.

Figure 12.2b Surface fire.

Wildfire Behavior

Wildfires are often started by lightning or by humans through carelessness or premeditated arson, and they can become very large and require significant resources to put them out (Table 12.1). When these fires spread toward urban areas, they can threaten homes and other structures and can turn deadly if people are unable to evacuate the areas in the path of the fire. Wildfires are a common occurrence in times of drought and can become major threats to communities that have sufficient fuel supplies surrounding them to sustain the fire. On a global basis, there are more than 4 million lightning strikes each day, and many of these start fires.

Types of Wildfires

There are three main types of wildfires, depending on the location of the burning material.

1. **Ground fires**. **Ground fires** consist of slow-burning material just below the surface and include plant roots and buried vegetative matter such as leaves or needles (**Figure 12.2a**). These fires can smolder below the surface so the damage they do is not readily seen. Ash from these and other fires can fill spaces in the soil and create a **hydrophobic** layer that impedes the downward movement of moisture. This condition contributes to erosion and other problems discussed later in the chapter.

2. **Surface fires.** **Surface fires** consume fuel lying on the surface, such as ground litter, limbs, fallen trees, and grasses (**Figure 12.2b**). The intensity of surface fires is quite variable depending on how dry the material is and how much fuel is present. Some of these surface fires can become very intense. Many of the wildfires in California are located in areas where fuels have dried out and fires are fanned by **Santa Ana winds**, which result from high pressure over the Great Basin of Nevada. Common from October of each year to the following March, they are produced by cool air in the eastern deserts of California being pushed westward over mountains. At the top, the cold, dry air descends, heating up and generating strong winds that are often associated with wildfires in southern California.

3. **Crown fires.** In forested areas, fires that begin on the surface often leap into the upper portions of trees to produce **crown fires** (**Figure 12.2c**). The ladder concept causes low-lying fires to climb up into the crowns of trees, where winds easily fan the flames from tree to tree (**Figure 12.3**). Unlike surface fires, which can be fought with fire personnel and equipment, crown fires are nearly impossible to control and the results are disastrous (**Figure 12.4**).

Fire Behavior Variables

Fire behavior is controlled by several variables. **Topography**, fuel types, and weather each play a major role in how a fire will spread.

Topography

Topography, or the lay of the land, can influence how rapidly a fire moves along the surface. The direction that a slope faces in terms of its exposure to sunlight is critical as the amount of vegetation can be much greater on those slopes receiving a high amount of sunlight. In the northern hemisphere, south-facing slopes produce thicker vegetation, which can become dried out in periods of low humidity by constant exposure to the sun. This is especially true from May to September.

If there are steep hillsides or canyons (Figure 12.4), any wind will be funneled upward and cause the fire to advance rapidly. As heated air from a fire rises, it will dry out vegetation at the higher elevations, thereby producing preheated, dry material that can subsequently be burned. A general rule is that the steeper the slope, the faster the fire advances and the more rapidly it spreads upslope. Convection of hot gases causes updrafts that fan the flames from below, causing the fire to spread faster.

Fuel Types

One reason we see large fires (**Table 12.2**) now is that for many years the prevailing philosophy was that all fires should be put out immediately. This forest management style resulted in an overall buildup of forest floor fuels, which served to generate large

Figure 12.2c Crown fire in Yellowstone National Park, 1988.

Photo by Jeff Henry, National Park Service.

crown fire
A fire that moves through the upper portions of trees; these are very difficult to extinguish.

topography
The general configuration of the land surface.

Figure 12.3 A progression of fuel heights produces a ladder effect to get the tops of trees burning.

Photo by Sara Jenkins.

Figure 12.4 This steep hillside caused the crown fire to spread rapidly.

TABLE 12.2	Large Wildfires Greater than 150,000 Acres in Alaska and the Lower 48 States		
Large Fires in Alaska—2000 to 2007			
Year	**Fire Name**	**Location**	**Size in Acres**
2004	Taylor Complex	Alaska Division of Forestry—AK	1,305,592
2004	Eagle Complex	BLM Upper Yukon Zone—AK	614,974
2004	Solstice Complex	BLM Upper Yukon Zone—AK	547,505
2004	Boundary	Alaska Division of Forestry—AK	537,098
2004	Central Complex	BLM Upper Yukon Zone—AK	451,162
2004	Pingo	BLM Upper Yukon Zone—AK	285,885
2004	Winter Trail	BLM Upper Yukon Zone—AK	279,865
2004	Chicken	BLM Tanana Zone—AK	257,720
2004	Wolf Creek	BLM Upper Yukon Zone—AK	197,067
2004	Camp Creek	Alaska Division of Forestry—AK	175,815
2000	Zitziana	BLM Tanana Zone—AK	164,387
2005	Chapman Creek	BLM Alaska Tanana Zone—AK	162,670
		TOTAL ACREAGE	4,979,740
Large Fires in the Lower 48 States 2000–2007			
Year	**Fire Name**	**Location**	**Size in Acres**
2006	East Amarillo Complex	Texas Forest Service—TX	907,245
2007	Murphy Complex	BLM Twin Falls District—ID	652,016
2002	Biscuit	Siskiyou National Forest—OR	499,570
2002	Rodeo/Chediski	BIA Fort Apache Agency—AZ	468,638
2007	Big Turnaround Complex	Okefenokee National Wildlife Refuge—GA	388,017

(continued)

fires once the material was ignited. In recent years, the U.S Forest Service, National Park Service, Bureau of Indian Affairs, and Bureau of Land Management, stewards of most of the forests and grasslands in the United States, have allowed large fires that were ignited naturally to burn while being monitored. These fires are considered to be in the wildland fire use (WFU) category. Such a designation was used in the summer of 2006, when a large fire on the north rim of Grand Canyon National Park in Arizona was started by a lightning strike. Of the more than 58,000 acres consumed by the fire, almost 19,000 acres (31 sq mi; 79 sq km) were considered as the wildland

TABLE 12.2	Large Wildfires Greater than 150,000 Acres in Alaska and the Lower 48 States (continued)		
Large Fires in the Lower 48 States 2000–2007			
Year	Fire Name	Location	Size in Acres
2007	Milford Flat Fire	SW Area, UT State Division of Forestry	363,052
2007	Cascade Complex	Boise National Forest—ID	302,376
2007	East Zone Complex	Payette National Forest—ID	300,022
2000	Valley Complex	Bitterroot National Forest—MT	292,070
2003	Cedar	Cleveland National Forest—CA	279,246
2005	Cave Creek Complex	Tonto National Forest—AZ	248,310
2006	Winters	BLM Winnemucca and Elko Districts—NV	238,458
2006	Derby Fire	Gallatin National Forest—MT	223,570
2006	Crystal	BLM Idaho Falls District—ID	220,042
2000	Clear Creek	Salmon-Challis National Forest—ID	216,961
2005	Clover	BLM Twin Falls District—ID	192,846
2000	Eastern Idaho Complex	BLM Idaho Falls District—ID	192,450
2006	Charleston Complex	BLM Elko District – NV	190,421
2000	SCF Wilderness	Salmon-Challis National Forest—ID	182,600
2005	Delamar	BLM Ely District—NV	170,046
2006	Day	Los Padres National Forest—CA	162,702
2000	Command 24	BLM Spokane District—WA	162,500
2002	McNally	Sequoia National Forest—CA	150,696
2006	Sheep	Elko Field Office, BLM—NV	150,270
		TOTAL ACREAGE	7,154,124

Source: National Interagency Fire Center.

fire use phase of the incident, while the remaining 39,000 acres (61 sq mi; 158 sq km) burned as a wildland fire.

Governmental fire agencies consider wildland fire use as the management of naturally ignited wildland fires in order to accomplish resource management goals for a given area. These goals include the reduction of likelihood of unwanted fires, the maintenance of natural ecosystems in a given area, and providing for the health and safety of firefighters and the public. The purpose of wildland use fires is the same as that of prescribed burns, which are discussed later in the chapter.

Figure 12.5 Pine trees destroyed by bark beetles. These trees become fuel for fast moving wildfires.

Figure 12.6 Results of a wildfire in a young forest that has experienced a decade of drought conditions. The 125-acre fire was started by a thrown radial tire along Interstate 40 outside Flagstaff, Arizona, in June 2006.

Figure 12.7 Fire on Wilson Mountain, north of Sedona, Arizona, June 2006.

Weather

Weather plays an obvious role in the spread of fires. Dry, windy conditions will greatly enhance the spread of wildfires, while moisture-laden storms will help quench a fire. Firefighters are aware of the amount of relative humidity in the air. A higher relative humidity means more moisture is present that can help impede the advance of a fire. As a general rule, relative humidity increases at night, which helps slow the growth of fires. Air temperature is also important as higher daytime temperatures will dry out the fuel.

Rapidly spreading fires are often accompanied by strong winds that tend to flatten out the flames and also transport glowing embers ahead of the fire, making it difficult to fight the fire along a well-defined line. Wind increases the rate of spreading of a fire.

Besides drying out fuel, drought can also stress trees and reduce their ability to ward off insects because there is less moisture in the trees themselves. This condition makes the trees susceptible to insect damage and diseases. In recent years, the pine bark beetle has killed thousands of acres of trees in North America, particularly conifers such as ponderosa pine and Douglas fir (**Figure 12.5**).

These standing dead trees are then fuel for raging wildfires that can rapidly move through an area, consuming healthy trees as well. Sometimes even without numbers of dead trees, healthy trees will burn, particularly if they are stressed from experiencing dry conditions for several years (**Figure 12.6**)

Firestorms

Very large wildfires are capable of producing their own weather by generating massive updrafts that can rise to 20 km (32,000 ft) or more. The upward movement of heated air causes surface winds to be drawn inward, generating a **firestorm**. These self-produced weather systems will have winds in front of the progressing fire and can also generate lightning that can produce more fires. Wind also increases the flow of oxygen to the fire, and the variable nature of wind can make it extremely difficult to forecast where a fire will move. Fire whirls are a phenomenon sometimes observed at locations of extreme heat. Fire whirls are small tornado-shaped currents that consist of fast, rotating flames. They are capable of carrying embers and other burning debris well above the ground surface and can deposit them ahead of the fire line. Some embers have been carried more than three miles (5 km) by wind and convection associated with a fire.

Fires have become larger in the past twenty years although the number of fires has decreased (see **Box 12.1**). Changes in climate have contributed to more dry fuel, which burns faster than wet trees.

Firefighting Techniques

In June 2006, a large wildfire burned in the scenic mountains three miles (5 km) north of Sedona, Arizona (**Figure 12.7**). Following accepted protocol, the first crew on the fire named it the Brins Fire, the name of the mesa near the initial fire point. This fire, started by a transient camper, consumed more than 4,200 acres (6.6 sq mi; 17 sq km)

Figure 12.8 Fixed-wing aircraft dropping fire retardant in rugged terrain.

Figure 12.9 Heavy-duty helicopter making a water drop on a fire line.

in very rugged terrain that required the use of aerial drops of retardant and water (**Figure 12.8, 12.9**). Several hundred residents in the threatened area were evacuated, while more than 1,000 firefighters took on the task of battling the fire (**Figure 12.10**).

Ground Crews

The attack on a fire uses several different techniques. The initial response is to send in firefighters to establish a fire line that controls the movement of the fire on the surface. Often hotshot crews are sent in to make the initial attack. These crews consisted of 20 to 25 highly trained and well conditioned firefighters. Personnel can come in by trucks or by walking if there are no roads. Helitack firefighting crews can be dropped by helicopter, or if the terrain is extremely rugged and the area is remote, smoke jumpers are called in. These specially trained firefighters parachute into the near vicinity of remote fires and begin the initial assault on the blaze.

Figure 12.10 Gila Hot Shot crew, Gila National Forest, New Mexico, heading off to fight the Brins Fire in northern Arizona.

Aircraft

Occasionally, ground crews are unsuccessful in their attempts to extinguish the fire and it climbs or ladders into the forest canopy. At this point, the firefighters coordinate their efforts with helicopter crews who drop water from the air. Should the fire be moving rapidly in the direction of major stands of fuel, large fixed-wing aircraft are called in to drop retardant just beyond the leading edge of the fire.

Retardant consists of a dry powder that contains various compounds of ammonium phosphate and water. The phosphate compounds are the retardant; water is simply the carrier for the chemicals. When the retardant coats the fuel, it prevents oxygen from coming in contact with the fuel, thus removing oxygen from the fire triangle. The mixture is also effective as a fire prevention substance even after the water has evaporated. Drops of the slurry coat everything in their path—trees, ground vegetation, and structures. Although it creates a sticky, red-colored surface, it can later be washed off.

firestorm
A widespread, intense fire that is sustained by strong winds and updrafts of hot air.

retardant
A chemical substance used to impede the spread of a fire.

BOX 12.1	The Human Effect

The occurrence of wildfires has changed in recent years with an increase in the population, especially in the West and Southwest of the United States (Box Figure 12.1.1). Although the number of fires has decreased in the past 20 years, the amount of acreage burned each year has shown a steady increase because of more human-caused fires. The average fire size, determined by dividing the number of fires into the total acreage destroyed, has increased from 32 acres per fire in the period 1988 to more than 97 acres per fire for the period 2003 to 2007—an increase of more than 200 percent!

In addition, the number of fires larger than 150,000 acres has increased in the past eight years (Table 12.2). In terms of being prepared for fires, we can often see storms coming that will alert us to the likelihood of fires starting in forests and grasslands. However, fire management officials have no indication where or when human-caused fires will take place, except perhaps during hunting or camping seasons.

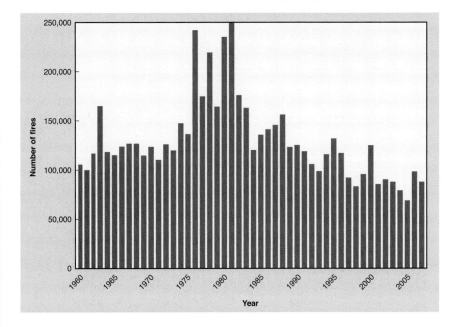

Box Figure 12.1.1 Number of wildland fires, 1960-2007. Although the number of fires has decreased significantly since 1980, the average size of fires has increased more than 200 percent.

Mapping

The growth of large fires (those exceeding 100 acres; 0.16 sq km) is mapped using global positioning system (GPS) and geographic information system (GIS) techniques. The techniques allow precise location of fire lines and provide the positions of crews working a fire. Very large fires are monitored by flight observers who survey the perimeter of a fire once a day and have continuous recording instrumentation on board that sends the coordinates via satellite to a computer. A map can then be produced providing an outline of the fire as it changes direction (**Figure 12.11**).

Evidence from the Geologic Record

Research done by Susan Rimmer of the University of Kentucky has disclosed the preservation of fossil charcoal in near-shore and terrestrial environments dating to the Late Silurian period. An increase in fire activity in the Late Devonian stretching into the Pennsylvanian occurred in Pennsylvania and West Virginia and points to fires that burned mostly surface vegetation. The sources of these fires are unknown. Major fires

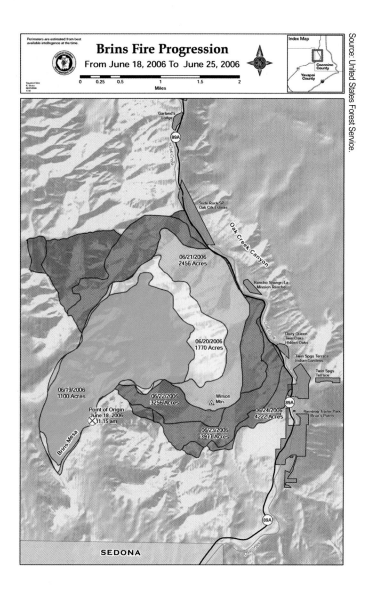

Figure 12.11 Map of progression of the Brins Fire, north of Sedona, Arizona.

resulted from the impact of the Chicxulub meteor in the Yucatan Peninsula at the Cretaceous-Tertiary boundary 65 million years ago. The global inferno that resulted produced enough soot to cover the Earth's surface. This soot is preserved in rock units in the Northern and Southern Hemispheres (Wolbach and others, 1990).

Obviously, major volcanic eruptions created extensive fires in the geologic past, but most of the record of these fires has been destroyed or buried beneath the volcanic deposits. During the past 1 million years, there has been a proliferation of plant life that has experienced burning. The ash and carbon that result from those fires is very mobile and not compacted, so it is seldom preserved.

Lessons from the Historic Record and the Human Toll

Wildfires have been a part of Earth's history, especially since combustible materials began to develop on Earth more than 350 million years ago. Fires that have damaged the largest amount of land are ones that burned in forested regions on the continents. However, numerous fires have occurred in populated areas and have destroyed homes, businesses, and the lives of numerous people.

Selected Historical Fires

Peshtigo Fire

The summer of 1871 in the upper Midwest was very dry, with drought conditions spreading across Wisconsin and Michigan. Forested areas had experienced occasional fires, and residents often set fires to clear their land for planting. Conditions were ripe for major conflagrations (fires) to occur.

On October 8, 1871, a massive low-pressure system began moving in from the west (**Figure 12.12**). Strong winds caused fires in the vicinity of Peshtigo (PESH-tee-goh) to quickly grow together (**Figure 12.13**), and by late evening they were encroaching on the town of approximately 1,700 people. The fire rapidly overtook the town and surrounding area and within a few days more than 1.2 million acres (1,875 sq mi; 4,850 sq km) had burned. Estimates of the loss of life ranged between 1,200 and 2,500 people and the final toll showed 12 towns in the region had been destroyed. This fire caused the greatest loss of human life from fire in the history of the United States. Unfortunately this entire event is seldom referred to as it occurred on the same day as, and within a few hours of, the Great Chicago Fire.

Chicago Fire

The Great Chicago Fire is well documented to have started in a barn belonging to Patrick O'Leary, although the story of a cow having knocked over a lantern is in dispute. (In 1999 the City Council of Chicago exonerated both the O'Learys and their cow, Naomi.) In 1871, Chicago was a bustling city of more than 330,000 residents and a major center of trade, banking, and farming. The same weather conditions that contributed to the Peshtigo fire caused the fire in Chicago to spread rapidly among the numerous wooden structures that were oftentimes closely spaced. Within a few hours, the fire had raged through the city and jumped the Chicago River to begin burning the downtown area (**Figure 12.14**). In less than 30 hours, more than three square miles (8 sq km) were turned to ash and almost 300 people had lost their lives.

Tillamook Burns

From 1933 to 1951, a series of large wildfires consumed more than 355,000 acres (555 sq mi; 1,440 sq km) of the Coast Range in western Oregon. The first fire began in August 1933, when a logger was dragging a downed tree along the ground. The

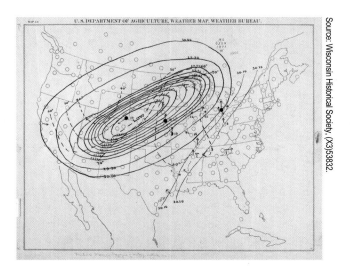

Figure 12.12 Weather conditions in the Midwest on October 8, 1871. A strong weather system produced extremely windy conditions throughout Wisconsin, Illinois, and adjoining states.

Source: Wisconsin Historical Society. (X3)53832.

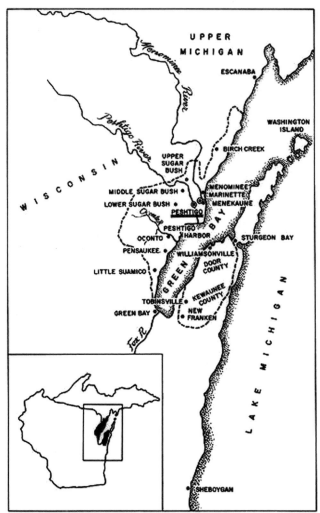

The district of the fire in Wisconsin and Upper Michigan, 1871. Broken lines indicate the approximate area devastated—some 1,280,000 acres. The insert map outlines Wisconsin and Upper Michigan and shows the estimated burned-over portions of the two states.

Figure 12.13 Area within solid line was burned by the Peshtigo fire, October 1871.

Chicago History Museum. Photo No. ICHi-26579.

Figure 12.14 Southwest corner of Dearborn and Monroe Streets, Chicago, October 1871. Photo by Jex Bardwell, courtesy of Chicago Historical Society.

steel cable ignited a log and the resulting fire spread to burn more than 240,000 acres (375 sq mi; 970 sq km). Numerous burned trees lay on the ground or were still standing. These helped fuel the second fire that took place in 1939 and burned 190,000 acres (300 sq mi; 770 sq km). Two additional fires, one in 1945 and the second in 1951, consumed more than 210,000 acres (330 sq mi; 850 sq km), some of which had been previously burned. Much of this area has now recovered because of the high rainfall in the Pacific Northwest that allows trees to grow rapidly.

Yellowstone

Yellowstone National Park, America's oldest national park, was ravaged by major wildfires in the summer and early fall of 1988. The park had a fire policy in place that allowed naturally occurring fires to burn on their own and not be suppressed unless property and human lives were endangered. The rationale was that nature could and would use fires to benefit the landscape. Up until 1988 there had been between 10 and 15 lightning-caused fires each year, which usually burned less than 100 acres each.

The park and surrounding area experienced a dry winter in 1987–1988. The lack of moisture produced large amounts of dry fuel, and stressed trees in the park became susceptible to insect infestation. These conditions, coupled with massive amounts of fuel that had built up from a lack of fires over the previous 80 or so years, allowed lightning to start several dozen fires within the park. In mid-July park officials realized they had to dispatch fire crews to fight these naturally produced fires—a move that was counter to their policy of letting such fires burn unchecked. The fires of Yellowstone National Park became the top priority of national firefighting agencies. More than 9,000 fire fighters from across the country battled the fires at a cost of more than $120 million.

In mid-September, rains finally arrived to help quench the fires, but not until significant snowfalls in November were the fires finally put out. The final toll showed more than 1.4 million acres (2,200 sq mi; 5,660 sq km) burned in the park and surrounding areas.

The Ravages of Fire on Communities

Fires in Oakland, California in October 1991

It started as a partially extinguished grass fire in the hills about three miles (5 km) northeast of downtown Oakland and within a similar distance from Berkeley, home of the University of California at Berkeley. The original fire was five acres and was considered under control in the evening of Saturday, October 19.

However, the fire was re-ignited by strong winds late Sunday morning and it quickly jumped two major highways. Frantic efforts by firefighters were for naught as several hundred homes and apartments were rapidly consumed. The fire eventually generated its own unpredictable winds that spread fire in all directions. Later in the evening of October 20, the winds stopped but the firestorm that had resulted proved too much for the army of firefighters battling the widespread blaze (**Figure 12.15**). The heroic efforts of many had been short-circuited by the lack of a water supply because of downed power lines that prevented pumps from delivering water to the hydrants. The fire ultimately killed 25 people and injured 150 others. Destroyed were 1,520 acres (2.4 sq mi; 6.1 sq km) along with 2,449 single-family dwellings and 437 apartment and condominium units. The economic loss was estimated at more than $1.5 billion.

Figure 12.15 Fire in the hills surrounding Oakland, California, October 1991.

Cerro Grande Fire in Los Alamos, New Mexico in May 2000

As part of a long-range plan to remove fuels from Bandelier National Monument, personnel of the National Park Service started a prescribed burn on the evening of May 4. Changes in local winds spread the fire beyond the containment lines and it was soon declared a wildfire. Within two days, strong winds turned it into a major fire (**Figure 12.16**). By May 10, winds were carrying embers more than a mile (2 km) beyond the defense lines that had been established. Because the towns of Los Alamos and White Rock lay in the immediate path of the fire, more than 18,000 residents of these towns were evacuated. Ground fuels located around homes ignited and caused many of the spot fires to consume these houses. More than 18,000 acres (28 sq mi; 73 sq km) had burned by May 10, destroying 235 homes and threatening the Los Alamos National Laboratory. The laboratory was spared major damage, but by the end, total losses amounted to 48,000 acres (75 sq mi; 194 sq km) of forest and grasslands and more than $1 billion of property. Fortunately, there was no loss of life in this fire.

Recent Major Fires Caused by Arson

Rodeo-Chediski Fire, East-Central Arizona in 2002

More than 469,000 acres (733 sq mi; 1,897 sq km) were burned in east-central Arizona when two separate fires converged in late June and early July 2002 (**Figure 12.17**). One fire, the Chediski, was caused when a lost hiker lighted a signal fire in an area that was extremely dry. The Rodeo fire was purposefully set by a seasonal firefighter seeking work. When the fires grew together they became the largest fire in the history of Arizona. More than 400 homes were destroyed. At its peak, more than 4,000 personnel, including four Type-1 **Incident Management Teams (IMT)**, which were coordinated by a national area command team, fought the inferno.

Hayman Fire in 2002

The Hayman fire, the largest in the history of Colorado, was started by a U.S. Forest Service employee who lit a fire in a fire ring in a nonburn zone. The fire eventually destroyed more than 130,000 acres (205 sq mi; 525 sq km) and included a one-day run that burned more than 60,000 acres (94 sq mi; 243 sq km).

Esperanza Fire in California in October 2006

An arsonist set a fire about 25 km (15 mi) west of Palm Springs, California, early on the morning of October 26, 2006. Santa Ana winds quickly spread the fire to more than 24,000 acres (38 sq mi; 97 sq km) in only 18 hours (**Figure 12.18**). Eventually it covered approximately 40,200 acres (63 sq mi; 163 sq km) before being contained. Five firefighters died when flames were blown over their position along a fire line. More than four dozen homes and outbuildings were destroyed. This fire had the greatest loss of life since the South Canyon fire in Colorado in 1994 (see **Box 12.2**).

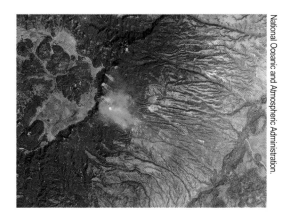

Figure 12.16 High-altitude photo of the Cerro Grande fire at its peak.

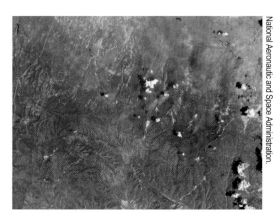

Figure 12.17 Rodeo-Chediski fire consumed more than 730 sq mi (1,900 sq km).

Incident Management Team (IMT)
A rapid response team that oversees the operations of large-scale firefighting.

Figure 12.18 Esperanza fire of October 2006.

| BOX 12.2 | The Personal Hazards of Firefighting |

Mann Gulch Fire, Montana, 1949

The Mann Gulch fire in the Helena National Forest of Montana overran 16 smoke jumpers who had para-chuted near the fire on August 5, 1949. While they were moving to a safer location, the fire blocked their route. Three survived: the foreman who ignited an escape fire into which he tried to move his crew, and two firefighters who found a route to safety. Considerable controversy has centered around the probable behavior of the fire and the actions of the crew members and their foreman.

South Canyon Fire, Glenwood Springs, Colorado, 1994

On July 2, 1994, lightning started a fire several miles west of Glenwood Springs, Colorado, in an area that was extremely dry because of a prolonged drought. High temperatures and low humidity contributed to the tinderbox conditions. Because of other fires in the vicinity, resources were not sent to this fire until it had burned for two days. By July 6 several small crews had been assigned to the fire, including two groups of smoke jumpers who had been dropped into the fire zone. During the mid-afternoon of July 6, winds up to 45 mph (72 kph) resulted in 200- to 300-foot (61 to 91m) flames causing the fire to jump a drainage area and rapidly overtake and kill 12 members of a hot shot crew along with two members of a helitack crew. Fortunately, 35 firefighters survived the onslaught. The combination of extremely steep terrain and cata-strophic weather conditions in an explosive fuel source, along with poorly established escape routes, led to the disaster. The entrapment of these firefighters in this 2,115 acre (3.3 sq mi; 8.6 sq km) fire marked the greatest loss of life in a wildfire in more than 40 years.

Two primary causes of the personnel disasters in the Mann Gulch fire and the South Canyon tragedies were that the firefighters were working downhill toward the fire and were inattentive to the weather and fire behavior.

The Mann Gulch and South Canyon fires bring home the fact that more than 920 personnel have died fighting wildfires in the United States (Box Figure 12.2.1). In nine instances, more than 10 firefighters died in single burnovers due to rapidly changing conditions and a lack of awareness of the terrain and the fuel conditions in which they were operating. The majority of health-related deaths were heart attacks suffered by fire personnel. In recent years, this statistic has increased as more volunteers, who are often not as well conditioned as professional fire crew members, are on the fire lines.

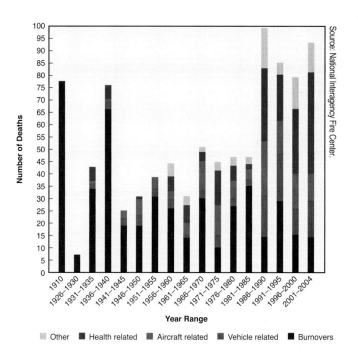

Source: National Interagency Fire Center.

Box Figure 12.2.1 Number and causes of deaths of wildland firefighters, 1910–2004.

Southern California Fires of October and November 2007

Tinderbox fire conditions occurred in October 2007 fueling wildfires that swept across more than 500,000 acres of southern California and destroyed more than 2,100 homes. As many as 15 major fires ranged from north of Santa Barbara to the United States–Mexico border. In San Diego County more than 350,000 homes were ordered evacuated, with more than 10,000 people being housed in Qualcomm Stadium, the home of the San Diego Chargers football team.

One fire that covered more than 36,000 acres was started by a 10-year-old boy playing with matches. That particular fire destroyed 21 homes and injured at least three people.

In November 2007 fires ravaged homes situated in canyons surrounding Malibu. Santa Ana winds drove the flames that destroyed more than 50 homes and burned almost 5,000 acres.

Minimizing Wildfire Hazards and Their Aftermath

Land Clearing

Areas of urban expansion often have an interface with forested land. Many municipalities have put plans in place that require **land clearing**, the clearing of vegetation around homes and other structures. This causes its own problem, however, as it increases soil erosion, and runoff can be up to several hundred times greater than the normal amount after surface fuels are cleared.

land clearing
The process of removing potential fire fuel.

Prescribed Burns

The continual buildup of surface fuels such as leaves, needles, branches, and low undergrowth and grasses provides the fuel for wildfires. Fire management agencies use prescribed burns to remove these surface fuels so as to lessen the likelihood of future uncontrolled fires. A **prescribed burn** is an intentionally-set fire that takes place under a set of defined parameters and predictable weather conditions. Prior to 2000 the term **controlled burn** was used to describe pre-set fires. Such fires are monitored and actively managed by fire personnel to prevent their getting out of control. Established control lines and some suppression techniques aid in keeping the fire within desired boundaries. In addition to burning material as it lies on the ground, fire managers will also often ignite slash piles consisting of small, rotten, or otherwise undesirable wood discarded during logging. There have been times, however, when slash has provided the fuel for devastating fires such as the fires in Michigan in the nineteenth century and the 1910 fires in northern Idaho.

prescribed burn
A controlled fire purposefully set to remove fuel from an area.

controlled burn
A fire set to remove fuel. Term is no longer used, as not all such fires were kept under control.

Prescribed burns reduce the fuel side of the fire triangle by removing ground litter and undergrowth. The resulting ash returns nutrients to the soil, helping to sustain the growth of new plants.

There are times, though, when a prescribed burn can turn disastrous, as in the case of the Cerro Grande fire in the summer of 2000. This fire, begun as a controlled burn, quickly got out of hand and inflicted major damage on the city of Los Alamos, New Mexico. Following this catastrophe, the term controlled burn is no longer used by fire management agencies.

Firebreaks and Burnouts

The spread of wildfires is slowed whenever firebreaks are established. **Firebreaks** are areas that have been cleared of surface and low-lying vegetation and ground litter. Clearing small areas involves actual removal of the fuel sources. However, a fire can generate airborne embers that can spread the fire past a narrow firebreak. Larger areas are cleared by setting prescribed burns on dozens or often hundreds of acres.

firebreak
A clearing that provides a gap across which fire should not burn.

burnout
An intentional fire that is lit to burn fuel that lies in the path of a spreading fire.

backburn
A controlled fire that burns back toward firefighters.

Incident Command System (ICS)
A system that is used to establish a chain of command for multiple agencies fighting fires.

public information officer (PIO)
A person who is designated to provide official information regarding disasters.

There are times when fire crews will conduct a **burnout**, a tactic that removes the nearby fuel before a fire gets to it. When a wildfire is threatening a community or if valuables are at risk, firefighters will sometimes deliberately set a back burn, a rather risky decision that is often a last resort. **Backburns** are carried out along a natural break, such as a roadway or near threatened buildings. The burn is drawn into the larger fire because of the convective heat rising from the larger fire.

Incident Command System

When personnel from a wide range of agencies began to be used in fighting wildfires, problems soon became apparent, especially in the ability of the various agencies to communicate with one another. Problems also developed over territorial control in terms of which group had jurisdiction in a given area.

As a result of fire suppression activity in southern California in the 1970s, the **Incident Command System (ICS)** was initiated. Depending on the magnitude of the fire, one agency assumes overall command of the incident and coordinates fire suppression activity with other responders. The federal government has 17 incident management teams (IMT) located in different regions throughout the United States. These teams consist of experts in all facets of disaster management and are drawn from a wide range of local, state, and federal agencies. They are located throughout the United States and can respond within a day or two to assume oversight of the firefighting when requested.

Informing the Public

The job of informing the public about wildfires is often undertaken by a **public information officer (PIO)** for the agency charged with direct oversight of the firefighting operations. The PIO is the contact point through whom other agencies and the general public learn about the progress of the fire and how it is expected to react in the short term. With so many agencies involved in a major fire, it is paramount to have a consistent message being delivered to the general population and other agencies that need information.

Results of Wildfires
Rejuvenation

Fires do have a useful purpose. They can help regenerate plants and other flora by opening seeds and by adding nutrients to the soil. This is evident in areas that have sustained severe burns when plants and trees rejuvenate themselves, often within a few months after the fire if moisture is present. In cold climates where there is little chemical breakdown, fire can speed up the process of enriching the soil with nutrients. Nitrogen is released from the burned plants and helps accelerate future plant growth. Large wildfires in Alaska are very beneficial to that state's high-latitude ecosystems because soils are slow to break down naturally in cold conditions.

Erosion

In the time following a large wildfire, the potential for major surface erosion is a primary concern. Steep slopes that were once covered by lush vegetation that could hold water and impede its rapid flow across the surface is absent. Rapid runoff removes top soil and cuts deep channels and rivulets into the surface (**Figure 12.19, 12.20**). In regions near highways, road cuts, and other steep slopes, mass wasting is accelerated. California experiences significant slides every year in burned areas.

Figure 12.19 Increased erosion and down cutting occurring in burnedout forest. Trees in the background are covered with more than 1 m of sediment. Gravel deposits are the result of reworked debris flows.

Figure 12.20 Outwash due to unimpeded surface water flow.

The **hydrophobic layer** causes more sheet erosion and flooding. Seasonal rains and snowfall will produce debris flows and landslides, along with flooding that can be catastrophic to low-lying areas. Replanting hillsides with grasses and seedlings helps alleviate the problem, but the landscape will not be the same for many years, especially in arid areas where plant life is slow to grow.

hydrophobic layer
A layer of material that cannot absorb water; usually forms in dry climates or in regions that have experienced intense fires.

Fiscal Resources

Another result of major fires is loss of revenue from the sale of timber. This means that less money is available to spend on cleanup and rehabilitation of the burned area. In addition, there are ecological issues, such as concerns about building temporary roads to harvest the timber that can be salvaged. After the timber is removed, these roads are often replanted with grasses and seedlings in order to help the area return to its previous condition. The result is a lack of fiscal resources to help a devastated area recover—to replant or reseed to encourage new growth and to help minimize the erosion and runoff problems that arise very soon after a fire.

Summary

Fires have existed for millions of years and will continue to be a natural hazard. Wildfires are unplanned, uncontrolled events that are caused by rapid oxidation of organic matter. Fuel, oxygen, and heat are needed to produce combustion. Removal of any one of these components of the fire triangle causes the fire to be extinguished.

Wildfires are usually caused by lightning or humans, either intentionally or accidentally. These fires can be ground, surface, or crown fires depending on where they occur in the fuel source. Factors that control fire behavior include surface topography, the types of fuels, and weather conditions, particularly wind speed.

Once a fire occurs, fire management personnel decide how to fight it. If it is located in an area that poses no threat to structures or people, the fire is usually allowed to burn itself out, although its progress is monitored. If fires pose a threat, then they are attacked by ground crews and aircraft. There have been many historical fires, including the Great Chicago Fire and those of Yellowstone National Park. Recently there has been an increase in the number of fires that affect communities, because urbanization has placed more structures and people closer to fuel sources. As a result, we need to be more aware of the effect fires have on developed areas and the impact major fires have on the landscape and environment. Preparation and planning will minimize the impact of fires in the future.

References and Suggested Readings

Arno, Stephen F. and Steven Allison-Bunnell. 2002. *Flames in Our Forest: Disaster or Renewal.* Washington, DC: Island Press.

Johnson, Edward A. and Kiyoko Miyanishi, eds. 2001. *Forest Fires—Behavior and Ecological Effects.* San Diego: Academic Press.

Macclean, John N. 2003. *Fire and Ashes—On the Front Lines of American Wildfires.* New York: Henry Holt.

Pyne, Stephen J. 1995. *World Fire—The Culture of Fire on Earth.* New York: Henry Holt.

Pyne, Stephen J. 1997. *Fire in America—A Cultural History of Wildland and Rural Fire.* Seattle: University of Washington Press.

Pyne, Stephen J. 2004. *Tending Fire—Coping with America's Wildland Fires.* Washington, DC: Island Press.

Pyne, Stephen, P. L. Andrews, and R. D. Laven. 1996. *Introduction to Wildland Fire.* 2d ed. New York: John Wiley.

Wolbach, Wendy S., Iain Gilmour, and Edward Anders. 1990. Major wildfires at the Cretacrous/Tertiary boundary. *In Global Catastrophes in Earth History—An Interdisciplinary Conference on Impacts, Volcanism, and Mass Mortality,* ed. Virgil L. Sharpton and Peter D. Ward, 391–400. Boulder, CO: Geological Society of America.

Wolf, Thomas J. 2003. *In Fire's Way—A Practical Guide to Life in the Wildfire Danger Zone.* Albuquerque: University of New Mexico Press.

Web Sites for Further Reference

http://earthobservatory.nasa.gov/NaturalHazards

http://gacc.nifc.gov/links/links.htm

http://landsat.gsfc.nasa.gov/images/image_index.html

http://water.usgs.gov/wid/index-hazards.html

http://www.fema.gov/hazard/index.shtm

http://www.ngdc.noaa.gov/seg/hazard/hazards.shtml

http://www.nifc.gov

http://www.usfa.dhs.gov

Questions for Thought

1. Explain the processes involved in the formation of a fire.

2. What are the three ways in which heat can be transferred?

3. Explain the fire triangle.

4. Why do fires spread more rapidly up a canyon than across a valley floor?

5. What defensive measures would you take to protect a summer resort home located in a forest against wildfires?

6. How do wildfires affect the soil layer and future runoff from rainfall?

7. How safe is your residence in terms of wildfire potential? What could be done to improve the situation?

8. What role do the Santa Ana winds of southern California play in the fires of that area?

9. Following a wildfire, what steps should be taken to begin restoring the ecology of the landscape?

10. How have the characteristics of wildfires changed over the past several decades?

Biological Hazards

13

Key Terms

acquired immunodeficiency
syndrome (AIDS)
anthrax
avian flu (H5N1)
bacteria
biological hazard (biohazard)
Black Plague (Black Death)
bubonic plague
cholera
epidemic
hepatitis
human immunodeficiency
virus (HIV)
influenza
leprosy (Hansen's disease)
malaria
microbe
pandemic
pathogen
prion disease
severe acute respiratory
syndrome (SARS)

(Continued)

smallpox	typhus	West Nile Virus (WNV)
tuberculosis (TB)	vector	yellow fever
typhoid fever	virus	

© Christopher Poliquin, 2010. Under license from Shutterstock, Inc.

Figure 13.1 The international biological hazard symbol.

Natural disasters, such as earthquakes, cyclonic storms, and floods, often generate conditions that produce widespread outbreaks of diseases. These disasters can greatly increase the risk of epidemics, especially in crowded situations such as refugee camps or temporary living areas for people displaced by natural disasters. Often poor sanitary conditions and the loss of a reliable water supply lead to explosive outbreaks of diseases. Biological hazards can quickly develop and have far-reaching, devastating effects at all levels of our population.

biological hazard (biohazard)
An organism or substance that is a possible threat to human health.

A **biological hazard** (also referred to as a **biohazard**) is any organism or substance that has the potential to threaten the health of humans or organisms (**Figure 13.1**). These can be in the form of a virus or bacterium. A different type of hazard is material produced by chemical waste or nuclear accidents; this can also become a primary cause of biological disasters.

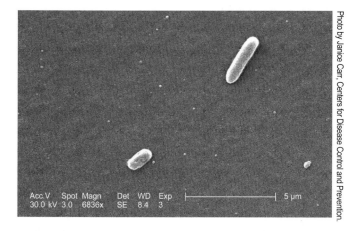

Photo by Janice Carr, Centers for Disease Control and Prevention.

Figure 13.2a Under a magnification of more than 6800 times, this scanning electron micrograph (SEM) depicted two *Escherichia coli* bacteria of the strain 0157:H7. Although most strains are harmless, and live in the intestines of healthy humans and animals, this strain produces a powerful toxin, which can cause severe illness.

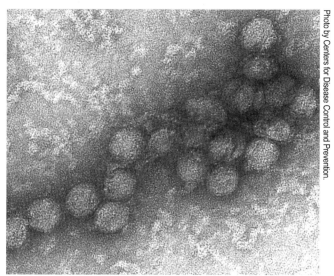

Photo by Centers for Disease Control and Prevention.

Figure 13.2b An electron micrograph of the West Nile virus.

Diseases that result from activity of bacteria, such as Escherichia coli (E. coli), or viruses, such as West Nile (**Figure 13.2a** and **b**), can adversely affect the well-being of any organism, including humans. **Bacteria** are microscopic creatures that usually consist of one cell, contain no chlorophyll, and reproduce by simple cell division. **Viruses** are ultramicroscopic organisms that contain either RNA or DNA, surrounded by a protein case. Viruses can only multiply within living cells.

Nature of Diseases

A **pathogen** is any agent that is capable of causing a disease. Most pathogens are microorganisms. Diseases have been in existence on Earth since the first **microbes**— both bacteria and viruses—were formed. Microbes infest a host, which serves as the source of energy and nutrition necessary for them to survive. Microbes have the ability to sustain their existence by not killing off all the hosts, which would terminate the particular disease—and the microbes. Many of Earth's 6.7 billion people live in areas that are constantly threatened by outbreaks of highly communicable diseases, such as malaria and tuberculosis (TB).

Many diseases are very resilient and have lasted for centuries. Malaria has affected humans for more than 50,000 years. Bubonic plague has been documented since before the first millennium and has persisted for almost 1,500 years. Several diseases, including TB, have become resistant to early treatment techniques and are now more difficult to combat.

Epidemics and Pandemics

Epidemics are the spreading of a highly contagious disease that affects the population for a short period of time. In the past, these diseases were usually contained within a region, but with today's mobile society, they can spread rapidly across countries and continents to produce a **pandemic** situation. This condition adversely affects a high percentage of the population. An example of an epidemic would be the spread of measles within a local community or a university student body. A pandemic would be much more widespread and create havoc for millions of people across continents. Examples of pandemics include the bubonic plague that overtook Europe in the mid-1300s or the Spanish influenza outbreak of 1918. In each case, tens of millions of people became infected and millions died. Diseases such as cancer or congestive heart failure, although killers of many people, are not contagious and hence do not produce epidemics or pandemics.

Historical Outbreaks

Several notable pandemics have affected large civilizations or the entire world. Among the earliest recorded was one that occurred in Athens in 430 B.C. through 426 B.C. As described by a Web site available from Indiana University:

> In the winter following the first year of the war, morale had fallen considerably in Athens. It was at the year's public funeral (held annually for men who had fallen in battle in the course of the year) that Pericles pronounced the famous funeral oration that is so often quoted as summing up the greatness of Periclean Athens (Thuc.2.34–46). Pericles' speech was an encomium on Athenian democracy and it provided the high point of Thucydides' account of the war. It is immediately and dramatically followed in his account by the description of the plague which struck the city in the following summer, as the Spartans again invaded Attica. Crowded together in the city as the result

bacteria
A tiny, single-celled organism that reproduces by cell division, having a shape similar to a rod, spiral, or sphere, and has no chlorophyll.

virus
Acellular, non-living infectious particles of either DNA or RNA associated with various protein coats; these only multiply in living cells.

pathogen
An agent, such as a bacterium or virus, that causes a disease.

microbe
A microorganism, especially one that can be a pathogen.

epidemic
The sudden occurrence of a highly-contagious disease or other event that is clearly in excess of normally expected numbers.

pandemic
A term applied to a highly-contagious disease that spreads throughout the world.

of Pericles' strategy, the Athenians fell victim to the virulent sickness that was spreading throughout the eastern Mediterranean. People died in large numbers, and no preventive measures or remedies were of any avail. It has been estimated that a quarter, and perhaps even a third, of the population was lost. The plague returned twice more, in 429 and 427/6, and Pericles himself died during this time, probably as a result of the disease.

Source: From http://www.indiana.edu/~ancmed/plague.htm. Used with permission.

In 2006, Manolis Papagrigorakis and others reported in the *International Journal of Infectious Diseases* that analysis of dental pulp extracted from the teeth in bodies in a mass grave in Athens, Greece, contained evidence that the people mostly likely died from typhoid fever.

The Antonine Plague, which lasted from A.D. 165 to A.D. 180, was mostly smallpox that had been brought back from the eastern Mediterranean. It is estimated that about 25 percent of those infected died—almost 5 million people. In a second outbreak (A.D. 251 to A.D. 260), as many as 5,000 people died each day in Rome and the surrounding area.

The bubonic plague first made its appearance when a pandemic outbreak occurred in A.D. 541. The Plague of Justinian began in Egypt and surfaced in Turkey the following year. At its worst in Constantinople, it killed 10,000 people each day and wiped out close to 40 percent of the population of the city. At the end of the outbreak, 25 percent of the entire population of the eastern Mediterranean region had died.

Other well-known pandemics include the Black Death outbreak of the mid-1300s, a series of influenza pandemics that extended from the late 1800s until the most recent worldwide outbreak in 1968–1969, and typhus, which had its first outbreaks in the late 1400s and recurred intermittently until the Second World War. More details on all these disease pandemics will be provided in the following sections that address past and present biological hazards.

Types of Diseases

Bubonic Plague

bubonic plague
A rare, but fast-spreading, bacterial infection caused by *Yersinia pestis;* transmitted in humans and rodents by infected flea piercings; symptoms include headaches and painful swelling of lymph nodes.

Black Plague (Black Death)
An epidemic of bubonic plague that killed more than 20 million people in Europe during the mid-fourteenth century.

Bubonic plague primarily affects rodents, but it is transmitted to humans by flea piercings (we incorrectly refer to these as bites, but fleas and mosquitoes do not have teeth). The disease is caused by a bacterium called *Yersinia pestis*, is highly communicable, and spreads very rapidly. Symptoms include a fever, chills, and painful swelling of the lymph glands called buboes—hence its name. Red spots develop on the skin and these eventually turn black from subdermal dried blood caused by internal bleeding, giving rise to the name **Black Plague**, or the **Black Death**. In some cases the bacterium enters the lungs, developing into pneumonic plague. This is easily spread through the air when infected humans breath and cough.

Following the pandemics that were recorded before A.D. 550, including the first recorded cases of bubonic plague, there was a lull in worldwide diseases until 1347,

---

BOX 13.1	**Ring Around the Rosies Rhyme**

Although typically considered to be just a myth, some feel that the song "Ring Around the Rosies" aptly describes the "black plague"; that is, "ring around the rosies" could refer to the red rash rings that those infected would get on their skin; "pocket full of posies" could refer to the fact that people would carry posies (flowers) in their pockets in the belief that this would keep the plague at bay; "ashes, ashes" might have originally been "Achoo, Achoo" to designate sneezing, another symptom; and "we all fall down" could refer to the many deaths that the disease caused.

Source: http://www.niehs.nih.gov/kids/lyrics/rosie.htm.

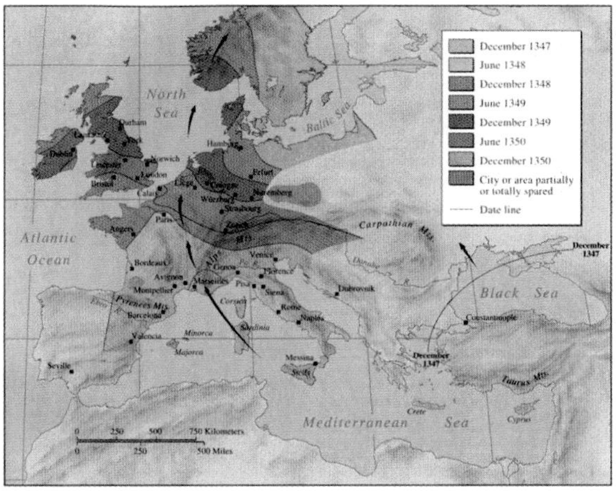

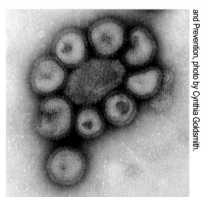

Source: http://www.odu.edu/~mcarhart/hist102/images.htm.

Figure 13.3 Spread of the Black Plague across Europe.

at which time the Black Plague resurfaced in China and found its way to Italy. Sailors on several Italian trade ships were dying from plague as the ships pulled into Sicily. Within a year, the plague had made its way through western Europe and into England (**Figure 13.3**). Here, it received the name "Black Death" (**Box 13.1**). Outbreaks slowed in the winter because fleas were dormant, but they had a resurgence each spring, when deaths increased dramatically. Although Ziegler (1991) chronicles an ongoing debate about the population of Europe and the number of people who died, it is fairly well established that by 1353, more than 20 million Europeans—almost one-third of Europe's population—had perished. This loss of life represented the estimated population growth that Europe had experienced in the previous 250 years.

Cholera

Cholera results from an intestinal infection by the bacterium *Vibrio cholerae* that could be present in drinking water or contaminated food. It is spread when people come in contact with feces from others who are infected, a problem in areas without proper sanitation facilities or safe water supplies. Extensive breakdown of infrastructure when cholera is present can produce epidemic conditions. Floods and earthquakes easily destroy a community's ability to handle its sanitation needs. There are documented cases of people contracting cholera from ingesting raw shellfish from coastal regions and brackish rivers, areas that are often tainted by raw sewage, especially when cyclonic storms hit populated coastal areas.

Symptoms of cholera range from almost nonexistent to severe, the latter affecting approximately 5 percent of those who become infected. Vomiting and extremely water-laden diarrhea create a rapid decrease in body fluids that brings on dehydration and shock. Death is immediate without treatment.

Table 13.1 shows pandemics and concentrated outbreaks of cholera that have occurred since the first outbreak in India in 1816–1826. By the late 1880s, most developed countries understood the need for a clean reliable water supply, which helped reduce the pervasive nature of the disease in most Western nations.

Influenza

Influenza viruses produce the flu, a disease that affects millions of people every year (**Figure 13.4**). Symptoms include a fever, headache, nasal congestion, muscle aches, a sore throat, and a general loss of energy and appetite. A case of the flu varies from being a mild illness to one that can be quite severe and lead to death in those with

cholera
An acute intestinal disease caused by the bacterium *Vibrio cholera;* often found in contaminated drinking water and food.

influenza
An acute viral infection that affects the respiratory system; this can occur as isolated, epidemic, or pandemic in scale; symptoms include fever, headache, nasal congestion, and general lethargy.

National Institutes of Health & Centers for Disease Control and Prevention, photo by Cynthia Goldsmith.

Figure 13.4 Three-dimensional structure of seasonal influenza virus from electron tomography.

TABLE 13.1	Major Outbreaks of Cholera Throughout the World Since 1816	
Dates	**Localities**	**Comments**
1816–1826	First in Bengal, to India in 1820, then to China and eastern Europe	
1829–1851	Began in Europe, then London, Canada, and New York in 1832; west coast of North America in 1834	
1852–1860	Mainly Russia	More than 1 million deaths, including composer Peter Tchaikovsky
1863–1875	Prevalent in Africa and Europe	
1866	North America	
1892	Hamburg, Germany	City water supply contaminated, killing more than 8,500
1899–1923	Russia	
1961–1966	First in Indonesia, then to Bangladesh, India, and Russia	
1994	Rwanda refugee camps	48,000 cases resulted in almost 24,000 deaths

low immunity, especially the young and the aged. An annual vaccination can greatly reduce a person's likelihood of getting the flu or at least make it less severe.

The Centers for Disease Control and Prevention (CDC) report that every year in the United States, on average,

- 5 to 20 percent of the population get the flu.
- More than 200,000 people are hospitalized from flu complications.
- About 36,000 people die from flu infections.

Pandemics of influenza have struck approximately three times every century since the 1500s. Recurrence intervals range between 10 and 50 years for these major outbreaks, beginning with the one that infected people in Africa and Europe in 1510. A more widespread episode occurred in 1889 and 1890, when influenza was reported in Russia in May 1889 and spread to North America by December of that year. In only five months, this pandemic had overtaken India and South America, and it finally reached Australia in the early spring of 1890.

In 1918 the incidence of influenza rose to a level never before seen on Earth. Although the strain began in the central United States, this episode became known as the "Spanish flu" because the spread of the disease received more press coverage in Spain, which was not involved in World War I and hence news in that country was not censored. The rapidity with which the pandemic spread is outlined at several different Web sites.

The supposed cause of this episode was from a mutated swine virus that affected soldiers at Camp Funston, Kansas, on March 11, 1918. Within two days, more than 500 people were sick, many of whom contracted pneumonia. Within a week, the flu had spread to every state in the United States and in the next two months the flu outbreak was worldwide. Interestingly, the disease usually peaked within a few weeks of its initial onset in a given area.

In the end, there were nearly 20 million cases in the United States, from which almost 1 million people died. The worldwide death toll has been estimated to range between 25 and 40 million people. Overall, most deaths occurred in people between 20 and 45 years old, generally considered the healthiest portion of a population.

Smallpox

Smallpox is an acute, contagious, and sometimes fatal disease caused by the variola virus (an orthopoxvirus), and marked by fever and a distinct, progressive skin rash. The only defense is prevention by vaccination, as no adequate treatment exists for the disease. Inflicted people develop raised bumps that appear on the skin covering the body. Historically, there were major episodes of outbreaks in the New World that killed several million people. In the twenty-first century, the United Nations reports that more than 300 million people worldwide have died from smallpox. The last documented case in the United States was in 1949 and the last case worldwide was in Africa in 1977. In 1980, the disease was declared eradicated following worldwide vaccination programs. However, in the aftermath of September 11, 2001, the CDC reports that the U.S. government is taking precautions to be ready to deal with a bioterrorist attack using smallpox as a weapon. Adequate vaccine supplies exist to inoculate everyone in the United States against the disease.

Typhus

The CDC reports that **typhus** generally occurs only in communities and populations in which body louse infestations are frequent (typically seen in refugee and prisoner populations, particularly during times of wars or famine; **Figure 13.5**). Typhus also occurs sporadically in cooler mountainous regions of Africa, South America, Asia, and Mexico, especially during the colder months when louse-infested clothing is not laundered and person-to-person spread of lice is more frequent.

The four main types of typhus are (1) *epidemic typhus*, (2) *Brill-Zinsser disease*, (3) *endemic* or *murine typhus*, and (4) *scrub typhus*. Scrub typhus was a major problem in World War I as it was associated with trench warfare that was prevalent in Europe. Typhus was a major cause of death during World War II.

The most severe of these is epidemic typhus, which is caused by *Rickettsia prowazekii*, a bacterium carried by human body lice as well as squirrel fleas and lice. If humans who carry *R. prowazeii* have blood sucked from them by lice, the lice become infected and excrete the bacillus. Uninfected humans who are then infected by these lice scratch themselves, working the lice feces into their skin.

Symptoms of epidemic typhus include chills and fever, headache, vomiting, and a rash that generally begins in the trunk area. If these conditions remain untreated, older adults who are infected can experience a death rate as high as 60 percent. Children usually recover well from epidemic typhus. Death rates for Brill-Zinsser disease, endemic or murine typhus, and scrub typhus are generally less than 1 percent. The best prevention for any form of typhus is to avoid the carrier insects and to use insect repellents and good hygiene.

Yellow Fever

Another viral disease that is transmitted between humans by mosquitoes is **yellow fever**. Symptoms are similar to other tropical diseases in that infected people experience fever, headache, muscle pain, jaundice, and nausea. One's pulse can slow down also, but all these symptoms generally vanish in a few days. There can be a sudden return of

smallpox
An acute viral disease that was once a major killer but has been eradicated; symptoms include headaches, vomiting, and fever, followed by a widespread skin rash that eventually permanently scars the skin.

typhus
An acute infectious disease with symptoms of high fever, severe headaches, and a skin eruption; common in wartime, famines, or catastrophes, it is spread by lice, ticks, or fleas.

yellow fever
Disease caused by a vector-transmitted virus; symptoms include high fever, headaches, jaundice, and often gastrointestinal hemorrhaging.

Centers for Disease Control and Prevention. Photo by James Gathany.

Figure 13.5 Body lice are parasitic insects that live on the body, and in the clothing or bedding of infested humans. They spread infections rapidly under crowded conditions where hygiene is poor, and there is frequent contact among people. Note the sensorial *setae*, or hairs that cover the louse's body, which pick up, and transmit information to the insect about changes in its environment such as temperature, and chemical queues. The dark mass inside the abdomen is a previously ingested blood meal. Information credit: Frank Collins, Ph.D., Centers for Disease Control and Prevention.

Figure 13.6 Major General William C. Gorgas, Surgeon General of the U.S. Army, eradicated yellow fever and reduced malaria in Panama.

anthrax
A serious disease usually acquired by ingestion of the bacterium *Bacillus anthracis* or its spores found in infected animals or their tissue; symptoms include coughing, chest pain, and general lethargy, which lead to death.

more serious conditions that include bleeding from the nose, eyes, mouth, and in the stomach. Kidney failure follows and usually death occurs within 10 to 14 days.

An epidemic of yellow fever struck the young United States in the late 1700s. From August to November 1793, between 4,000 and 5,000 people perished from the disease that at the time was unexplained. Philadelphia, then the new nation's capital, was the hardest hit in terms of deaths. The government came to a halt because so many workers were either afflicted with the disease or left the city to seek a safer area. A recurrence in 1798 killed many who were involved with forming the new government of the United States.

Historically, yellow fever was a major disease in equatorial regions and was one of the key impedances to the construction of the Panama Canal in the early 1900s. The research of William C. Gorgas (**Figure 13.6**), who also played a key role in the quest to eliminate malaria in the region, was paramount in controlling outbreaks of yellow fever. Gorgas contracted yellow fever as a young soldier and became immune to it in his later life. He also contracted typhoid fever in 1898.

Agricultural and forestry workers in South America and Africa are most affected by yellow fever as they are exposed to mosquitoes in their workplace, fields, and forests. Moist savanna regions in Central and West Africa are primary areas when the rainy season produces conditions that foster mosquito larva growth.

Yellow fever is now very rarely a concern to travelers but they must have a vaccination to prevent the disease. A single vaccination provides immunity for 10 years or more, with inoculation of a booster dose required every additional 10 years. If people are able to avoid mosquito piercings through the use of protective clothing, mosquito nets, and insect repellent, they can minimize the chance of contracting the disease. As the world's climate becomes warmer, more localities will experience increased infestation of mosquitos. This could result in an increase in yellow fever and other mosquito-borne diseases, especially in developing nations.

Anthrax

Anthrax is caused by a bacterium, *Bacillus anthracis*, which forms spores. Spores are cells that can lie dormant for decades but come to life when the right conditions exist. Anthrax most commonly occurs in wild and domestic lower vertebrates (cattle, sheep, goats, camels, antelopes, and other herbivores), but it can also occur in humans when they are exposed to infected animals or tissue from infected animals.

There are three types of anthrax:

1. Skin (or cutaneous), contracted through the skin by handling infected animals
2. Lung, contracted by breathing airborne spores
3. Digestive (gastrointestinal), which enters the body through ingestion of the spores that resided in undercooked meat from infected animals

Anthrax is not known to be a communicable disease, so someone who has the disease cannot pass it directly to another human.

In 2001, anthrax was used as a terrorist weapon when letters that contained anthrax were sent through the postal system. Of the 22 cases that were treated, five people died. This caused a national alert but was a relatively short-lived episode even though it disrupted many lives. Unfortunately, the chances are good that it will be used again in the future as a weapon of bioterrorism.

The CDC assigns biological agents with possible bioterrorism abilities to three groups:

1. Category A items pose the highest threat of producing major effects on public health, which includes anthrax.
2. Category B items can spread over large areas or require public awareness.
3. Category C items need a high level of planning to protect the health of the public.

Individual symptoms are quite varied depending on the type of anthrax. Cutaneous anthrax begins with small sores that develop into a blister that becomes a skin ulcer with a black center. This form of anthrax is fairly treatable if caught in the early stages. Antibiotics can cure most cases, but even if untreated, approximately 80 percent of the infected people do not die.

Gastrointestinal anthrax typically begins with nausea, loss of appetite, bloody diarrhea, and a fever. This is followed by stomach pains. This form of anthrax can result in death rates of 25 percent to more than 50 percent of the infected people.

Anthrax resulting from airborne spores is the most severe and can kill the majority of those infected. Anthrax spores that are inhaled first produce cold or flu-like symptoms that degenerate into a cough, chest discomfort, shortness of breath, muscle aches, and general lethargy. Because these symptoms are very similar to those of the flu, it is necessary to have them checked if there is any suspicion that anthrax might be the underlying cause of the illness.

The symptoms of the three types of anthrax can appear over a wide range of time. Symptoms related to all three types can become evident within seven days of coming in contact with the spores. Unfortunately, inhalation anthrax can take up to six weeks for the symptoms to become evident.

Treatment is generally effective if it is started soon after diagnosis. Antibiotics are used in all three cases. For someone who has been exposed to anthrax but is not yet displaying the symptoms, antibiotics along with inoculation with the anthrax vaccine will arrest the problem. If a person has been infected, the usual treatment involves a 60-day prescription of antibiotics, but final success of the treatment is dependent on which type of anthrax is involved and how quickly treatment was initiated.

At the present time, the only prevention for the general public is to avoid contact with the spores that produce anthrax. A vaccine that prevents anthrax does exist but is not generally available. People who could be exposed to the disease through their work or research have the opportunity to be vaccinated, along with those who might be exposed through an attack that might involve the use of anthrax as a weapon.

HIV

Human immunodeficiency virus (HIV) produces acquired immunodeficiency syndrome (AIDS) in humans. This disease was first diagnosed in the 1981 in the United States (although AIDS-like diseases had killed people 30 years earlier) and has spread to become a global problem.

By killing or damaging cells of the body's immune system, HIV progressively destroys the body's ability to fight infections and certain cancers. People diagnosed with AIDS may become infected with life-threatening diseases called opportunistic infections, which are caused by microbes such as viruses, bacteria, or fungi that usually do not make healthy people sick.

More than 40 million new cases of HIV and AIDS occur worldwide each year and the mortality rate is about 3 million per year. The greatest occurrence of HIV and AIDS is in eastern and southern Africa, where estimates place infection rates as high as 25 percent of the population. This is a blood-borne disease that is transmitted through unprotected sexual contact with infected partners. It is also possible to become infected through the use of nonsterile needles or, on extremely rare occasions, by tainted blood received through transfusions. HIV can be transmitted by women to their babies during pregnancy or birth. Approximately one-quarter to one-third of all untreated pregnant women infected with HIV will pass the infection to their infants. HIV can also be spread to babies through the breast milk of mothers infected with the virus.

The incidence rate of AIDS is highest among homosexual men, but education programs in high-incidence countries have helped stem the spreading rate. Recent data, however, show an increase in the Americas and in Asian countries.

human immunodeficiency virus (HIV)
An RNA (ribonucleic acid) virus that causes an immune system failure and can lead to AIDS.

acquired immunodeficiency syndrome (AIDS)
A serious disease that results from an infection with HIV; it is spread through direct contact with contaminated bodily fluids.

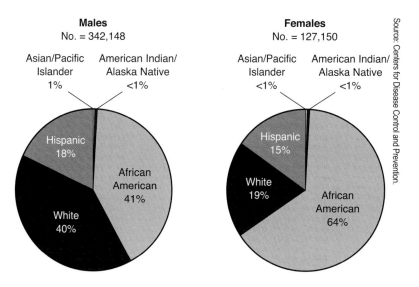

Source: Centers for Disease Control and Prevention.

Males
No. = 342,148

Asian/Pacific Islander 1%
American Indian/ Alaska Native <1%
Hispanic 18%
African American 41%
White 40%

Females
No. = 127,150

Asian/Pacific Islander <1%
American Indian/ Alaska Native <1%
Hispanic 15%
White 19%
African American 64%

Figure 13.7 African Americans account for about half of all people living with HIV/ AIDS within each sex category. According to information from 33 states, during 2005, among men, 41 percent living with HIV/AIDS are African American, and among women, 64 percent living with HIV/AIDS are African American.

As reported by the National Institutes of Health, more than 900,000 cases of AIDS have been reported in the United States since 1981. As many as 950,000 Americans may be infected with HIV, one-quarter of whom are unaware of their infection. The epidemic is growing most rapidly among minority populations. According to the CDC, AIDS affects nearly seven times more African Americans and three times more Hispanics than whites. In recent years, an increasing number of African American women and children are being affected by HIV/AIDS. In 2005, 49 percent of people in the United States who contracted HIV and AIDS were African Americans, and these diseases are the leading cause of death for that ethnic group (**Figure 13.7**).

In early 1999, a group of scientists reported in the journal *Nature* that they had linked HIV to chimpanzees in west equatorial Africa. Their evidence pointed to strains of HIV-1 coming in contact with humans through the blood of chimpanzees that had been hunted as a food source.

Symptoms of an HIV infection are not evident when someone is first infected. Within one or two months, flu-like symptoms appear and may include a fever, fatigue, headaches, and swollen lymph nodes. Because these generally disappear within a few weeks, they are usually mistaken for some other viral infection. It is during this period that people are highly infectious, especially through unprotected sexual contact.

Medical workers have noticed that the more severe or persistent symptoms may take 10 years or more to be evident, at which time the disease is well-established in the individual. The time it takes for the symptoms to be clearly defined varies with each person.

Much progress has been made toward the treatment of HIV and AIDS during the past decade. Advances and new drug treatments have improved the survival rate. The most current information concerning HIV and AIDS will be found on appropriate Internet sites sponsored by major medical institutions such as the CDC and National Institutes of Health (NIH).

Hepatitis

hepatitis
A liver disease caused by a virus; four different forms exist; symptoms include fever, jaundice, fatigue, liver enlargement, and abdominal pain.

Hepatitis is a liver disease caused by a virus. Several forms of hepatitis (designated A, B, C, D, and E) exist. Hepatitis A affects individuals through the presence of the Hepatitis A virus (HAV). There are no long-term effects and once someone gets it, he or she is immune for life. Hepatitis A is passed from person to person through contact with fecal material of infected people. It is critical to main clean surroundings if you have to work with ill patients.

The most widespread of the five types of hepatitis is Hepatitis B (HBV). Its symptoms include jaundice, fatigue, abdominal pain, nausea, and pain in the joints. One-third of the world's population is infected with HBV, making it the number one infectious disease at the present time (and of all time). Death from chronic liver disease occurs in approximately 25 percent of all cases, usually as a result of liver cancer or cirrhosis (scarring) of the liver. Most deaths occur in underdeveloped countries.

HBV is spread by contact with the blood of infected people, which enters the bloodstream of the noninfected person. Unprotected sexual contact with infected people and the sharing of needles and other drug paraphernalia will put the uninfected at severe risk of contracting Hepatitis B. HBV can produce a lifelong infection and can cause cirrhosis of the liver, liver cancer, liver failure, and death. A vaccine has existed since 1982 and is recommended for people between birth and age 18 if they have any likelihood of contracting the disease.

The CDC reports the following data for the incidence of HBV in the United States:

- The number of new infections per year has declined from an average of 260,000 in the 1980s to about 60,000 in 2004.
- The highest rate of disease occurs in 20- to 49-year-olds.
- The greatest decline has been observed among children and adolescents due to routine hepatitis B vaccination.
- An estimated 1.25 million are chronically infected Americans, of whom 20 to 30 percent acquired their infection in childhood.

Hepatitis C is contracted through contact with blood of an infected person. A primary means is by shared use of needles and other sharp objects related to drug use. Hepatitis D needs HBV present to exist. Hepatitis E, which is transmitted in similar ways to HAV, rarely occurs in the United States.

Leprosy

Leprosy (Hansen's disease) is a chronic, infectious disease that attacks the skin and peripheral nerves but can have many other manifestations. Two forms of the disease exist: paucibacillary Hansen's disease is the milder form that forms one or more pigmented skin lesions under the skin. The multibacillary form is recognized by symmetric skin lesions, nodules, thickened dermis, and nasal mucosa which produce nasal congestion and nose bleeds. The source of leprosy is *Mycobacterium leprae*, a rod-shaped bacterium that spreads very slowly and mainly infects the skin, nerves, and mucous membranes.

leprosy (Hansen's disease)
A chronic, mildly infectious disease caused by *Mycobacterium leprae*, which affects the nervous system, skin, and nasal regions; it is characterized by skin ulcerations and nodules.

It is uncertain how the disease is spread but researchers believe that humans contract the bacillus through respiratory droplets. The CDC reported that in 2002 there were 763,917 new cases detected worldwide, of which only 96 were in the United States. The World Health Organization (WHO) reported that in 2002 more than 90 percent of the cases occurred in Brazil, Madagascar, Mozambique, Nepal, and Tanzania.

Hansen's disease is spread by close contact with infected people. In the countries that have the majority of cases, drug therapy has not been implemented to a level that arrests its spread. Although there has been a steady decrease in outbreaks, Hansen's disease is still prevalent in localized areas. There are between 1 and 2 million people worldwide who are permanently disabled because of this disease. Once a person is infected, the disease can be treated with antibiotics, which rid the carrier of the causative bacteria.

The National Leprosarium of the United States, a facility that treated people inflicted with leprosy, existed in Carville, Louisiana (see **Figure 13.8**). Between 1894

National Park Service.

Figure 13.8 This historic image shows a Carville, Louisiana, Leprosarium patient pointing his walking stick towards a stone marker paying tribute to those 125 patients that had died while at the hospital between 1894 and 1922.

and 1922 there were 125 patients who died and were buried on the grounds. Once people entered the facility they never left. Because of the nature of the disease, the facility was self-contained. It had its own food-processing buildings, laundry, treatment centers, and recreational facilities.

Malaria

malaria
An infectious disease caused by the bacterium *Plasmodium falciparum,* which is transmitted by the female *Anopheles* mosquito; symptoms include high fever, chills, and sweating.

vector
An agent (for example, a flea or mosquito) that can spread parasites, a virus or bacteria.

Since the beginning of history, malaria has killed half of the men, women, and children who have lived on the planet. It has outperformed all wars, all famines and all other epidemics. Until World War II it still accounted for 50 percent of the business at most cemeteries. (Nikiforuk, 1991)

Malaria (derived from Italian: *mala aira,* meaning bad air) was so named because it was first thought to have been caused by bad air. A serious disease that has affected humans for more than 50,000 years, malaria has the cyclic symptoms of body chills, muscular pain, flu-like illness, and a high fever. Different strains exist, and one type can cause more serious problems, damaging vital organs including the heart, kidneys, lungs, or brain, thereby producing death. The bacterium *Plasmodium falciparum* that causes malaria cannot sustain its growth cycle in the female *Anopheles* mosquito (**Figure 13.9**) if temperatures are lower than 20°C (68°F). If those conditions exist, transmission of the disease ceases. The female *Anopheles* is the reservoir for malaria and is the **vector** that infects people by means of puncturing the skin. Male mosquitoes feed on nectar from flowers and pose no threat of transmitting the disease.

Most often found in Central and South America, Africa, and South Asia, malaria is a serious disease in those regions. This was especially true in the 1800s, when the British Army had large numbers of soldiers stationed throughout the tropics. The British government recognized that if the disease, and more importantly, its cause and source, could be identified, Great Britain would continue to dominate the regions it controlled in South Asia.

Sir Ronald Ross, a former major and surgeon in the British Army (**Figure 13.10**), studied the malaria parasite in birds and recognized its connection with mosquitoes, although he did not see that malaria was constrained to being transmitted by the female *Anopheles* mosquito. An infected mosquito carrying various protozoans belonging to the genus *Plasmodium* can easily pass along the disease. In 1880, Charles L. A. Laveran, a French army surgeon stationed in Algeria, was the first to notice parasites in the blood of a patient suffering from malaria. For his discovery and continuing work, Laveran was awarded the Nobel Prize in Physiology or Medicine in 1907.

Centers for Disease Control and Prevention. Photo by James Gathany.

Figure 13.9 This female *Anopheles* mosquito is taking a blood meal from a human host by pumpig the ingested blood through her "labrum", which is visible here as a thin red, "needle-like" structure between the mosquito's head and the host's skin.

In rare instances, the disease can be transmitted if people come in contact with infected blood. It is also possible for an infected pregnant woman to pass malaria to her fetus. However, it is a myth that malaria can simply be passed to anyone who comes in close contact with a carrier.

Today, malaria infects between 350 and 500 million people each year and causes between 1 and 3 million deaths annually, mostly among young children in sub-Saharan Africa. No vaccine is currently available for malaria, so preventive drugs must be taken continuously to reduce the risk of infection. Unfortunately, these prophylactic drug treatments are too expensive for most people living in the endemic areas of the tropical and subtropical regions of the world.

Malaria was prevalent in Central and equatorial South America and created a major challenge during the early stages of construction of the Panama Canal. In 1906, more than 26,000 workers were involved with the project. Malaria and yellow fever hospitalized more than 80 percent of them during some stage of their employment. By the time the workforce had swelled to more than 50,000 people in 1912, fewer than 12 percent of them were hospitalized. Through the efforts of William C. Gorgas, malaria was significantly reduced as a biological threat in the region.

Centers for Disease Control and Prevention.

Figure 13.10 Sir Ronald Ross

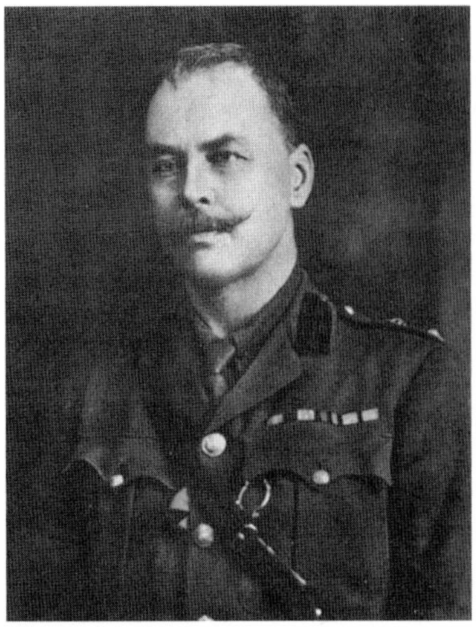

Because warmer environments exist closer to the equator, malaria can be transmitted any time during the year, and its movement through the population will be much more intense. The greatest degree of transmission occurs on the African continent south of the Sahara Desert (**Figure 13.11**).

As reported by the CDC, although the best breeding conditions exist for mosquitoes within subtropical and tropical regions, malaria will *not* be transmitted

- At high elevations
- During cooler seasons in some areas; this will change as a result of global warming
- In deserts (excluding the oases)
- In some islands in the Pacific Ocean, which have no local *Anopheles* species capable of transmitting malaria
- In some countries where transmission has been interrupted through successful eradication

Malaria is less likely to occur in developed countries where public health measures are better established. Although the United States and western Europe have eliminated malaria, these areas still harbor *Anopheles* mosquitoes that can transmit malaria, and reintroduction of the disease is a constant risk. One interesting side note is that people who have sickle cell anemia do not contract malaria due to mutations of their red corpuscles.

Distribution of Malaria

Source: Centers for Disease Control and Prevention.

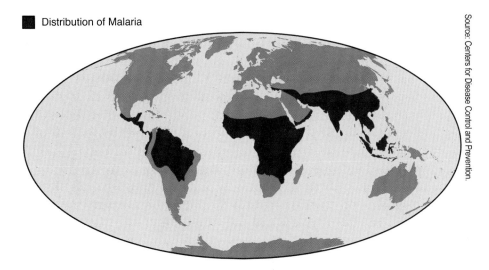

Figure 13.11 Worldwide distribution of malaria is shown in red. Note that the regions center about the tropics where warm and moist conditions allow the virus to grow and be spread by mosquitoes.

Tuberculosis

Captain among these Men of Death.

John Bunyan, 1660, author of *The Pilgrim's Progress*

---

tuberculosis (TB)

An infection caused by *Mycobacterium tuberculosis* and passed among humans by inhaled, airborne droplets; symptoms include coughing of fluids, fever, and chest pains.

From the preceding quotation, we see that tuberculosis was a disease of epic proportions, even 350 years ago. **Tuberculosis (TB)** is a pulmonary or lung disease that results from a bacterial infection produced by *Mycobacterium tuberculosis*. However, it can spread to other parts of the body. TB is commonly transmitted to other people on respiratory droplets when the infected person coughs or laughs.

Two forms of TB exist: latent and active. If a person has latent TB, the bacteria are being carried around by the person but are not transmittable. Approximately 10 percent of latent TB cases transform into active TB. Active TB spreads through the body of the carrier and can become transmitted to others if the carrier's lungs are infected. If not treated, active TB will kill more than half its victims.

The World Health Organization reported in 2004 that almost 15 million people were inflicted with active TB, and there were almost 9 million new cases, all of which contributed to 1.7 million deaths that year (Table 13.2). Most deaths occur in developing countries, but there is a significant increase in other parts of the world, primarily among those with HIV/AIDS because of the reduced ability of their immune systems to counteract the disease and the occurrence of strains that are resistant to drug treatment.

Tuberculosis and HIV form a lethal combination, as each disease hastens the spread of the other in an infected person. People who are HIV-positive have very weak

TABLE 13.2	Data Related to Incidence of TB Worldwide						
Estimated Incidence, Prevalence, and TB Mortality, 2004							
	Incidence[a]				Prevalence[a]		TB Mortality
	All forms		Smear-positive[b]				
WHO region	number (thousands) (percent of global total)	per 100000 pop	number (thousands)	per 100000 pop	number (thousands)	per 100000 pop	number (thousands)	per 100000 pop
Africa	2,573 (29)	356	1,098	152	3,741	518	587	81
The Americas	363 (4)	41	161	18	466	53	52	5.9
Eastern Mediterranean	645 (7)	122	289	55	1,090	206	142	27
Europe	445 (5)	50	199	23	575	65	69	7.8
South-East Asia	2,967 (33)	182	1,327	81	4,965	304	535	33
Western Pacific	1,925 (22)	111	865	50	3,765	216	307	18
Global	**8,918 (100)**	**140**	**3,939**	**62**	**14,602**	**229**	**1,693**	**27**

[a]Incidence: new cases arising in given period; prevalence: the number of cases that exist in the population at a given point in time.

[b]Smear-positive cases are those confirmed by smear microscopy and are the most infectious cases. Pop indicates population.

Source: World Health Organization.

immune systems that make them prime candidates to be infected with TB and hence become very ill. TB is one of the leading causes of death among those who are HIV-positive, accounting for about one in eight of all AIDS deaths worldwide.

Medical cures did not exist for TB until about 50 years ago. The use of antibiotic drugs reduced the onset of TB and reduced its severity for those infected. Treatment usually takes six to nine months but it is generally successful. As often happens, some strains of TB have developed resistance to certain drugs, thereby making them ineffective combatants, and new measures of fighting the disease must be adopted.

Typhoid Fever

The bacterium *Salmonella typhi* is the cause of **typhoid fever**, which is is common worldwide as it affects more than 21 million people annually. Approximately 400 cases are diagnosed each year in the United States, three-quarters of which are attributable to international travelers. The bacteria travel in the blood and intestinal tract and are shed through fecal material.

typhoid fever
An illness spread by contamination of water, food, or milk supplies with *Salmonella typhi;* symptoms include fever, diarrhea, stomach aches, and rash.

Although typhoid fever is rare in developed countries, it is still very prevalent throughout Latin America, Asia, and Africa. *S. typhi* is transmitted through unclean drinking water and food that has been handled by an infected person. Symptoms usually include a high temperature, headache, weakness, stomach pains, and a loss of appetite. A small number of cases are accompanied by a rash of flat, rose-colored spots. Immediate medical attention is critical.

An interesting case in history involves Mary Mallon, a cook who worked in New York City between 1900 and 1907. She became the first person to be recognized as a healthy individual who was clearly documented to be a carrier of typhoid fever. As a cook, she ended up infecting 47 people with typhoid fever through her dessert dish—peaches and ice cream. "Typhoid Mary," as she was named, continually denied her involvement with transmitting the disease and also refused to stop work as a cook. This only made the situation worse. She ended up infecting family members and, as she cared for them, continued to infect others. She died from pneumonia at age 69 but her autopsy revealed live typhoid bacteria in her body. She was cremated and interred in the Bronx section of New York City.

Prevention of typhoid includes being careful of what you eat and drink, and washing your hands with soap and water after using the bathroom is important. Vaccination is helpful but it is not completely effective in preventing the disease.

For studying several of the diseases discussed in this chapter, five medical researchers who examined their causes and effects on humans won Nobel Prizes for physiology or medicine (Table 13.3). Obviously, their work had wide-ranging and

TABLE 13.3	Nobel Prizes for Physiology or Medicine Awarded for Research into Various Major Diseases		
Year	Name	Country	Achievement
1902	Sir Ronald Ross	Great Britain	Recognizing the role of mosquitoes in spreading malaria
1907	Charles L. A. Laveran	France	Discovering the role of parasites in blood as related to the transmission of malaria
1928	Charles Nicolle	France	Researching the causes of typhus
1948	Paul Mueller	Switzerland	Recognizing that DDT killed mosquitoes
1951	Max Theiler	South Africa	Developing a vaccine for yellow fever

beneficial effects on humanity as did the contributions of countless other scientists and medical personnel who, with less recognition, have greatly expanded our current knowledge of these and other diseases.

Numerous other diseases exist that could easily generate pandemics. Many millions of people worldwide are inflicted with a multitude of diseases that include the Ebola virus, syphilis and other sexually transmitted infections (STIs), whooping cough, measles, and others. Only time will tell if any of these become universally prevalent.

Newer Diseases

No one knows what the future holds in terms of diseases—both those known and unknown. Until the early 1980s, we did not know about AIDS, but for almost three decades it has been a concern for the medical community worldwide. West Nile virus, avian flu, and severe acute respiratory syndrome (SARS) have surfaced in recent years as new global challenges. Undoubtedly there will be new diseases and mutated strains of existent diseases that make it onto the scene. The challenge will be how to control them and if possible to find preventive measures to lessen the illnesses and the resulting death tolls.

West Nile Virus

West Nile virus (WNV)
A type of virus that infects mosquitoes and birds, but can be transmitted to humans and other animals; symptoms include fever and other flu-like symptoms that may develop into encephalitis and meningitis.

West Nile virus (WNV) appears to be a seasonal epidemic that is present in the summer and early fall in North America. It is caused by skin punctures from infected mosquitoes, which get the disease from drawing blood from infected birds. Symptoms of WNV include a high fever, headache, disorientation, and muscle weakness. Fewer than 1 percent of those infected with the disease develop a severe case of WNV. About 20 percent experience a mild case; the remaining 80 percent show no symptoms at all.

Prevention is similar to other mosquito-borne diseases: use insect repellent, keep arms and legs covered with long clothing, and eliminate possible mosquito breeding sites, such as standing bodies of water. The disease has been on the decrease in the United States in recent years, with only 722 cases reported through December 2009. Approximately 4 percent of those infected with WNV die each year.

Avian Flu

avian flu (H5N1)
An acute viral disease of chickens and other birds (except pigeons) capable of being transmitted to humans; first noticed in the Far East, its symptoms include fever and lack of energy.

In the winter of 2003 and spring 2004, the virus associated with avian flu (H5N1) was discovered in birds in Vietnam. Fear increased that new strains of this flu would evolve and become co-joined with human flu viruses, thereby producing a flu that could prove extremely contagious and lethal to humans. Researchers felt that a pandemic could quickly spread across the globe. Unlike regular influenza, avian flu works its way into the deeper parts of the lungs. This makes it very lethal, but it also inhibits its spread from one person to another. Nations were put on alert to combat the potential spread of the avian flu after cases were diagnosed in several Eastern European countries as well as China and Russia. In October 2005 the virus had spread to more than eight countries, but only 67 people had died. This was far fewer than had died at a similar stage of previous flu epidemics. Fortunately, very few human-to-human transmissions were noted. The current phase of pandemic alert set by the World Health Organization (WHO) is level 3, described as "no or very limited human-to-human transmission."

In the spring of 2006, the WHO reported the first evidence that avian flu virus had mutated and spread among family members in Indonesia. Fortunately, the strain died out after killing seven of eight people in the family. Obviously the potential exists for avian flu to reappear and cause a great deal of concern among health workers. To date, fewer than 200 people have died worldwide from avian flu, and health officials have not recognized a sustained human-to-human transmission that would cause major concern.

Prion Diseases

Prion diseases are part of a class of transmissible spongiform encephalopathies (TSE) that are characterized by a malformed protein molecule that produces clumps in the brain. Bovine spongiform encephalopathy (BSE or mad-cow disease) is a TSE, as is scrapie, a nervous system disease in goats and sheep. Human consumption of a diseased animal can lead to the spread of TSE among humans. Once contracted, prion diseases are nearly always fatal.

Included in the human category of TSEs is kuru, a rare brain disorder that reached epidemic proportions in the mountains of New Guinea in the 1960s. As a result of rituals involving mortuary cannibalism among some of the indigenous tribes, kuru spread to a large percentage of the population, but it has now essentially disappeared due to government intervention.

prion disease
An infliction that attacks the brain and nervous system, and disrupts normal protein activity within the neural cells.

SARS

Severe acute respiratory syndrome (SARS) is caused by a virus belonging to the genus *Coronavirus*. One epidemic of SARS occurred in Asia (predominantly mainland China) and lasted from November 2002 to July 2003. More than 8,000 cases were documented, resulting in 774 deaths. Concern spread worldwide that a pandemic could result, but rapid response and preventive measures precluded the pandemic. The disease has not been eradicated and could reemerge without warning. Because it is an atypical form of pneumonia, it can rapidly spread among humans and have wide-ranging effects.

severe acute respiratory syndrome (SARS)
A respiratory disease of unknown source that first occurred in mainland China in 2003; symptoms include fever and coughing or difficulty breathing; it is sometimes fatal.

Examples from the Geologic Record

Major die-offs of animals have most likely occurred in the geologic past but the organisms that could have produced such deaths have themselves been destroyed. Large-scale mass extinctions resulted from major wildfires, impacts of extraterrestrial objects, major volcanic eruptions, and global climate changes that lowered temperatures. Undoubtedly, individual animals experienced torn flesh through predation of other organisms. Unless these wounds were immediately fatal, the animal would heal itself and move along to die by other means.

Lessons from the Historic Record and the Human Toll

Many serious diseases have existed in the past and continue to do so now. Throughout human history, countless millions of people have died from diseases, most of them related to pandemic outbreaks. Because the sources of some diseases are difficult to eradicate, such as wiping out all mosquitoes worldwide, it will be virtually impossible to completely eliminate many of these maladies.

However, there is a glimmer of hope. Smallpox is an example of a success story. A disease that was once a dreaded scourge on Earth, it has been put into the extinct category, as it has been eradicated after a very concerted, worldwide effort. The CDC reported that the last recorded case in the United States was in 1949 and the last one in the world was in Somalia in 1977. Although the disease is eradicated, samples of the virus are kept under guard at the Centers for Disease Control for future research and the production of a smallpox vaccine, if needed.

Minimizing Biological Hazards

One means of reducing the spread of insect-borne diseases is the use of insecticides. Perhaps the most widely used has been DDT (dichloro-diphenyl-trichloroethane), first synthesized in 1874 by a German chemistry student as a thesis project. In 1939, its ability to kill insects was discovered by the Swiss scientist Paul Müller (Nobel Prize in physiology or medicine in 1948). Several countries used it during World War II to rid areas of lice that carried typhus. Following the war, it was used in numerous countries to kill mosquitoes that carried malaria. However, the use of DDT had profound effects on many forms of wildlife, including the embryos of bald eagles and peregrine falcons. Its use in the United States was banned in 1972. However, it is still used in many tropical regions to curtail the spread of mosquitos.

An example of the eradication of a disease in a developed country occurred in 1951 when malaria was officially declared no longer a threat in the United States. This happened after a four-year plan had been carried out to spray millions of homes and their surroundings to kill the carriers. As late as 1947, 15,000 cases of malaria were reported in the United States; by 1951 there were none.

Safe water sources and clean living and sanitary environments will help minimize many diseases that are associated with poverty. Diseases can move with people who are sick whenever they are trying to avoid a pending natural disaster, such as an imminent volcanic eruption. We must all be vigilant to prevent the spread of diseases that can become a challenge to our very existence. Global warming will enhance the spread of diseases that thrive in warmer temperatures. Malaria and yellow fever will spread when mosquitoes are able to expand their habitats across the globe.

Summary

Throughout history, humanity has dealt with biological hazards that have a profound effect on the health and well-being of millions of people. Following major natural disasters, such as earthquakes or floods, diseases spread rapidly among people and communities where existent defenses against them have been compromised. Epidemics prevail in local or regional settings but are eventually controlled. On a global scale, diseases can move across oceans to other continents generating pandemic situations. Since the first century A.D., diseases have killed millions in every part of the globe. Bubonic plague spread as a result of increased trade routes with the Far East; influenza spread in the early twentieth century with increased travel opportunities, and HIV/AIDS has afflicted millions as a result of unsafe sex practices among people of many different countries and ethnicities. Much research was done in the late nineteenth and early twentieth century to curb the spread of malaria, yellow fever, and tuberculosis but these diseases are still prevalent in certain regions of the world.

Careful planning and widespread education are necessary to blunt the impact of these diseases, particularly in developing countries and regions that do not have the economic means to solve the problems themselves. The reduction of smallpox, once a dreaded disease, to the realm of "eradicated" is evidence that a worldwide effort can produce remarkable benefits. The challenges of the future lie in controlling relatively new diseases, such as avian flu, prion diseases, and SARS, to the point that they do not generate pandemics requiring the energy and attention of millions of people to prevent them from affecting the global economy and well-being of Earth's inhabitants.

References and Suggested Readings

Ewald, Paul W. 1994. *Evolution of Infectious Disease.* New York: Oxford University Press.

Giesecke, Johan. 2002. *Modern Infectious Disease Epidemiology.* 2d ed. London, England: Arnold.

Herlihy, David. 1997. *The Black Death and the Transformation of the West.* Cambridge, MA: Harvard University Press.

Hurster, Madeline M. 1997. *Communicable and Non-Communicable Disease Basics—A Primer.* Westport, CT: Bergin and Garvey.

Lashley, Felissa R. and Jerry D. Durham, eds. 2002. *Emerging Infectious Diseases—Trends and Issues.* New York: Springer.

Nikiforuk, Andrew. 1991. *The Fourth Horseman—A Short History of Epidemics, Plagues, Famines and Other Scourges.* New York: M. Evans.

Rifkind, David and Geraldine L. Freeman. 2005. *The Nobel Prize Winning Discoveries in Infectious Diseases.* Amsterdam, Holland: Elsevier Academic Press.

Ziegler, Philip. 1991. *The Black Death.* Wolfeboro Falls, NH: Alan Sutton.

Web Sites for Further Reference

http://www.cdc.gov/
http://www.cdc.gov/ncidod/dbmd/diseaseinfo/
http://www.cyndislist.com/medical.htm
http://www.infoplease.com/ipa/A0903696.html
http://www.mla-hhss.org/histdis.htm

Questions for Thought

1. What is a biological hazard and what form can it take?

2. Distinguish between a bacterium and a virus.

3. Explain the difference between an epidemic and a pandemic.

4. Why was the Black Plague so named?

5. List the major diseases that are spread by insects.

6. How does a warming climate affect the potential spread of diseases?

7. Why is the "Spanish flu" outbreak in 1981 so named, and where did it begin?

8. List major diseases that are readily passed along from one person to another by means of direct contact.

Human and Natural Interactions on the Environment*

14

© Debbie Franzen, 2010. Under license from Shutterstock, Inc.

© Just ASC, 2010. Under license from Shutterstock, Inc.

© Tonis Valing, 2010. Under license from Shutterstock, Inc.

© Lane V. Erickson, 2010. Under license from Shutterstock, Inc.

© egd, 2010. Under license from Shutterstock, Inc.

Collage of human and natural environments that affect the space in which humans, animals, and plants exist on Earth.

Key Terms

biodiversity
biogeochemical cycle
biomagnification
carbon cycle
carbon/oxygen cycle
desert
ecology
ecosystem
nitrogen cycle
phosphorus cycle

* This chapter was written with W. Sylvester Allred, Department of Biological Sciences, Northern Arizona University.

The preceding chapters have discussed how natural catastrophes on Earth affect the landscape and the people who inhabit those areas. This closing chapter will address how environments above, on, and below the surface have been changed by humans and nature itself.

Human Population Growth

In Chapter 1 we learned that global population has been estimated at eight million inhabitants 10,000 years ago. With the domestication of animals and the development of agriculture that allowed people to move away from hunter-gatherer societies, the number of people on Earth began to rise. Growth rates are estimated to have been fairly low for the next 11,000 years, but by the Middle Ages (1100 to 1500) almost 500 million people were living on the planet. A setback occurred between 1348 and 1350 when bubonic plague (the Black Plague) struck Europe, killing tens of millions of people (refer to Chapter 13). The world population rebounded in the 1400s and by 1800 it was slightly less than one billion.

The United Nations Population report *The World at Six Billion* provides the following facts about changes in the world population over the past two hundred years:

- It reached one billion in 1804;
- It took 123 years to reach 2 billion in 1927;
- 33 years to reach 3 billion in 1960;
- 14 years to reach 4 billion in 1974 ;
- 13 years to reach 5 billion in 1987; and
- 12 years to reach 6 billion in 1999.

This report estimated that global population would reach 7 billion in 2013. As of December, 2009, global population is estimated at 6.8 billion (**Figure 14.1** and U.S. Census population clock). Obviously the United Nations estimate will be eclipsed prior to 2013. There has been a decline in the annual growth increment from about 2 percent in 1970 to 1.1 percent in the early twenty-first century. However, it is still a positive growth rate. The amount of growth in 2004 was almost 76 million people—equivalent to adding the population of Egypt to the planet each year!

Contributors to the exponential growth in population worldwide include increased industrialization in Europe in the eighteenth century; increased food sources and improved nutrition; the eradication of certain diseases; and development of medi-

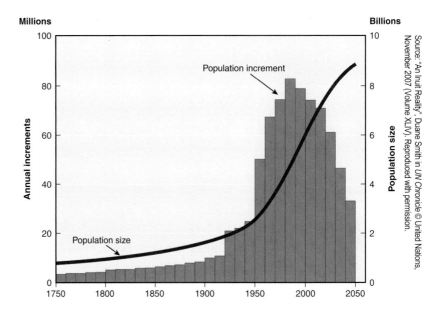

Figure 14.1 The dark line shows global population as measured on the right hand vertical axis. Annual incremental changes are shown by the rectangles which are scaled on the left.

TABLE 14.1	Average Life Expectancy, Measured in Years. Entries contain the average number of years to be lived by a group of people born in the same year, if mortality at each age remains constant in the future. The entry includes *total population* as well as the *male* and *female* components. Life expectancy at birth is also a measure of overall quality of life in a country and summarizes the mortality at all ages. Data represent estimates made in 2009.

Rank	Country	Life Expectancy (in years)
1	Macau	84.36
2	Andorra	82.51
3	Japan	82.12
4	Singapore	81.98
5	San Marino	81.97
6	Hong Kong	81.86
7	Australia	81.63
8	Canada	81.23
9	France	80.98
10	Sweden	80.86
36	United Kingdom	79.01
50	United States	78.11

Source: Central Intelligence Agency, *World Fact Book.*

cines to improve longevity. Unfortunately diseases such as cholera, influenza, malaria, and tuberculosis continue to kill hundreds of millions of people each year, mainly in underdeveloped countries that lack sanitation and good health services. Childhood diseases also take a huge toll in poor nations, where infant mortality exceeds 10 percent of the live births.

Average life expectancy varies widely throughout the world. Many countries in Africa and Asia have life expectancies of less than 50 years. The more-developed countries have the highest expectancies (Table 14.1). Note the position of the United States in the ranking, well below other countries that have considerably higher longevities.

The Effects of Overpopulation

Land Use

The continued exponential population growth worldwide has placed additional demands on Earth's resources, both those that occur naturally such as water and oil, and on resources that are produced, such as agricultural goods. Destruction of the landscape has had a profound effect on the environment. Our natural resources have

Figure 14.2 Large crowds often gather to celebrate important events. These events give an indication of how many people live in areas that covered much more space than this gathering spot in Ontario, Canada.

desert
A region that receives less than 25 cm (10 in) of annual precipitation.

not increased, only our ability to find them. Essentially no new land is created (tens of thousands of years are often necessary for volcanic material to form productive soils).

Urbanization has placed a higher percentage of people into compact cities (**Figure 14.2**) . In the United States only 3 percent of the area of the country is categorized as urban (60 million acres). Often these municipalities are built at the expense of farmland, areas once used to produce crops and raise animals.

The Department of Agriculture (USDA) reports that 97 percent (2.2 billion acres) of the land in the United States is classified as agricultural, forest, or other use land. In the United States the population explosion in the Southwest has transformed once **desert** areas into bedroom communities. In recent years the overbuilding in cities such as Phoenix and Las Vegas has resulted in thousands of unsold homes and a significant decrease in home values due to a glut of housing. Parts in southern California and southern Arizona that were once rich agricultural areas have been reshaped into suburbs populated by large numbers of people.

Agriculture

The USDA reports the following major uses of agriculture land in 2002:

- Cropland, 442 million acres (20 percent of the land area)
- Grassland pasture and range, 587 million acres (26 percent)
- Forest-use land (total forest land exclusive of forested areas in parks and other special uses), 651 million acres (19 percent)
- Special uses (parks, wilderness, wildlife, and related uses), 297 million acres (13 percent)
- Urban land, 60 million acres (3 percent)
- Miscellaneous other land (deserts, wetlands, and barren land), 228 million acres (10 percent).

Proportions vary across regions of the United States due to differences in climate, geographic setting, and population densities. For example, the Northeast has 12 percent of its area in cropland, compared with 58 percent in the Corn Belt. However, nearly 60 percent of the Northeast is in forest, compared with only 2 percent in the Northern Plains. The land in Alaska, the largest state in the Union, skews the data due to large amounts of forest and very little cropland, so it is not included in these values.

Small changes have occurred in land use, in percentage terms, between 1997 and 2002, for the 48 contiguous States. As reported by USDA, the largest acreage change from 1997 to 2002 was a 13-million-acre decrease in cropland (a drop of 3 percent). This continues a long downward trend from 1978, when cropland totaled 470 million acres. Total cropland area in 2002 was at 442 million acres, its lowest point since in 1945. Cropland has been relatively constant from 1945 to 1997, ranging between 442 and 471 million acres and averaging about 463 million acres (**Figure 14.3**).

While land defined as forest use has been on a declining trend from 612 million acres in 1964, forest-use lands increased from 552 to 559 million acres between 1997 and 2002. Forest-use land excludes forested areas in special uses, estimated at 98 million acres in 2002.

The USDA reports an estimated 60 million acres in 2002 as urban use, compared to an estimated 66 million acres in 1997. This decline in the estimates is due to changes in the way urban land area is measured and provides little information about changes in urbanized area from 1997 to 2002. Rural residential land, a new land use category introduced in 1997, was estimated to be about 94 million acres in 2002, up from the 1997 estimate of 83 million acres (**Figure 14.4**).

Figure 14.3 The amount of crop land has remained relatively constant over the past sixty years.

Figure 14.4 An increase in the urban population has led to the construction of many new homes in areas that were once used for agricultural purposes.

Factors Contributing to Land Use Change

Land has shifted into crop production from other nonurban uses in response to rising commodity prices. However, land-use changes are gradual, due to conversion costs.

Between 1945 and 2000, the population of the United States more than doubled from 133 million to 281 million people. In 2000, there were 106 million households, a quarter of which were one-person households. More land was converted to urban uses, especially for homes (see Figure 14.4). New residential uses also require land for schools, office buildings, shopping sites, and other commercial and industrial uses. The amount of land converted to urban use rose steadily from 15 million acres in 1945 to an estimated 60 million acres in 2002. These increases came mostly from pasture, range, and forest land, a shift that decreased the amount of land available for farming.

Effects of Human Population Growth on the Environment

Rapid increases in global population during the past two centuries have affected cycles related to the environment and ecosystems. These alterations and disruptions have changed the amounts of important elements and components of the environment, such as carbon, oxygen, nitrogen, phosphorus, and water. The disruption and pollution of the atmosphere and hydrosphere have long-ranging effects on ecosystems and biodiversity.

biodiversity
All the diversity of life on Earth, ranging from single cell organisms to the most complex ecosystems.

biogeochemical cycle
Natural processes that recycle nutrients in various chemical forms from the environment, to organisms, and then back to the environment. Examples are the carbon, oxygen, nitrogen, phosphorus, and hydrologic cycles.

ecosystem
A community of plants, animals and other organisms that interact together within their given setting.

carbon cycle
The movement of carbon between the biosphere and the nonliving environment. Carbon moves from the atmosphere into living organisms that then move it back into the atmosphere.

The Biogeochemical Cycles

Besides water cycling through the biosphere in plants and animals, other components, such as carbon and nitrogen, have high concentrations in the biosphere. Cycles that include the interactions between the biosphere and other reservoirs involve biological processes including respiration, photosynthesis, and decomposition (decay), which are referred to as biogeochemical cycles. A **biogeochemical cycle** is a pathway by which a chemical element or molecule moves through both biotic and abiotic components of an **ecosystem** (**Figure 14.5**). All chemical elements occurring in organisms are part of biogeochemical cycles and, in addition to being a part of living organisms, these chemical elements also move through the hydrosphere, atmosphere, and lithosphere. All the chemicals, nutrients, or elements, such as carbon, nitrogen, oxygen, and phosphorus, used by living organisms are recycled. An important example is the **carbon cycle**, which is important in regulating our global climate by acting as a greenhouse gas within the atmosphere.

Figure 14.5 A mountain stream ecosystem contains trees, soil, moisture, and the atmosphere, all of which interact with one another to provide a dynamic, living environment.

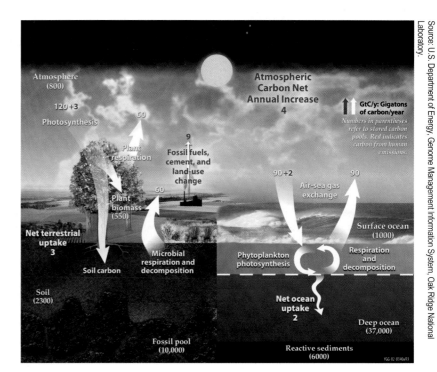

Source: U.S. Department of Energy, Genome Management Information System, Oak Ridge National Laboratory.

Figure 14.6 The carbon cycle shows where carbon is contained in the geosphere, hydrosphere, biosphere, and atmosphere. Each year more than 3 gigatons of carbon are added to the cycle.

Carbon Cycle

Carbon (C) is the basic building block of life. As the fourth most abundant element in the universe (after hydrogen, helium, and oxygen), carbon occurs in all organic substances, including DNA, bones, coal, and oil. Carbon moves through the hydrosphere, geosphere, atmosphere, and biosphere, each of which serves as a reservoir of carbon and carbon dioxide (CO_2). The carbon cycle (**Figure 14.6**) involves the biogeochemical movement of carbon as it shifts between living organisms, the atmosphere, water environments, and even solid rock. The least amount of carbon is contained in the atmosphere, while the largest amount is found in the lithosphere.

Carbon dioxide occurs in all spheres on Earth and, as a gas, is readily mobile. Our increased use of fossil fuels has increased the amount of CO_2 in the atmosphere, with some of gas being taken up in the hydrosphere and biosphere. Marine and fresh water organisms use CO_2 in their life cycles and can increase the amount of CO_2 in the geosphere. Coral reefs form by corals extracting dissolved CO_2 from sea water and creating their colonies in warm water environments.

Oxygen Cycle Reservoirs and Flux

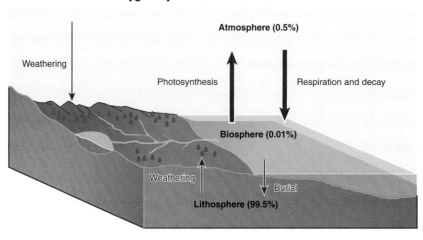

Figure 14.7 Storage and movement of oxygen occur between the atmosphere, biosphere, hydrosphere, and lithosphere. Photolysis is the chemical decomposition of the atmosphere caused by sunlight.

Generally there is a dynamic equilibrium between and among the four reservoirs of carbon. However, these are often changed by natural processes that contribute to the production and hence increased abundance of CO_2. Volcanic eruptions and wildfires are two natural hazards that upset the carbon equilibrium. These two examples are discussed earlier in the text.

Carbon/Oxygen Cycle

Carbon (C) and oxygen (O_2) are elements that are required of all living organisms. In the earliest stages of Earth's formation, no free O_2 existed in the atmosphere, and very little was present in the oceans. Approximately one billion years after Earth formed, bacteria began to produce O_2 and the amounts of this gas increased significantly. Oxygen is the second most abundant gas in the atmosphere and the most common element in Earth's crust by weight (refer to Chapter 2). As an anion, oxygen can actively combined with other cations and is a key constituent in numerous organic compounds. Oxygen is very prevalent in the lithosphere and is often freed up through chemical reactions to move into the atmosphere, hydrosphere, and biosphere (**Figure 14.7**). Additional amounts of oxygen and carbon are introduced to the atmosphere and hydrosphere by volcanic activity (refer to Chapter 4).

carbon/oxygen cycle
This cycle is connected to the movement of carbon dioxide through the biosphere, lithosphere, and atmosphere as carbon dioxide is used by different organisms.

Nitrogen Cycle

Nitrogen (N_2) is the predominant gas in the atmosphere, comprising 78 percent of the total gases. Although it is a key part of proteins and other organic substances, nitrogen is difficult for most living organisms to assimilate. Soil bacteria must alter N_2, which is then taken up by plants such as peanuts, peas, beans, soybeans, and alfalfa. These bacteria, which are found in the roots of these plants, generate a chemical reaction that allows N_2 to be transformed into ammonia (NH_3) and nitrate compounds (those containing NO_3). Animals and other organisms that use these plants as a food source then take in the N_2 and expel it as organic waste. Waste products rich in N_2 are used as fertilizers which return the N_2 to the soil and the **nitrogen cycle** begins again (**Figure 14.8**)

nitrogen cycle
The movement of nitrogen through the biosphere and the atmosphere.

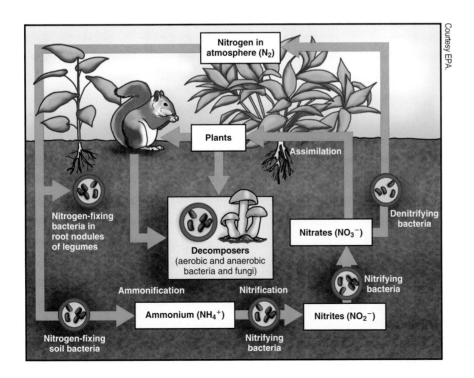

Figure 14.8 The Nitrogen Cycle.

Phosphorus Cycle

Phosphorus (P) is an important nutrient for plants and animals. It moves through the biosphere, hydrosphere, and geosphere but is lacking from the atmosphere, because it is in a liquid state at normal temperature and pressure. Phosphorus is part of DNA molecules and is a major component in animals and humans, as it makes up a significant portions of bones and teeth. Phosphorus is also a component of ATP, the energy molecule for all life forms. Due to its relatively low mobility, the **phosphorus cycle** is the slowest-acting of the ones described here.

phosphorus cycle
The movement of phosphorus through the biosphere and the geosphere.

Water Cycle and Water Resources

The water cycle, which is also referred to as the hydrologic cycle, shows where water is located and how it moves through the atmosphere, hydrosphere, lithosphere, and biosphere (**Figure 14.9**). The movement of water is related to the three states or conditions in which water can exist—as a liquid, solid, or a gas. More than 75 percent of Earth's surface is covered by water in either a liquid or solid form. Initially, water formed on the surface through its being expelled by volcanic eruptions from water-bearing magmas and also by comets from outer space bringing frozen ice that later melted.

The United States Environmental Protection Agency (EPA) was formed in 1970 to "protect human health and to safeguard the natural environment—air, water and land—upon which life depends (part of the EPA Mission statement). The EPA sets the standards for the amounts of potentially harmful substances that can affect human life and the natural environment.

Federal legislation has addressed the need for guidelines to establish and maintain water and air standards. The Clean Water Act is a U.S. federal law that regulates the discharge of pollutants into the nation's surface waters, including lakes, rivers, streams, wetlands, and coastal areas. Passed in 1972 and amended in 1977 and 1987, the Clean Water Act was originally known as the Federal Water Pollution Control Act. The Clean Water Act is administered by the EPA, which sets water quality standards, handles enforcement, and helps state and local governments develop their own pollution control plans.

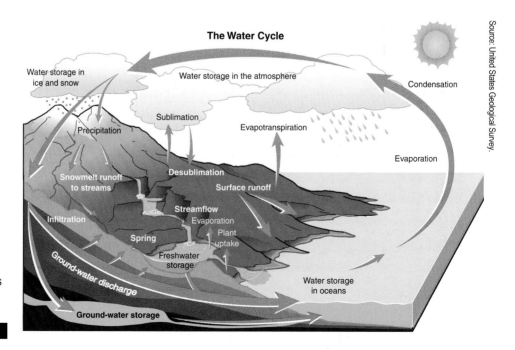

The Water Cycle

Water storage in ice and snow

Water storage in the atmosphere

Condensation

Sublimation

Evapotranspiration

Precipitation

Evaporation

Snowmelt runoff to streams

Desublimation

Surface runoff

Infiltration

Streamflow

Evaporation

Plant uptake

Spring

Freshwater storage

Ground-water discharge

Water storage in oceans

Ground-water storage

Source: United States Geological Survey.

Figure 14.9 The water cycle (also referred to as the hydrological cycle) shows how water in its various phases moves or rests on Earth.

Water Pollution

Water is essential for plant and animal growth. Water must be clean enough for its consumer to use it without adversely affecting the health and well-being of the plant or animal.

Most individuals do not think about where the water they drink comes from or where the water that goes down the drains winds up. If asked, their responses range from—"the city provides my water" to "I don't know, it just flows when I open the faucet." Only when we are without water do we consider where it is and ask if it is drinkable.

"Is it drinkable?" becomes an important question. Because water is the universal solvent, most chemicals will eventually become incorporated. None of us would intentionally drink water with harmful chemicals such as arsenic, cadmium, or lead, in it. Thus we trust that our water suppliers, be it a city or a bottler of water, have some standards in what can and cannot be in the water we drink. The EPA sets water standards for the nation, and the 50 states can set the standards to be the same or be even more strict.

When you flush anything down the drain, for example, some expired prescription drugs, the water treatment plant may or may not remove those drugs before the treated water is discharged into a watercourse. Downstream another city takes in that water and treats it to destroy harmful bacteria and viruses, but what about the other substances in the water? Where would you like to live? Upstream or downstream of such a community?

An interesting debate has been going on in Flagstaff, Arizona, over the past few years. A ski resort wants to use partially treated water, which the city of Flagstaff is willing to sell, to make snow. Would you consider catching snowflakes when the snow is being made or having a snowball fight with that snow? What if you fell down while snowboarding and accidently swallowed a mouthful of that snow?

Much of the freshwater we have on the planet is used in agricultural endeavors to feed and nourish our ever-increasing population. Pumps run 24/7/365 to move water from deep aquifers to the surface for irrigation (**Figure 14.10**). These aquifers, filled with fossil water (water laid down during the last Ice Age), are not recharging at anywhere near their depletion rate. Thus the water is being mined in a permanent sense. In some areas of our country where there is a preponderance of agricultural activities and associated irrigation, saltwater encroachment is occurring in the aquifers.

The pollution of water in aquifers beneath cities and military installations occurs as a result of leaks and accidental spills into the ground, which eventually enters the water table. For example, petroleum pipelines connecting refineries and distribution centers around the country occasionally break. Certainly if the leak is detected mediation (clean up) measures are instigated. But what about undetected leaks, or leaks from years ago when regulations were lacking or ignored, many times out of ignorance. Today we have monitoring wells around sites of known hazardous leaks. These wells are regularly sampled to insure that the chemical plume's vicinity to ground water is known. But again these are the sites we know about.

When the Cuyahoga River in Cleveland, Ohio caught on fire in June 1969, a wakeup call was sounded to the nation and the world. The call was simple—we cannot keep using our water supplies as a place to dump. The mentality of "out of sight, out of mind" had to stop. We are still working on this one.

Runoff from agricultural and livestock feeding facilities represents another source of water pollution. Fertilizers, insecticides, fungicides, and herbicides that are sprayed on

Figure 14.10 Irrigation of fields uses large amounts of ground and surface water. The water must be of good quality to provide the moisture needed to raise crops and other agricultural products.

© Daniel Alvarez, 2010. Under license from Shutterstock, Inc.

© Homa, 2010. Under license from Shutterstock, Inc.

Figure 14.11 Surface drainage often enters the seashore without any treatment of the materials in the water.

fields to improve crop yields find their way into streams, rivers, lakes, and eventually oceans. Fertilizers, while enriching the soil, can end up in the water supply and alter the biota by accelerating the growth of certain algal components. Bacteria feed on the decaying algae, depleting the dissolved oxygen levels, causing fish kills with their associated ramification to the environment. It takes one domino falling into another to cause the remaining ones in line to fall.

Usually not mentioned but certainly on the list of water pollution is thermal pollution. Downstream outflows from power generating stations, which use river water to cool the machinery of power production, alter temperature regimes, which alter habitat and thus the biota. Some organisms flourish while others die off in the heated water.

As stated above, oceans are the final receiver of continental runoff (**Figure 14.11**). Oceans and lakes seem vast and bottomless, but in some areas, usually at the mouths of rivers, they may be virtual cauldrons of toxins. These toxins may be suspended in the water column or accumulate in the sediments. Ocean food chains are linked to the water column and the sediments.

Organisms feeding in these areas accumulate toxic compounds that are then later accumulated at higher levels in larger organisms (biomagnification). Biomagnification was introduced by Rachel Carson in her book, *Silent Spring*. Considering that many species of ocean life are consumed by humans around the world, and depending on where in the world we are, ocean life is sometimes a staple in the diet and sometimes a delicacy. Either way, the toxins placed by us, either accidently or by intent, return to us in an often harmful way.

Air Pollution

Clean air is necessary for life to survive and thrive on the planet. We have seen that the atmosphere near the surface consists of 78 percent nitrogen, 21 percent oxygen, and about 1 percent of many lesser gases. Oxygen is the component critical for animal life. However, when the air (and atmosphere) is contaminated with pollutants such as nitrous and sulfurous oxides and dangerous hydrocarbon volatiles, the air is rendered harmful (**Figure 14.12**). Pollution can come from numerous sources and can include particulate matter such as dust and ash from volcanoes and wildfires. Anthropogenic sources include coal-fired power plants, factories, vehicles that operate on fossil fuels, and even dry-cleaning establishments.

© manfredxy, 2010. Under license from Shutterstock, Inc.

Figure 14.12 Pollutants from industrial smokestacks enter the atmosphere without any reduction in the particulates or gases generated through the manufacturing process.

A secondary effect of air pollutants is the formation of haze that can reduce visibility around large cities and in wilderness areas and in national parks. In northern Arizona a large coal-fired generating station is located only 25 km (15 miles) from the northeastern boundary of Grand Canyon National Park. Until several years ago, a similar type of station located to the west of the park contributed significantly to lower visibility in the regions of national parks in northern Arizona and southern Utah. The Mohave Generating Station in Laughlin, Nevada, was shut down in December 2005 rather than have its operators face the expense of installing scrubbers and other pollution control equipment to reduce its emissions.

The Clean Air Act, originally passed in 1963, has been amended several times to insure healthy air in the United States. The EPA is entrusted to insure that clean air standards are met by monitoring the amounts of air pollutants across the country.

Soil Contamination

The introduction of hazardous materials either on the surface or into pore spaces in soil can produce harmful conditions for humans, animals, and plant life. Before the EPA was established in 1970, numerous instances of dumping of hazardous chemicals and other materials caused the soil and ground water to be filled with substances that had a dangerous effect of life. There was no concern or knowledge that materials placed in dump areas would produce negative effects on future generations.

Dangerous contaminants quickly work their way into the food chain as plants take up the material into their structure and then animals ingest the plants. Eating of the animals by humans and other animals then passes the substances along, where they can have a very deleterious effect on life.

Heavy minerals and liquids can contaminate the soil and lead to life-threatening conditions (**Figure 14.13**). The cleaning up of contaminated areas involves either: (a) treating the soil in place with chemical and other procedures that clean the soil, (b) leaving the soil in place and containing it to a small area, thereby preventing the contaminants from reaching animals, humans, or plants, or (3) removing the soil and treating it or disposing of it in a way that removes the harmful material.

Worldwide there are millions of sites that contain some amount of contamination. The proliferation of harmful chemical and the thoughtless manner in which industrial wastes were handled in the early and middle twentieth century have created numerous harmful conditions. Two such examples in the United States are the incident at Love Canal in the state of New York and the mining of uranium in the western United States.

Figure 14.13 Chemical containers eventually rust and decay, releasing harmful chemicals and other toxins into the water and soil. Cleanup operations are expensive and often not totally effective in reducing the hazardous conditions.

© Sasha Radosavljevich, 2010. Under license from Shutterstock, Inc.

The Love Canal Environmental Crisis

To quote Eckardt Beck, former Administrator of Region 2 of the EPA, "Love Canal is one of the most appalling environmental tragedies in American history." Construction of the canal first began in the 1890s as the plan of William T. Love, a developer who had the idea of constructing a canal between the upper and lower Niagara Rivers in New York. The idea was to generate inexpensive hydroelectric power for the homes and industries in the area. In the early twentieth century digging on the canal began but was abandoned shortly thereafter. The excavated area was left as an eyesore, which became a dumping ground for chemical waste in the 1920s and 1930s. In 1953 a chemical company sold the land, which had been buried in soil, to the City of Niagara Falls for one dollar.

Several years later, as the region began to expand, the area was developed as homes and a school were constructed on the land that rested on decaying chemical containers and other hazardous waste. The winter of 1976 and spring of 1977 brought record precipitation to the region, and the percolating water was enough to release toxic chemicals that had been stored in the subsurface. Within months residents began developing major health problems that were traced to a myriad of extremely harmful chemicals. Although the affected area only covered about 16 acres, the 21,000 tons of buried toxic waste led to miscarriages and birth defects in unusually high proportions.

Assistance soon came to the area. Intervention on the part of the State of New York and the federal government through a Presidential Declaration by President Carter in 1978 provided funds to initiate clean up the area. This event resulted in the establishment of the Comprehensive Environmental Response, Compensation, and Liability Act (CERCLA), or the Superfund Act. More than 200 homes were purchased, and implementation of a major environmental restoration project led to restoration of the area.

Uranium Mining in the West

Uranium is a naturally occurring element that is processed to provide fuel for nuclear power plants. In the 1950s and 1960s uranium was refined to use in the manufacturing of nuclear bombs as part of the arms race with the Soviet Union. The main source for uranium in the United States is the western states, where the geology provided numerous sources from deposits found in sedimentary rocks. The mining and refinement of the source material created large piles of mine tailings that contained small amounts of uranium and other dangerous elements.

Very little attention was paid to the massive piles of waste material. There are many of these piles that have leached harmful substances into the environment, especially into the hydrosphere. The city of Moab, Utah, was an active center for mining, as well as Grand Junction, Colorado. Each of these cities lies along the Colorado River. A major cleanup site is located within a few kilometers of Moab as a massive pile of material is located on the banks of the Colorado River. The EPA has designated this location as a Superfund site. An estimated 16 million tons of mine tailings are being moved about 50 km north to the town of Crescent Junction, Utah, in order to get the toxic materials away from the river. Completion of the project is estimated for the year 2022. In the meantime, the waters of the Colorado River will continue to carry some of the toxic materials downstream.

The cost of hauling away the Moab tailings could exceed $1 billion, according to the latest estimate by the U.S. Energy Department, the agency managing the cleanup. Congress ordered updated cost projections based on a cleanup timetable that is nearly a decade shorter than DOE's. The department had been planning to spend about $30 million a year for the next twenty years to remove the leftover uranium waste piled up on the banks of the Colorado River outside of Moab. But securing annual funding is likely to be much tougher when the annual costs are between $79 million and $103 million from 2010 to 2019, as the DOE estimates. The total cost would be between $844.2 million and $1.1 billion, under the projections DOE submitted in a report to Congress on July 1, 2008. (Reprinted by permission of the *Salt Lake Tribune*.)

Ecosystems Effects and Impacts

ecology
The study of the interaction between organisms and their environments.

Ecology is the study of the interactions between organisms and their environment. Organisms such as plants (producers), consumers (squirrels, foxes), and decomposers (fungi, bacteria) are the biotic components. Examples of the abiotic components are water, temperature, soil nutrients, pH of the substrate, elevation, latitude, aspect, slope, and amount of solar radiation received. If both the biotic and abiotic components are combined and studied, this discipline is referred to as ecosystem ecology.

A grassland ecosystem, for example, would include all of the grasses, insects, birds, rodents, earthworms as well as the soil nutrients, water availability, soil and ambient temperatures, and parent rock (**Figure 14.14**).

All of the ecosystems on the planet form the biosphere. With few exceptions, ecosystems are named after the dominant vegetation that occurs within the system. Some examples of ecosystems are: grasslands, savannahs, tundra, tropical rain forests, deciduous forests, deserts, coral reefs, oceans, and estuaries.

Each ecosystem has its own flora and fauna. Some of these are endemic, while others are found across several ecosystems. Ecosystems suffer from the effects caused by humans, directly and indirectly. An example of this would be the loss of a section of tropical rain forest. When the trees are removed, such as by burning or logging activities,

© Larsek, 2010. Under license from Shutterstock, Inc.

Figure 14.14 Grasslands provide habitats and food for many different animals.

animals and plants that used the trees as their habitat are immediately without a residence. The nutrients that were stored in the trees are removed entirely from that ecosystem, The soil, which is already very poor in nutrients, will not be replenished. When the seasonal rains come, there is no vegetation to intercept its flow, and soil erosion becomes rampant. If the surface has significant slope, landslides can result. Next the unchecked water with the eroded soil enters streams and rivers, causing sediment loads, which interfere and greatly reduce primary production of alga and aquatic vegetation. This loss of photosynthetic activity causes collapse of food chains within the associated aquatic habitats.

Deserts are another ecosystem that has been greatly affected over the past century by humans. Deserts, although they may on first glance seem to be horrible places to survive, do have their own very unique life forms that have evolved through adaptations to endure the hot, dry days and cold evenings (**Figure 14.15**). Deserts, especially in the southwestern United States, have suffered dramatic changes due to development of cities. Houses, roads, buildings, and infrastructure now rise up from the same place that cacti and other desert flora once flourished. Furthermore, the desert temperatures, coupled with atmospheric pollution from cars, trucks, and planes, plus the dust from unpaved desert roads, create a brown pall over many southwestern cities.

Acidification, the process of lowering the pH of terrestrial and aquatic environments, begins with the combination of gases and water vapor in the atmosphere. Examples of three gases released into the atmosphere are CO_2, SO_2, and NO_2. Volcanoes, internal combustion engines, and various industries produce these oxides. Geological activities such as volcanoes emit oxides of sulfur, and carbon dioxide is released from catastrophic forest fires. Once these gases are released and combined with water vapor the resulting acidic precipitation falls onto terrestrial and aquatic habitats. At first the buffering effects of various ions of the soil and water prevent the radical shift of pH. However, over time the acidification process gains momentum, resulting in losses of vegetation resulting in the loss of food and cover for the various organisms. In some lakes, acidification has rendered the water sterile of all life. If this sounds like the "Domino Theory"—it is. To paraphrase John Muir, everything is tied together in an ecosystem and therefore changing one aspect changes everything else within the system.

Since 75 percent of Earth's surface is covered by oceans and lakes, it is logical to some of our species (*Homo sapiens*) that the oceans are limitless with respect to being able to swallow the sewage, garbage, sludge, and other pollutants so frequently and directly shoved into the water. There is an indirect absorption by oceans of agricultural runoff from rivers that are fed by surface runoff, which are fed by farms. This agricultural "cocktail" contains fertilizers, herbicides, fungicides, insecticides, and in some cases fecal materials from feedlots. The oceans and their associated shores and intertidal zones are the largest ecosystem and as such touch every human in some way. The most obvious way is the utilization of the numerous foods that are harvested from the oceans. The pollutants, for example mercury, that have entered the oceans do come back to shore in the fish we eat. This is because of a process called **biomagnification**. Small amounts of a residual pollutant that are absorbed in the lower links of a food chain are increased in the tissues of the larger members of the food chain.

Figure 14.15 Desert ecosystems are fragile due to the low amounts of moisture they receive. Although viewed as inhospitable to humans, they thrive with plant and animal life that has adapted to the conditions.

biomagnification
The increase in amounts of toxins in animals higher in the food chain as a result of ingesting organisms that contain toxic materials.

Changes to Biodiversity

Biodiversity is a combination of two terms—biological and diversity. The term means all of the diversity of life, from the unique gene combinations of individual species to the many ecosystems across the Earth where these species live. There is an intrinsic value bestowed on the megafauna and megaflora, as it should be. However, much of the biodiversity on Earth is hidden because it is either microscopic or in places that are difficult to explore. Consider the soil and its depths and the ooze of oceans and lakes and the aphotic zone of the abysses of the oceans. There is so much yet to be discovered, and much to be lost if we continue down the paths of ignorance and exploitation.

No one knows about the biodiversity that may have already been lost due to following the above two paths. The majority of humans have a soft spot for Bambi and her allies (we can see her), but not that many have reasons to care about unseen down-in-the-ground microbes. These buried "biodiverse" organisms have unique adaptations that permit them to survive in otherwise uninhabitable places. "Buried" might also refer to species that live in or on another species, such as parasites. The presence of parasites often generates a negative response, yet these organisms are part of a larger ecosystem. Epiphytes, plants that attach and live on other plants, but do not derive anything except a place to attach, are also in a way buried since they are not as visible (**Figure 14.16**). There are bromeliads that live high on the trunks of tropical rain forest trees. Bromeliads collect water around their stems and leaves, creating little pools (**Figure 14.17**).

There are tiny frogs that spend their entire life in the pool, from egg to adult. So who cares if they are lost; after all, out of sight out of mind, as the saying goes. We do assign labels to some species like endangered, threatened, species of concern—and once a label is appended to that species some protection is provided. Good examples would be bald eagles and peregrine falcons; populations of both have recovered sufficiently to have their labels removed.

Causes of loss of biodiversity are legend. A common way for biodiversity to be reduced in an area is remove the habitat or alter it so significantly that there is no chance for certain species to hang on. The Dusky seaside sparrow fits into this category. Between flooding and then draining its habitat in Florida and with the addition of numerous pesticides in its environment, the "duskies" departed, going extinct in the late 1980s. The dodo, which was thought to have only existed on Mauritius in the Indian Ocean became extinct in the late seventeenth century (**Figure 14.18**). On rare occasions a species is "rediscovered in the wild." An example is the Banggai crow (*Corvus unicolor*), thought extinct from the early 1900s, was rediscovered in 2007 on Peleng Island in eastern Indonesia.

Another reason for losses of endemic biodiversity, organisms that only live in a particular area, is the invasion of non-native biota, meaning organisms that did not evolve within the same ecosystem and therefore have no natural controls on their spread or population levels. There is a National Invasive Species Information Center in the United States. Some examples of invasive species follow:

- The majestic American chestnut tree grew to great heights and diameters in the eastern United States until the chestnut blight (a fungus) was accidentally released in the United States on some imported wood from Asia. A few American chestnut trees remain, perhaps because of immunity to the blight.
- The European gypsy moth, brought into the United States in the 1860 for silk production is a notorious defoliator of hardwood trees across North America.
- The European starling was introduced into the United States in the 1890 for the sole purpose of having all birds mentioned in any of William Shakespeare's writings to have a habitat in this country. These birds have displaced many native songbirds and have spread to all 48 contiguous states and Alaska. Of the original 100 starlings released in New York's Central Park, hundreds of millions of these birds have resulted.
- Kudzu, a member of the bean family, was introduced into the United States around 1880 from Japan. Kudzu was originally touted as a good plant to control erosion. It found the southern United States to its liking and took off, covering everything in its path. Kudzu has grown around roads by traveling across electrical

Figure 14.16 Spanish moss hanging from a live oak tree in the southern United States is an example of a epiphyte.

Figure 14.17 This bromeliad, found in a tropical rainforest in South America, collects its water in the leaves and sends it down to the root system.

Figure 14.18 Artist's rendering of the extinct dodo.

lines from one side of the road to another. As you can see, some organisms came in accidentally, other introduced on purpose, but either way each organism was not native and has disrupted native species as a consequence. This is one reason the United States Customs Service asked arriving travelers into this country if they have any plants or animals in their possession.

You have seen the bumper sticker—"Extinctions Are Forever!" This is one slogan that speaks for itself and is hard to argue with.

Summary

Human inhabitants on Earth have increased in number because of improved agricultural output and the domestication of animals. A steady growth occurred from several million people about 10,000 years ago to about 500 million in the Middle Ages. Diseases, including the Black Plague, reduced the population, but it rebounded to reach one billion by the early 1800s. A rapid increase followed and the global population now is almost 7 billion. Improved food and nutrition, along with the reduction of some diseases, led to this increase.

Land use changed as more areas became urbanized. This reduces the areas available for farming and forests. The encroachment of humans into once pristine, fertile land has produced changes to the biogeochemical cycles and the carbon, nitrogen, and phosphorus cycles, in addition to the water cycle, are continually threatened.

The safety and effectiveness of these cycles on Earth has been reduced by pollution of the water, air, and soil. Many examples of environmental crises exist. One case involved the Love Canal crisis of the late 1970s, in which the United States government intervened to restore an improved living environment in New York. A second, more wide ranging crisis is centered on the mining of uranium and the resulting tailings in the western United States. The placement of tailings along river drainages created myriad problems downstream from the deposits. Recovery is ongoing, but slow.

Ecology is the study of the interactions between organisms and their environment. Organisms such as plants, consumers, and decomposers are the biotic components. Examples of the abiotic components are water, temperature, soil nutrients, pH of the substrate, elevation, latitude, aspect, slope, and amount of solar radiation received. If both the biotic and abiotic components are combined and studied, this discipline is referred to as ecosystem ecology. Examples of ecosystems include grasslands and deserts, each of which contains its own flora and fauna. The biodiversity of any ecosystem encompasses the range of life that exists in the system, much of which is not seen because many parts of an ecosystem thrive underground.

Every day Earth loses some species through extinction. In the past several centuries well known cases of the loss of a species have occurred, including the dodo bird. Less obvious are the demise of smaller flora and fauna that are affected by the encroachment of humans into the habitats that exist in deserts and other fragile settings. Invasive species that did not pre-exist in an ecosystem, but were introduced later, can upset the balance of naturally occurring flora and fauna. These newly introduced organisms often overtake an area. The introduction of non-native species, although often done by humans, can be caused by nature through disasters such as landslides and storms.

References and Suggested Readings

Barker, Barry. 2003. *Environmental Studies: Concepts, Connections, and Controversies*: Dubuque, IA: Kendall/Hunt Publishing Company.

Botkin, Daniel B. and Edward A. Keller. 2009. *Environmental Science: Earth as a Living Planet*: Hoboken, NJ. John Wiley & Sons.

Campbell, Neil A., Jane B. Reece, and Eric J. Simon. 2007. *Essential Biology with Physiology*, 2nd ed. San Francisco, CA: Pearson Publishing Company.

Keller, Edward A. 2011. *Environmental Geology*, 9th ed.: Upper Saddle River, NJ: Prentice Hall.

Levkov, Jerome. 2008. *As the Earth Turns: Perspective on Environmental Issues in the 21st Century*: Dubuque, IA: Kendall/Hunt Publishing Company.

Mader, Sylvia. 2010. *Essentials of Biology*, 2nd ed. Dubuque, IA: McGraw-Hill Publishing Company.

Pruitt, Nancy L. and Larry S. Underwood. 2006. *BioInquiry: Making Connections in Biology*, 3rd ed.: Hoboken, NJ. John Wiley & Sons.

Web Sites for Further Reference

http://biology.usgs.gov
http://www.census.gov/main/www/popclock.html
http://www.epa.gov/
http://www.epa.gov/superfund/students/wastsite/soilspil.htm
https://www.cia.gov/library/publications/the-world-factbook/rankorder/2102rank.html
http://www.ers.usda.gov/data.agproductivity/
http://www.epa.gov/maia/html/nitrogen.html

Questions for Thought

1. Explain in general terms how global population has changed over the past 12,000 years.

2. Cite data to show how rapidly the world's population has changed in the past two centuries.

3. What conditions have caused the population to increase during the past two centuries?

4. Explain how changes in land use in the United States have occurred in the past fifty years.

5. What is the biogeochemical cycle? (and give two examples)

6. What role do the oceans play in the water cycle?

7. Research an example of water pollution in an area near your hometown.

8. Explain the effects of coal-fired power plants on the atmosphere.

9. Although grassland ecosystems tend to occur in regions with a lower amount of rainfall, explain the effect the grasses have on animal life in the area.

10. Explain the concept of extinction in terms of a plant or animal.

11. How widespread is water pollution?

Glossary

acquired immune deficiency syndrome (AIDS) A serious disease that results from an infections with HIV; it is spread through direct contact with contaminated bodily fluids.

acre foot The amount of water that covers one acre to a depth of one foot; equivalent to 325,872 gallons of water.

active A term applied to a volcano that has erupted in recorded history or is currently erupting ash pyroclastic material that measures less than 2 mm in diameter.

advisory An announcement that is issued when a tsunami could result from a large earthquake that occurs near a coastal region.

air mass A large body of air of considerable depth which are approximately homogeneous horizontally. At the same level, the air has nearly uniform physical properties, especially temperature and moisture.

angle of repose The natural angle of a slope that forms in a pile of unconsolidated materials.

anthrax A serious disease usually acquired by ingestion of the bacterium *Bacillus anthracis* or its spores found in infected animals or their tissue; symptoms include coughing, chest pain, and general lethargy, which lead to death.

anticyclone An area of high atmospheric pressure having clockwise circulation in the northern hemisphere and counterclockwise motion in the southern hemisphere.

ash A pyroclastic material that has an average particle size less than 2 mm.

asteroid A rocky or metallic body ranging in size from a few centimeters to almost 1000 km in size.

asteroid belt A region around the Sun lying between Mars and Jupiter that is the source of asteroids.

asthenosphere The uppermost layer of the mantle, located below the lithosphere. This zone of soft (plastic), easily deformed rock exists at depths of 100 kilometers to as deep as 700 kilometers.

astronomical unit (AU) The distance of the Earth from the Sun, averaging 150 million km or 93 million miles; used to describe large distances between bodies in space.

atmosphere The air surrounding the Earth, from sea level to outer space.

avian flu (H5N1) An acute viral disease of chickens and other birds (except pigeons) capable of being transmitted to humans; first noticed in the Far East, its symptoms include fever and lack of energy.

backburn A controlled fire that burns back toward firefighters.

backshore zone The part of the beach extending landward from the high tide level to the area reached only during storms.

backwash The flow of water down the beach face toward the ocean from a previously broken wave.

bacteria A tiny, single-celled organism that reproduces by cell division, having a shape similar to a rod, spiral, or sphere, and has no chlorophyll.

barrier island A long, usually narrow accumulation of sand, that is separated from the mainland by open water (lagoons, bays, and estuaries) or by salt marshes.

Barringer Crater The best preserved impact crater on Earth; located in northern Arizona, it has a diameter of almost 1 km and is about 180 m in depth.

basalt An extrusive, fine-grained, dark volcanic rock that contains less than 50 percent silica by weight and a relatively high amount of iron and magnesium.

beach An aggregation of unconsolidated sediment, usually sand, that covers the shore.

beach drift The movement of sand along a zigzag path along the beach parallel to shore due to successive waves on the beach.

beach nourishment A soft stabilization technique primarily accomplished by adding sediment to the coastline.

berm A low, incipient, nearly horizontal or landward-sloping area, or the landward side of a beach, usually composed of sand deposited by wave action.

Big Bang Theory The idea that the Universe formed from an initial point mass that exploded about 12 to 15 billion years ago and moved outward in an every-expanding fashion.

biodiversity All the diversity of live on Earth, ranging from single cell organisms to the most complex ecosystems.

biogeochemical cycle Natural processes that recycle nutrients in various chemical forms from the environment, to organisms, and then back to the environment. Examples are the carbon, oxygen, nitrogen, phosphorus, and hydrologic cycles.

biogeochemical cycle Natural processes that recycle nutrients in various chemical forms from the environment, to organisms, and then back to the environment. Examples are the carbon, oxygen, nitrogen, phosphorus, and hydrologic cycles.

biological hazard (biohazard) An organism or substance that is a possible threat to human health.

biomagnification The increase in amounts of toxins in animals higher in the food chain as a result of ingesting organisms that contain toxic materials.

biosphere The living and dead organisms found near the Earth's surface in parts of the lithosphere, atmosphere, and hydrosphere. The part of the global carbon cycle that includes living organisms and biogenic organic matter.

Black Plague (Black Death) An epidemic of bubonic plague that killed more than 20 million people in Europe during the mid-fourteenth century.

blizzard A severe winter storm that has the following conditions that are expected to prevail for a period of 3 hours or longer; sustained wind or frequent gusts to 35 miles an hour or greater and significant falling and/or blowing snow (i.e., reducing visibility frequently to less than ¼ mile).

block A large angular fragment of lava measuring more than 64 mm in diameter.

body wave A seismic wave that is transmitted through the Earth. P-waves and S-waves are body waves.

bomb A large lava fragment larger than 64 mm in diameter that becomes rounded as it is thrown into the air.

break When a wave steepness exceeds 1/7, the wave becomes too steep to support itself and it breaks, or spills forward, releasing energy and forming whitecaps often observed in choppy waters or the surf area along a beach.

breaker A wave in which the water at the top and leading edge falls forward producing foam.

breakwater A structure built offshore and parallel to shore that protects a harbor or shore from the full impact of waves.

brush fire A fire that burns primarily brush and material low to the ground.

bubonic plague A rare, but fast-spreading, bacterial infection caused by *Yersinia pestis*; transmitted in humans and rodents by infected flea piercings; symptoms include headaches and painful swelling of lymph nodes.

burnout An intentional fire that is lit to burn fuel that lies in the path of a spreading fire.

caldera A large basin-like depression that is many times larger than a volcanic vent.

carbon cycle The movement of carbon between the biosphere and the nonliving environment. Carbon moves from the atmosphere into living organisms that then move it back into the atmosphere.

carbon/oxygen cycle This cycle is connected to the movement of carbon dioxide through the biosphere, lithosphere, and atmosphere as carbon dioxide is used by different organisms.

carrying capacity The maximum population size that can be regularly sustained by an environment.

carton sequestration The storage or removal of carbon from the environment or the reducing or elimination of its presence.

cause A controlling factor that makes a slope vulnerable to failure.

cellulose The primary substance that makes cell walls in plant tissue.

cholera An acute intestinal disease caused by the bacterium *Vibrio cholera*; often found in contaminated drinking water and food.

cinder cone A conical-shaped hill crated by the build up of cinders (lapilli) and other pyroclastic material around a vent.

circular orbital motion The movement of a particle by moving in a circle.

climate The long-term average weather, usually taken over a period of years or decades, for a particular region and time period.

climate change The long-term fluctuations in temperature, precipitation, wind, and other aspects of the Earth's climate.

closed system A system in which no matter or energy can leave or enter from the outside.

coastal submergence The permanent flooding of coastal areas by global sea level rise or land subsidence.

coastline A unique boundary where the geosphere, atmosphere, and hydrosphere meet and the systems interact.

cold front A mass of cold air moving toward a mass of warm air. Strong winds and rain typically accompany a cold front.

collapse The sudden sinking of the surface into a subsurface void.

comet A body of icy and rocky material that moves around the sun; most comets originate from the Oort Cloud.

composite volcano A volcano that forms from alternating layers of lava and pyroclastic debris; also known as a stratovolcano.

conduction Heat transfer directly from atom to atom in solids.

controlled burn A fire set to remove fuel. Term is no longer used, as not all such fires were kept under control.

convection (1) (Physics) heat transfer in gas or liquid by the circulation of currents from one region to another; also fluid motion caused by an external force such as gravity. (2) (Meteorology) The phenomenon occurring where large masses of warm air, heated by contact with a warm land surface and usually containing appreciable amounts of moisture, rise upward from the surface of the earth.

convection cell Within the geosphere it is the movement of the asthenosphere where heated material from close to the Earth's core becomes less dense and rises toward the solid lithosphere. At the lithosphere-asthenosphere boundary heated asthenosphere material begins to move horizontally until it cools and eventually sinks down lower into the mantle, where it is heated and rises up again, repeating the cycle.

convergent plate boundary An area where two or more lithospheric plates are coming together, such as along the west coast of South Africa.

core The innermost layer of the Earth, made up of mostly of iron and nickel. The core is divided into a liquid outer core and a solid inner core. The core is the densest of the Earth's layers.

Coriolis Effect An imaginary force that appears to be exerted on an object moving within a rotating system. The apparent force is simply the acceleration of the object caused by the rotation. Along the equator, there will be no such rotation.

crest The highest point of a wave.

crown fire A fire that moves through the upper portions of trees; these are very difficult to extinguish.

crust The rocky, relatively low density, outermost layer of the Earth.

cyclone A rotating mass of low pressure in the atmosphere that covers a large area; the warm air mass rotates counterclockwise in the northern hemisphere and clockwise in the southern hemisphere. The term is strictly applied to large low pressure storms in the Indian Ocean and South Asia.

cyclonic storm A generic term that covers many types of weather disturbances that are typified by low atmospheric pressure and rotating, inwardly directed winds.

decompressional (or pressure) melting The melting of hot rocks in the subsurface caused by a reduction in overlying pressure as the material rises toward the surface.

deep-focus earthquake An earthquake that has its focus located between a depth of 300 kilometers and roughly 700 kilometers.

desert A region that receives less than 25 cm (10 in) of annual precipitation.

discharge The volume of water flowing through a stream channel in a given period of time, usually measured as cubic feet per second or cubic meters per second.

dissolution The natural dissolving of a substance by chemical reactions.

divergent plate boundary An area where two or more lithospheric plates move apart from each other, such as along a mid-oceanic ridge.

Doppler radar Radar that can measure radial velocity, the instantaneous component of motion parallel to the radar beam (i.e., toward or away from the radar antenna).

dormant A term applied to a volcano that is not currently erupting but has the likelihood to do so in the future.

drainage basin An area that drains water to a given point or feature, such as a lake.

drawdown A lowering of water level at the beach when the trough of a tsunami comes ashore.

driving force A force that produces down-slope movement caused by gravity.

dust Pyroclastic material that is silt- or clay-size (less than 1/256 mm).

dynamic equilibrium The state in which the action of multiple forces produces a steady balance, resulting in no change over time.

Earth system Composed of the geosphere, hydrosphere, atmosphere, and biosphere, and all of their components, continuously interacting as a whole.

Earth system science Study of our planet as a system composed of numerous interconnecting subsystems governed by natural laws.

ebb tide That period of tide between a high water and the succeeding low water; falling tide.

ecology The study of the interaction between organisms and their environments.

ecosystem A community of plants, animals and other organisms that interact together within their given setting.

El Niño An irregular variation of ocean current that from January to March flows off the west coast of South America, carrying warm, low-salinity, nutrient-poor water to the south. It is associated with the Southern Oscillation.

emergent coastline Is a coastline which has experienced a fall in sea level, because of global sea level change, local land uplift, or isostatic rebound.

epicenter The point on Earth's surface directly above the focus. This is the position that is reported for the occurrence of an earthquake as latitude and longitude value can be assigned to the point.

epidemic The sudden occurrence of a highly-contagious disease or other event that is clearly in excess of normally expected numbers.

eustatic Global changes in sea level.

exponential growth The growth in which some quantity, such as population size, increases by a constant percentage of the whole during each year or other time period; when the increase in quantity over time is plotted, this type of growth yields a curve shaped like the letter J.

extinct A term applied to a volcano that no longer is expected to erupt.

eye The central core of a cyclonic storm, normally relatively small and lacking clouds, moisture, and wind.

eye wall The boundary between the eye of a cyclonic storm and the inner most band of clouds.

Factor of Safety (FoS) The ration of resisting force to driving force.

Ferrel cell In the general circulation of the atmosphere, the name given to the middle latitude cell produced by sinking motion near 30 degrees and rising motion near 60 degrees latitude.

fetch The distance over which the wind blows across open water.

fire triangle Consists of three parts that cause fire to occur; heat, fuel, and oxygen.

firebreak A clearing that provides a gap across which fire should not burn.

firestorm A widespread, intense fire that is sustained by strong winds and updrafts of hot air.

floodplain The flat area alongside a stream that becomes flooded when water exceeds the banks of the stream.

flood stage The point in time when a body of water, such as a river, rises to a level that causes damage to the adjacent areas.

flood tide The incoming or rising tide; the period between low water and the succeeding high water.

focus The point in the subsurface where an earthquake first originates due to breaking and movement along a fault plane; sometimes referred to as a hypocenter.

forest fire A fire that occurs in a forest or stand of trees.

frequency Occurrence of specific events. Used in interpreting the past record of events to predict occurrences of that event in the future.

front A boundary or transition zone between two air or water masses of different properties.

fuel The part of the fire triangle that actually burns.

geologic time scale A relative time scale based upon fossil content. Geological time is divided into eons, eras, periods, and epochs.

geosphere The soils, sediments, and rock layers of the Earth including the crust, both continental and beneath the ocean floors.

global warming The gradual increase in global temperatures caused by the emission of gases that trap the sun's heat in the Earth's atmosphere (greenhouse effect). Gases that contribute to global warming include carbon dioxide, methane, nitrous oxides, chlorofluorocarbons (CFCs), and halocarbons (the replacements for CFCs). The carbon dioxide emissions are primarily caused by the use of fossil fuels for energy.

glucose An organic substance with the chemical formula $C_6H_{12}O_6$.

gravitational energy The force of attraction between objects due to their mass and is produced when an object falls from higher lower elevations.

greenhouse effect The heating that occurs when gases such as carbon dioxide trap heat escaping from the Earth and radiate it back to the surface.

greenhouse gases Atmospheric gases, primarily carbon dioxide, methane, and nitrous oxide restricting some heat-energy from escaping directly back into space.

groin Solid structures built at an angle from a shore to reduce erosion from long shore currents, and tides.

ground fire Fire that moves along the ground and rises several feet above the surface.

hail Solid, spherical ice precipitation that has resulted from repeated cycling through the freezing level within a cumulonimbus cloud.

headland A steep-faced irregularity of the coast that extends out into the ocean.

heat The part of the fire triangle that provides the energy to create fire.

heat transfer Heat moving from a hot body to a cold one through processes of conduction, convection, or radiation, or any combination of these.

heat wave A prolonged period of excessively hot weather and high humidity.

hepatitis A liver disease caused by a virus; four different forms exist; symptoms include fever, jaundice, fatigue, liver enlargement, and abdominal pain.

human immunodeficiency virus (HIV) A RNA (ribonucleic acid) virus that causes an immune system failure and can lead to AIDS.

hurricane Term applied to cyclonic storms that occur in the North Atlantic Ocean or eastern Pacific Ocean. Minimum wind velocity is 74 miles per hour. Refer to cyclone and typhoon.

hydrologic cycle The cyclic transfer of water in the hydrosphere by water movement from the oceans to the atmosphere and to the Earth and return to the atmosphere through various stages or processes such as precipitation, interception, runoff, infiltration, percolation, storage, evaporation, and transportation.

hydrophobic Incapable of absorbing water.

hydrophobic layer A layer of material that is cannot absorb water; usually forms in dry climates or in regions that have experienced intense fires.

hydrosphere The part of the Earth composed of water including clouds, oceans, seas, ice caps, glaciers, lakes, rivers, underground water supplies, and atmospheric water vapor.

hypothesis A tentative explanation to explain the cause, or why, of the phenomenon being studied.

ice-jam flood A flood, usually in the spring, that results from broken pieces of river ice blocking the flow of a river, thereby flooding areas adjacent to the river.

igneous rock A rock formed when molten rock (magma) has cooled and solidified (crystallized). Igneous rocks can be intrusive (plutonic) and extrusive (volcanic).

impact energy Cosmic impacts with a larger body convert their energy of motion (kinetic energy) to heat.

Incident Command System (ICS) A system that is used to establish a chain of command for multiple agencies fighting fires.

Incident Management Team (IMT) A rapid response team that oversees the operations of large-scale firefighting.

influenza An acute viral infection that affects the respiratory system; this can occur as isolated, epidemic, or pandemic in scale; symptoms include fever, headache, nasal congestion, and general lethargy.

information bulletin Notification to scientists and researchers that an earthquake has occurred, but not necessarily a tsunami was generated.

inner core The solid central art of Earth's core.

intermediate-focus earthquake An earthquake that has its focus located between a depth of 70 kilometers and 300 kilometers.

island arc An arc-shaped chain of volcanic islands produced where an oceanic plate is sinking (subducting) beneath one another.

isotope A different form of an element that has the same atomic number but a different atomic weight due to a change in the number of neutrons present.

jet stream A high-speed, meandering wind current, generally moving from a westerly direction at speeds often exceeding 400 kilometers (250 miles) per hour at altitudes of 15 to 25 kilometers (10 to 15 miles).

jetty A structure extending into the ocean to influence the current or tide in order to protech harbors, shores, and banks.

kinetic energy The energy inherent in a substance because of its motion, expressed as a function of its velocity and mass, or $MV^2/2$.

Kuiper Belt A region beyond the orbits of Neptune and Pluto that is a source of short-period comets.

La Niña A phenomenon characterized by unusually cold ocean temperatures in the eastern Equatorial Pacific, compared to El Niño, which is characterized by unusually warm ocean temperatures in the eastern Equatorial Pacific.

lahar A volcanic mudflow or landslide that contains unconsolidated pyroclastic material.

land clearing The process of removing potential fire fuel.

landslide A general term used to describe the down-slope movement of material under force of gravity.

lapilli Pyroclastic material that ranges in size from 2 mm to 64 mm; sometimes referred to as cinders.

latent heat of condensation Heat released when water vapor absorbs heat to be transformed to water.

latent heat of fusion Heat released when water freezes to form ice.

latent heat of vaporization Heat stored in water vapor as it changes states from a liquid to a vapor.

lava Molten rock that flows onto the surface and cools.

lava dome A dome-shaped mountain formed by very viscous lava flows.

law (or principle) The unvarying sequence of a set of naturally occurring events.

leprosy (Hansen's disease) A chronic, mildly infectious disease caused by *Mycobacterium leprae*, which affects the nervous system, skin, and nasal regions; it is characterized by skin ulcerations and nodules.

lightning A visible electrical discharge produced by a thunderstorm. The discharge may occur within or between clouds, between the cloud and air, between a cloud and the ground or between the ground and a cloud.

liquefaction A condition that exists when an overabundance of liquid, usually water, is present.

lithosphere The outer layer of solid rock that includes the crust and uppermost mantle. This layer, up to 100 kilometers (60 miles) thick, forms the Earth's tectonic plates. Tectonic plates float above the more dense, flowing layer of mantle called the asthenosphere.

lithospheric plate A series of rigid slabs of lithosphere (16 major ones at present) that make up the Earth's outer shell. These plates float on top of a softer, more plastic layer in the Earth's mantle known as the asthenosphere.

local magnitude A term that describes the size of an earthquake in an area near the epicenter; see Richter scale.

longshore current A current that flows parallel to the shore just inside the surf zone. It is also called the littoral current.

longshore drift The net movement of sediment parallel to the shore.

magma Molten rock that lies below the surface.

magnitude The size or scale of an event, such as an earthquake.

malaria An infectious disease caused by the bacterium *Plasmodium falciparum*, which is transmited by the female *Anopheles* mosquito; symptoms include high fever, chills, and sweating.

mantle The layer of the Earth below the crust and above the core. The uppermost part of the mantle is rigid and, along with the crust, forms the 'plates' of plate tectonics. The mantle is made up of dense, iron and magnesium rich (ultramafic) rock such as periodotite.

mass movement (mass wasting) A term used more by geologists to describe the down-slope movement of material.

mesosphere (1) (Geosphere) the division of the mantle between the asthenosphere and the outer core; (2) (Atmosphere) the division of the atmosphere above the stratosphere that begins about 50 kilometers (31 miles) in altitude and extends to about 80 kilometers (50 miles).

metamorphic rock A rock that has been altered physically, chemically, and mineralogically in response to strong changes in temperature, pressure, shearing stress, or by chemical action of fluids.

meteor A rapidly moving body that passes into the atmosphere and begins to burn up, producing what is sometimes referred to as a "shooting star".

meteorite A meteor that hits Earth's surface.

meteoroid A relatively small object that moves through space; most range in size from dust to less than 10 m in size; composed of rock, metal, or ice.

microbe A microorganism, especially one that can be a pathogen.

mineral Any naturally occurring inorganic substance found in the earth's crust as a crystalline solid.

mitigation The process of making less severe or intense; measures taken to reduce adverse impacts on humans or the environment.

moment magnitude A measure of an earthquake that is based on the area affected, the strength of the rocks involved, and the amount of movement along the primary fault.

natural catastrophe A massive natural disaster often affecting a large region and requiring significant amounts of time and money for recovery.

natural disaster The loss of life, injuries, or property damage as a result of a natural event or process, usually within a more local geographic area.

natural hazard An event or phenomenon that could have a negative impact on people and their property resulting from natural processes in the Earth's environment.

neap tide A tide that occurs when the difference between high and low tide is least; the lowest level of high tide. Neap tide comes twice a month, in the first and third quarters of the moon. Contrast with spring tide.

nitrogen cycle The movement of nitrogen through the biosphere and the atmosphere.

nor'easter A strong low pressure system with winds from the northeast that affects the mid-Atlantic and New England states between September and April. These weather events are notorious for producing heavy snow, copious rainfall, and tremendous waves that crash onto Atlantic beaches, often causing beach erosion and structural damage.

normal fault A plane along which movement has occurred such that the upper block overlying the fault has moved down relative to the lower block.

normal force A force that is acting perpendicular to a surface.

occluded front A composite of two fronts, formed as a cold front overtakes a warm or quasi-stationary front. Two types of occlusions can form depending on the relative coldness of the air behind the cold front to the air ahead of the warm or stationary front. A cold occlusion results when the coldest air is behind the cold front and a warm occlusion results when the coldest air is ahead of the warm front.

oceanic crust That part of the Earth's crust of the geosphere underlying the ocean basins. It is composed of basalt and has a thickness of about 5 km.

oceanic ridge An uplifting of the ocean floor that occurs when convection currents beneath the ocean bed force magma up where two tectonic plates meet at the divergent boundary. The ocean ridges of the world are connected and form a single global ridge system that is part of every ocean and also by far the longest mountain range on Earth.

oceanic trench Deep, linear steep-sided depression on the ocean floor caused by the subduction of oceanic crustal plate beneath either other oceanic or continental crustal plates.

offshore zone The portion of beach that extends seaward from the low tide level.

Oort Cloud A far-reaching, spherically shaped body of material located between 50,000 and 100,000 AU from the Sun; the major source of comets.

outer core The liquid outer layer of the core that lies directly beneath the mantle.

overwash A deposit of marine-derived sediments landward of a barrier system, often formed during large storms; transport of sediment landward of the active beach by coastal flooding during a tsunami, hurricane, or other event with extreme wave action.

oxidation A process in which oxygen combines with a substance.

oxygen A basic element that allows oxidation to take place; a necessary part of the fire triangle.

pandemic A term applied to a highly-contagious disease that spreads throughout the world.

pathogen An agent, such as a bacterium or virus, that causes a disease.

phosphorus cycle The movement of phosphorous through the biosphere and the geosphere.

phreatomagmatic eruption Term used to describe volcanic eruptions that involve magma and water coming in contact; the resulting steam caused extremely explosive activity.

plate A slab of rigid lithosphere (crust and uppermost mantle) that moves over the asthenosphere.

plate boundary According to the theory of plate tectonics, the locations where the rigid plates that comprise the crust of the earth meet. As the plates meet, the boundaries can be classified as divergent (places where the plates are moving apart, as at the mid-ocean ridges of the Atlantic Ocean), convergent (places where the plates are colliding, as at the Himalayas Mountains), and transform (places where the plates are sliding past each other, as the San Andreas fault in California).

plate tectonics The theory that the Earth's lithosphere consists of large, rigid plates that move horizontally in response to the flow of the asthenosphere beneath them, and that interactions among the plates at their borders (boundaries) cause most major geologic activity, including the creation of oceans, continents, mountains, volcanoes, and earthquakes.

plunging breaker Forms on shorelines with more steep offshore slopes and have a curling crest that moves over an air pocket.

potential energy Energy available in a substance because of position (e.g., water held behind a dam) or chemical composition (hydrocarbons). This form of energy can be converted to other, more useful forms (for example, hydroelectric energy from falling water).

precession The wobble that occurs when a spinning object slows down.

prescribed burn A controlled fire purposefully set to remove fuel from an area.

prion disease An infliction that attacks the brain and nervous system, and disrupts normal protein activity within the neural cells.

public information officer (PIO) A person who is designated to provide official information regarding disasters.

P-wave The primary wave, which is a compressional wave; this is the fastest moving of all the seismic waves, and arrives first at a recording station. P-waves are a type of body wave.

pyroclastic Related to material that is thrown out by a volcanic eruption.

pyrolysis The process in which a series of reactions break down complex chemical compounds into simpler ones.

radiant heat Produces an increase in temperature caused by radiation.

radiation Energy emitted in the form of electromagnetic waves. Radiation has differing characteristics depending upon the wavelength. Because the radiation from the sun is relatively energetic, it has a short wavelength (ultra-violet, visible, and near infrared) while energy radiated from the Earth's surface and the atmosphere has a longer wavelength (e.g., infrared radiation) because the Earth is cooler than the Sun.

radioactive decay Natural spontaneous decay of the nucleus of an atom where alpha or beta and/or gamma rays are released at a fixed rate.

rapid-onset hazards Hazards that develop with little warning and strike rapidly. They expend their energy very quickly, such as volcanic eruptions, earthquakes, floods, landslides, thunderstorms, or lightning.

recurrence interval A measure of the elapsed time between events of a similar size, such as earthquakes, floods, or large storms.

regolith The layer of varied material that overlies unaltered bedrock and includes unconsolidated and fragmental particles.

relative humidity The percentage of moisture present in the air as measured against the amount it can hold at a given temperature and pressure to be saturated.

resisting force A force that tends to prevent downhill movement.

retardant A chemical substance used to impede the spread of a fire.

reverse fault A plane along which movement has occurred such that the upper block overlying the fault has moved up relative to the lower block.

Richter scale A measurement scale developed by Charles Richter to determine the size of earthquakes in California. This is commonly used to describe many types of earthquakes but it is more correctly used for small, localized events.

rift valley A depression formed on the surface caused by the extension of two adjacent blocks or masses or rock.

rip current A movement of water back into ocean through narrow zones through the surf zone.

river A large stream.

rock Naturally occurring aggregate of minerals. Rocks are classified by mineral and chemical composition; the texture of the constituent particles; and also by the processes that formed them. Rocks are thus separated into igneous, sedimentary, and metamorphic rocks.

rock cycle A sequence of events in which rocks are formed, destroyed, and reformed by geological processes. Provides a way of viewing the interrelationships of internal and external processes and how the three rock groups relate to each other.

rogue wave A large solitary wave caused by constructive wave interference that usually occurs unexpectedly amid waves of smaller size.

runoff Water that flows across a surface and into a stream or other body of water.

run up The height to which water comes up onto the shoreline, either from natural wave action or the occurrence of a tsunami.

Santa Ana winds Seasonal winds that original from high-pressure over the Great Basin of Nevada, pushing dry air from east to west across southern California.

scientific method A systematic way of studying and learning about a problem by developing knowledge through making empirical observations, proposing hypotheses to explain those observations, and testing those hypotheses in valid and reliable ways.

sea arch An opening through a headland caused by erosion.

sea stack An isolated rock island that is detached from a headland by wave action.

seawall A massive structure built along the shore to prevent erosion and damage by wave action.

sedimentary rock A sedimentary rock is formed from pre-existing rocks or pieces of once-living organisms. They form from deposits that accumulate on the Earth's surface. Sedimentary rocks often have distinctive layering or bedding.

seismic gap An area in a faulted region where no seismic activity has occurred in recent time. These areas are undergoing stress buildup and could be the site of future earthquakes.

seismic sea wave A large ocean wave that is produced by a major disturbance in the ocean, such as an earthquake or a landslide.

seismogram The written or electric record of ground motion detected by a seismometer.

seismograph A device that records the ground motion of an earthquake.

seismometer An instrument used to detect very small movements in the ground. This movement is amplified electronically to generate seismograms. Sometimes called a seismograph.

severe acute respiratory syndrome (SARS) A respiratory disease of unknown source that first occurred in mainland China in 2003; symptoms include fever and coughing or difficulty breathing; it is sometimes fatal.

shallow-focus earthquake An earthquake that has its focus located between the surface and a depth of 70 kilometers.

shear force A force that acts parallel to a surface.

shear strength The internal force of an object that acts to counter a force acting parallel to a plane.

shield volcano A broad volcano that has gentle slopes consisting of low viscosity basaltic lava flows.

silicate Refers to the chemical unit silicon tetrahedron, SiO_4, the fundamental building block of silicate minerals. Silicate minerals represent about one third of all minerals and hence make up most rocks we see at the Earth's surface.

silicic Rich in silicon dioxide, which is the chemical composition of the mineral quartz.

sinkhole A circular depression on the surface that forms from the collapse of material into an underlying void.

slide The movement of material along a curved or flat plane.

slope failure The down-slope movement of an unstable area.

slow-onset hazard A hazard that takes years to develop such as drought, insect infestations, disease epidemics, and global warming and climate change.

slump A mass movement in which generally unconsolidated material moves down a hillside along a curve, rotational subsurface plane.

smallpox An acute viral disease that was once a major killer but has been eradicated; symptoms include headaches, vomiting, and fever, followed by a widespread skin rash that eventually permanently scars the skin.

southern oscillation The shift between El Niño conditions and La Niña conditions in the areas around the equatorial Pacific Ocean.

spilling breaker Forms on shorelines with gentle offshore slopes and are characterized by turbulent crests spilling down the front slope of the wave.

spring tide The highest high and the lowest low tide during the lunar month. The exceptionally high and low tides that occur at the time of the new moon or the full moon when the sun, moon, and earth are approximately aligned. Contrast with neap tide.

storm An atmospheric disturbance manifested in strong winds accompanied by rain, snow, hail, or other precipitation and often by thunder and lightning.

storm surge An abnormal rise in sea level associated with intense storms, such as cyclones, typhoons, and hurricanes.

stratosphere The level of the atmosphere above the troposphere, extending to about 50-55 kilometers above the Earth's surface.

stream A body of water that flows downhill under the influence of gravity and lies within a defined channel.

stress The force being applied to a surface; forces can be compressional, extensional, or shearing.

strike-slip fault A fault in which the motion of the two adjacent blocks is horizontal, with little if any vertical movement.

subduction A process of one crustal plate sliding down and below another crustal plate as the two converge.

subduction zone Also called a convergent plate boundary. An area where two plates meet and one is pulled beneath the other.

sublimation The process that changes a solid into a gas, bypassing the liquid phase.

submarine landslide The collapse of land material either underwater or from the land that slides into water, producing a massive wave.

submergent coastline Is a coastline which has experienced a rise in sea level, due to a global sea level change or local land subsidence.

subsidence A relatively slow drop in the ground surface caused by the removal of rock or water located under the surface.

surf zone The nearshore zone of breaking waves.

surface fire A fire that moves along the ground and consumes ground litter.

surface wave A seismic wave that moves along a surface or boundary.

surging breaker Forms when the offshore slopes abruptly and the wave energy is compressed into shorter distance and the wave surges forward right at the shoreline.

swash A turbulent sheet of water that rushes up the slope of the beach following the breaking of a wave at shore.

S-wave The secondary wave, which is a shear wave that moves material back and forth in a plane perpendicular to the direction the wave is traveling. S-waves do not travel through liquids. S-waves are a type of body wave.

swell A wave of uniform wavelength moving away from a storm center. They can travel great distances before the energy of the wave is released by breaking and crashing onto the coast.

talus The natural pile of material that builds up at the base of a cliff or hillside by material falling from above.

temperature of combustion The temperature at which a specific type of material will ignite.

tephra A general term for all types of pyroclastic material produced by a volcano.

theory A comprehensive explanation of a given set of data that has been repeatedly confirmed by observation and experimentation and has gained general acceptance within the scientific community.

thermal energy The amount of energy in a system that is related to the temperature of its constituents; for hurricanes this is heat originally taken from the oceans.

thrust fault A reverse fault having a fault plane dipping less than 30 degrees and usually less than 15 degrees.

thunderstorm A local storm produced by a cumulonimbus cloud and accompanied by lightning and thunder.

tide The periodic rising and falling of the water that results from the gravitational attraction of the moon and sun acting on the rotating earth.

topography The general configuration of the land surface.

tornado A rotating column of air usually accompanied by a funnel-shaped downward extension of a cumulonimbus cloud and having a vortex several hundred yards in diameter whirling destructively at speeds of up to 600 kilometers per hour (350 miles per hour).

tornado warning A warning is issued when a tornado is indicated by the WSR-88D radar or sighted by spotters; therefore, people in the affected area should seek safe shelter immediately.

tornado watch A watch is issued by the National Weather Service when conditions are favorable for the development of tornadoes in and close to the watch area. Their size can vary depending on the weather situation.

trade winds A system of relatively constant low level winds that occur in the tropics. The trade winds blow from the northeast to the equator in the Northern Hemisphere and from the southeast to the equator in the southern hemisphere.

transform fault A strike-slip fault with side to side horizontal movement that offsets segments of an oceanic ridge.

transform plate boundary An area where two plates meet and are moving side to side past each other.

tributary A stream that flows into another larger stream.

trigger An event that disturbs the equilibrium of a slope, causing movement to occur.

tropical cyclone A low-pressure system having a warm center that developed over tropical (sometimes subtropical) water and has an organized circulation pattern. The magnitude of its winds defines it as a disturbance, depression, storm, or hurricane/typhoon.

tropical depression A slow-forming cyclonic storm with sustained surface winds of 38 miles.

tropical disturbance The beginning stage of a cyclonic storm lasting at least 24 hours, originating in the tropics or subtropics; clouds and moisture become organized and a vertically rotating wind mass creates atmospheric instability.

tropical storm A cyclonic storm with sustained surface winds between 39 miles per hour to 73 miles per hour. At this level of activity the system is assigned a name to identify and track it.

troposphere The lowest part of the atmosphere that is in contact with the surface of the Earth. It ranges in altitude above the surface up to 10 or 12 kilometers.

trough The low spot between two successive waves.

tsunami A series of giant, long wavelength waves produced by the displacement of large amounts of ocean water by an earthquake, volcanic eruption, landslide, or meteorite impact.

tuberculosis (TB) An infection caused by *Mycobacterium tuberculosis* and passed among humans by inhaled, airborne droplets; symptoms include coughing of fluids, fever, and chest pains.

typhoid fever An illness spread by contamination of water, food, or milk supplies with *Salmonella typhi*; symptoms include fever, diarrhea, stomach aches, and rash.

typhoon A cyclonic storm that forms in the central and western Pacific Ocean. Refer to cyclone and hurricane.

typhus An acute infectious disease with symptoms of high fever, severe headaches, and a skin eruption; common in wartime, famines, or catastrophes, it is spread by lice, ticks, or fleas.

undercutting A process whereby a slope or hillside has supporting material removed by erosion.

updraft A small-scale current of rising air. If the air is sufficiently moist, then the moisture condenses to become a cumulus cloud or an individual tower of a towering cumulus.

upwelling In ocean dynamics, the upward motion of sub-surface water toward the surface of the ocean. This is often a source of cold, nutrient-rich water. Strong upwelling occurs along the equator where easterly winds are present. Upwelling also can occur along coastlines, and is important to fisheries and birds in California and Peru.

vector An agent (for example, a flea or mosquito) that can spread parasites, a virus or bacteria.

virus A cellular, non-living infectious particles of either DNA or RNA associated with various protein coats; these only multiply in living cells.

viscosity A measure of the internal resistance of a substance to flow; a lower viscosity means the material flows easily.

volcanic arc Arcuate chain of volcanoes formed above a subducting plate. The arc forms where the downgoing descending plate becomes hot enough to release water and gases that rise into the overlying mantle and cause it to melt.

volcanic explosivity index (VEI) A measure of the intensity of a volcanic eruption, with a value of 0 representing the quietest and 8 being the most explosive.

warning The highest level that is given when a tsunami has been formed and its arrival is highly possible.

watch An alert that a tsunami could potentially occur in an area.

watershed See drainage basin.

wave Energy in motion that is the result of some disturbance that moves over or through a medium with speeds determined by the properties of the medium. Ocean waves are usually generated by wind blowing across the water surface.

wave base Depth equal to one-half the wavelength where there is no movement associated with surface waves.

wave height The vertical distance between the crest and adjacent trough of a wave.

wave period The time it takes for one full wavelength to pass a given point.

wave refraction The process by which the part of a wave in shallow water is slowed down, causing it to bend and approach nearly parallel to shore.

wave speed The velocity of propogation of a wave through a liquid, relative to the rate of movement of the liquid through which the disturbance is propagated.

wave steepness The measured ratio of wave height to wave length.

wave-cut platform A gently sloping surface produced by wave erosion, extending far into the sea or lake from the base of the wave cut cliff.

wavelength The distance separating two adjacent crests (or troughs) on a wave form.

weather The composite condition of the near earth atmosphere, which includes temperature, barometric pressure, wind, humidity, clouds, and precipitation. Weather variations over a long period create the climate.

weathering A process that includes two surface or near-surface processes that work in concert to decompose rocks. Both processes occur in place. No movement is involved in weathering. Chemical weathering involves a chemical change in at lest some of the minerals within a rock. Mechanical weathering involves physically breaking rocks into fragments without changing the chemical make-up of the minerals within it.

West Nile virus (WNV) A type of virus that infects mosquitoes and birds, but can be transmitted to human and other animals; symptoms include fever and other flu-like symptoms that may develop into encephalitis and meningitis.

wildfire An uncontrolled fire in a natural setting.

wind The horizontal motion of the air past a given point. Winds begin with differences in air pressures. Pressure that's higher at one place than another sets up a force pushing from the high toward the low pressure. The greater the difference in pressures, the stronger the force.

yellow fever Disease caused by a vector-transmitted virus; symptoms include high fever, headaches, jaundice, and often gastrointestinal hemorrhaging.

Index